Lecture Notes in Mechanical Engineering

Lecture Notes in Mechanical Engineering (LNME) publishes the latest developments in Mechanical Engineering - quickly, informally and with high quality. Original research reported in proceedings and post-proceedings represents the core of LNME. Volumes published in LNME embrace all aspects, subfields and new challenges of mechanical engineering. Topics in the series include:

- Engineering Design
- Machinery and Machine Elements
- Mechanical Structures and Stress Analysis
- Automotive Engineering
- Engine Technology
- Aerospace Technology and Astronautics
- Nanotechnology and Microengineering
- Control, Robotics, Mechatronics
- MEMS
- Theoretical and Applied Mechanics
- Dynamical Systems, Control
- Fluid Mechanics
- Engineering Thermodynamics, Heat and Mass Transfer
- Manufacturing
- Precision Engineering, Instrumentation, Measurement
- Materials Engineering
- Tribology and Surface Technology

To submit a proposal or request further information, please contact the Springer Editor in your country:

China: Li Shen at li.shen@springer.com
India: Dr. Akash Chakraborty at akash.chakraborty@springernature.com
Rest of Asia, Australia, New Zealand: Swati Meherishi at
swati.meherishi@springer.com
All other countries: Dr. Leontina Di Cecco at Leontina.dicecco@springer.com

To submit a proposal for a monograph, please check our Springer Tracts in Mechanical Engineering at http://www.springer.com/series/11693 or contact Leontina.dicecco@springer.com

Indexed by SCOPUS. The books of the series are submitted for indexing to Web of Science.

More information about this series at http://www.springer.com/series/11236

Magdalena Diering · Michał Wieczorowski ·
Christopher A. Brown
Editors

Advances in Manufacturing II

Volume 5 - Metrology and Measurement Systems

 Springer

Editors
Magdalena Diering
Poznan University of Technology
Poznan, Poland

Christopher A. Brown
Worcester Polytechnic Institute
Worcester, MA, USA

Michał Wieczorowski
Poznan University of Technology
Poznan, Poland

ISSN 2195-4356   ISSN 2195-4364   (electronic)
Lecture Notes in Mechanical Engineering
ISBN 978-3-030-18681-4   ISBN 978-3-030-18682-1   (eBook)
https://doi.org/10.1007/978-3-030-18682-1

This Springer imprint is published by the registered company Springer Nature Switzerland AG
The registered company address is: Gewerbestrasse 11, 6330 Cham, Switzerland

Preface

This volume of Lecture Notes in Mechanical Engineering contains selected papers presented at the 6th International Scientific-Technical Conference Manufacturing 2019, held in Poznan, Poland, on May 19–22, 2019. The conference was organized by the Faculty of Mechanical Engineering and Management, Poznan University of Technology, Poland, under the scientific auspices of the Committee on Machine Building and Committee on Production Engineering of the Polish Academy of Sciences.

The aim of the conference was to present the latest achievements in mechanical engineering and to provide an occasion for discussion and exchange of views and opinions. The main conference topics were:

- quality engineering and management
- production engineering and management
- mechanical engineering
- metrology and measurement systems
- solutions for Industry 4.0.

The organizers received 293 contributions from 36 countries around the world. After a thorough peer review process, the committee accepted for conference proceedings 167 papers, prepared by 491 authors from 23 countries (acceptance rate of about 57%). Extended versions of selected best papers will be published in the scientific journals: *Flexible Services and Manufacturing Journal, Research in Engineering Design, Management and Production Engineering Review* and *Archives of Mechanical Technology and Materials*.

The book **Advances in Manufacturing II** is organized into five volumes that correspond with the main conference topic mentioned above.

The book **Advances in Manufacturing II - Volume 5 - Metrology and Measurement Systems** is a collection of material that was created as a result of an international forum for the dissemination and exchange of academics' and practitioners' knowledge of and experience with scientific information on metrology and measurement systems. This crosses many disciplines, including surface metrology, biology, chemistry, civil engineering, food science, material science, mechanical engineering, manufacturing, metrology, nanotechnology, physics, tribology, quality engineering, computer sciences and other. Editors, reviewers and experts from various areas of metrology and measurement systems in manufacturing ensure that the book makes a significant contribution to the development of the mentioned area and is attractive not only to academics, but to industry and business as well. The book consists of 22 chapters, prepared by 69 authors from 7 countries.

We would like to thank the members of the International Program Committee for their hard work during the review process.

We acknowledge all that contributed to the staging of Manufacturing 2019: authors, committees and sponsors. Their involvement and hard work were crucial to the success of the Manufacturing 2019 conference.

May 2019
Magdalena Diering
Michał Wieczorowski
Christopher Brown

Organization

Steering Committee

General Chair

Adam Hamrol — Poznan University of Technology, Poland

Chairs

Olaf Ciszak — Poznan University of Technology, Poland
Stanisław Legutko — Poznan University of Technology, Poland

Scientific Committee

Stanisław Adamczak, Poland
Michal Balog, Slovakia
Zbigniew Banaszak, Poland
Myriam Elena Baron, Argentina
Stefan Berczyński, Poland
Johan Berglund, Sweden
Wojciech Bonenberg, Poland
Christopher A. Brown, USA
Anna Burduk, Poland
Somnath Chattopadhyaya, India
Shin-Guang Chen, Taiwan
Danut Chira, Romania
Edward Chlebus, Poland
Damir Ciglar, Croatia
Marcela Contreras, Mexico
Nadežda Cuboňová, Slovakia

Jens J. Dahlgaard, Sweden
María de los Angeles Cervantes Rosas, Mexico
Andrzej Demenko, Poland
Magdalena Diering, Poland
Ewa Dostatni, Poland
Jan Duda, Poland
Davor Dujak, Croatia
Milan Edl, Czech Republic
Sabahudin Ekinovic, Bosnia and Herzegovina
Mosè Gallo, Italy
Bartosz Gapiński, Poland
Józef Gawlik, Poland
Hans Georg Gemuenden, Norway
Boštjan Gomišček, UEA

Marta Grabowska, Poland
Wit Grzesik, Poland
Michal Hatala, Slovakia
Sandra Heffernan, New Zealand
Christoph Herrmann, Germany
Ivan Hudec, Slovakia
Vitalii Ivanov, Ukraine
Andrzej Jardzioch, Poland
Mieczyslaw Jurczyk, Poland
Wojciech Kacalak, Poland
Lyudmila Kalafatova, Ukraine
Anna Karwasz, Poland
Mourad Keddam, Algeria
Sławomir Kłos, Poland
Ryszard Knosala, Poland
Janusz Kowal, Poland
Drazan Kozak, Croatia
Agnieszka Kujawińska, Poland
Janos Kundrak, Hungary
Maciej Kupczyk, Poland
Ivan Kuric, Slovakia
Oleksandr Liaposhchenko, Ukraine
Piotr Łebkowski, Poland
José Mendes Machado, Portugal
Aleksandar Makedonski, Bulgaria
Ilija Mamuzic, Croatia
Krzysztof Marchelek, Poland
Tadeusz Markowski, Poland
Edison Perozo Martinez, Colombia
Thomas Mathia, France
Józef Matuszek, Poland
Adam Mazurkiewicz, Poland
Andrzej Milecki, Poland
Mirosław Pajor, Poland
Ivan Pavlenko, Ukraine
Dragan Perakovic, Croatia

Alejandro Pereira Dominguez, Spain
Marko Periša, Croatia
Emilio Picasso, Argentina
Jan Pitel, Slovakia
Alla Polyanska, Ukraine
Włodzimierz Przybylski, Poland
Luis Paulo Reis, Portugal
Álvaro Rocha, Portugal
Rajkumar Roy, UK
Iwan Samardzic, Croatia
Krzysztof Santarek, Poland
Jarosław Sęp, Poland
Bożena Skołud, Poland
Jerzy Sładek, Poland
Roman Staniek, Poland
Beata Starzyńska, Poland
Tomasz Sterzyński, Poland
Tomasz Stręk, Poland
Antun Stoić, Croatia
Manuel Francisco Suarez Barraza,
 Mexico
Marek Szostak, Poland
Rafał Talar, Poland
Franciszek Tomaszewski, Poland
María Estela Torres Jaquez, Mexico
Justyna Trojanowska, Poland
Stefan Trzcieliński, Poland
Maria Leonilde R. Varela, Portugal
Sachin D. Waigaonkar, India
Edmund Weiss, Poland
Michał Wieczorowski, Poland
Ralf Woll, Germany
Magdalena Wyrwicka, Poland
Jozef Zajac, Slovakia
Jan Żurek, Poland

Program Committee

Available on http://manufacturing.put.poznan.pl/en/.

Special Sessions

Collaborative Manufacturing and Management in the Context of Industry 4.0

Special Session Organizing Committee

Leonilde Varela	University of Minho, Portugal
Justyna Trojanowska	Poznan University of Technology, Poland
Vijaya Kumar Manupati	Mechanical Engineering Department, NIT Warangal
José Machado	University of Minho, Portugal
Eric Costa	Solent University, UK
Sara Bragança	Solent University, UK

Intelligent Manufacturing Systems

Special Session Organizing Committee

Ivan Pavlenko	Sumy State University, Ukraine
Sławomir Luściński	Kielce University of Technology, Poland

Tooling and Fixtures: Design, Optimization, Verification

Special Session Organizing Committee

Vitalii Ivanov	Sumy State University, Ukraine
Yiming Rong	Southern University of Science and Technology, China

Advanced Manufacturing Technologies

Special Session Organizing Committee

Jozef Jurko	TU Košice, Slovak Republic
Michal Balog	TU Košice, Slovak Republic
Tadeusz E. Zaborowski	TU Poznań, Poland

*The Changing Face of Production Engineering and Management
in a Contemporary Business Landscape*

Special Session Organizing Committee

Damjan Maletič	University of Maribor, Faculty of Organizational Sciences, Enterprise engineering Laboratory, Slovenia
Matjaž Maletič	University of Maribor, Faculty of Organizational Sciences, Enterprise engineering Laboratory, Slovenia
Tomaž Kern	University of Maribor, Faculty of Organizational Sciences, Enterprise engineering Laboratory, Slovenia

Enabling Tools and Education for Industry 4.0

Special Session Organizing Committee

Dorota Stadnicka	Politechnika Rzeszowska, Poland
Dario Antonelli	Politecnico di Torino, Italy
Katarzyna Antosz	Politechnika Rzeszowska, Poland

Staff for the Industry of the Future

Special Session Organizing Committee

Magdalena Wyrwicka	Poznan University of Technology, Poland
Anna Vaňová	Faculty of Economics, Matej Bel University, Slovakia
Maciej Szafrański	Poznan University of Technology, Poland
Magdalena Graczyk-Kucharska	Poznan University of Technology, Poland

*Advances in Manufacturing, Properties, and Surface Integrity
of Construction Materials*

Special Session Organizing Committee

Szymon Wojciechowski	Poznan University of Technology, Poland
Grzegorz M. Królczyk	Opole University of Technology, Poland
Sergei Hloch	Technical University of Kosice, Slovakia

Materials Engineering

Special Session Organizing Committee

Monika Dobrzyńska-Mizera	Poznan University of Technology, Poland
Monika Knitter	Institute of Materials Technology, Poznan University of Technology, Poland
Robert Sika	Institute of Materials Technology, Poznan University of Technology, Poland
Dariusz Bartkowski	Institute of Materials Technology, Poznan University of Technology, Poland
Waldemar Matysiak	Institute of Materials Technology, Poznan University of Technology, Poland
Anna Zawadzka	Institute of Materials Technology, Poznan University of Technology, Poland

Advanced Mechanics of Systems, Materials and Structures

Special Session Organizing Committee

Hubert Jopek	Institute of Applied Mechanics, Poznan University of Technology, Poland
Paweł Fritzkowski	Institute of Applied Mechanics, Poznan University of Technology, Poland
Jakub Grabski	Institute of Applied Mechanics, Poznan University of Technology, Poland
Krzysztof Sowiński	Institute of Applied Mechanics, Poznan University of Technology, Poland
Agata Matuszewska	Institute of Applied Mechanics, Poznan University of Technology, Poland

Virtual and Augmented Reality in Manufacturing

Special Session Organizing Committee

Filip Górski	Poznan University of Technology, Poland
Paweł Buń	Poznan University of Technology, Poland
Damian Grajewski	Poznan University of Technology, Poland
Jorge Martin-Gutierrez	Universidad de la Laguna, Spain
Letizia Neira	Universidad Autónoma de Nuevo León, Mexico
Eduardo Gonzalez Mendivil	Tecnologico de Monterrey, Mexico

Statistical Comparison of Original and Replicated Surfaces

Milena Kubišová[1(✉)], Vladimír Pata[1], Libuše Sýkorová[1], and Mária Franková[2]

[1] Faculty of Technology, Tomas Bata University in Zlín,
Vavrečkova 275, 760 01 Zlín, Czech Republic
mkubisova@utb.cz
[2] Faculty of Mechanical Engineering, Technical University of Kosice,
Masiarska 74, 04001 Kosice, Slovakia

Abstract. This article is focused on the statistical comparison of the original and replicated surface. Surface replication has been used for a long time, but recently a material with the commercial name Dentacryl™ has been used for the replication. However, this material has some disadvantages that outweigh the benefits. The article focuses on the use of Siloflex© dental impression material that is very sensitive to the detailed control. It also uses the non-contacting profile of Talysurf CLI 500, where initial parameters of the roughness (Ra, Rz, Rp, Rv and Rt) were recorded both on the original surface and on the replicated surface. These parameters are then compared using statistical methods such as K- means Clustering, PCA, and Factor Analysis.

Keywords: Talysurf CLI 500 · Statistical comparison · Replica · Surface · Comparison of the surfaces

1 Introduction

The article focuses on surface replication research. SILOFLEX© marks were used to produce the replica. Even though this substance is primarily intended for dental purposes, it was used for its ability to scan micrometric surface details. The replica is composed of the basic material Putty© used for reinforcement and Stomaflex Light© for the replica printing itself. This combination proved to be the most stable [1, 2].

A variety of depths and elevations can be found on the surface of the products, measuring up to thousands of micrometres. Surface properties have a significant effect on whether the product is "smooth" or "rough". Smoothness and roughness parameters are mathematically matched to formulas and graphs to eliminate the error of measured data.

Surface quality means its structure and roughness. Determined by machining methods, appearance, depth of tool traces, and type of machined material. The roughness parameters are evaluated on the actual profiles, which are obtained as a perpendicular or perpendicular intersection: oblique, a plane with the rear surface.

All of the above is necessary to determine what the parts will have the sliding properties of how the mould will be injected. It is not always possible to place the sensing surface under the measuring device because it is too large or tangible, so it is

M. Diering et al. (Eds.): *Advances in Manufacturing II* - Volume 5, LNME, pp. 1–10, 2019.
https://doi.org/10.1007/978-3-030-18682-1_1

possible to print the surface and evaluate the printout separately. The imprint in this work is called a "replica" [1, 3].

The main part of the chapter is research on the preparation of surface replicas, determination of how to print the surface and how to evaluate the replica. This technology is very little explored regarding product surface quality control. It can bring new directions and ways to evaluate. For the production of replicas are used impression compounds used mainly in dentistry. Therefore, in the field of dentistry, we are looking not only for the necessary inspiration for the choice of material suitable for surface sensing but also for knowledge in the field of preparation and application of impression materials [1, 4].

The current state of knowledge is divided into two diametrically different categories. The first one is a scientific category that deals mainly with the surface quality sensing methodology, but less with its mathematical assessment. The issue of surface replica creation in scientific practice is not adequately addressed at all.

In contrast, a second, practical category requires the assessment of the quality of the surfaces that are difficult to remove or are not removable at all (weight, dimension) and their replacement by replicas.

However, a comparison of the quality of the replica and the replicated surface is not adequately described in the literature [1, 2, 5].

From the point of view of mathematical evaluation of surfaces and replicas, in practice, ISO 4287, 4288 and 25178-2 are used, which is inadequate.

There is no way to compare originals and replicas in mathematics; there is no way to compare several parameters at a time; there is no way to determine a systematic error.

Furthermore, it was necessary to find a suitable type of substance for replicating surfaces of moulds that are patterned (not polished).

We can say that, to date, Dentakryl, or other resin-based resin impregnation materials, are used. However, these were entirely inappropriate for purposes and will speak later.

Another problem we were dealing with was finding suitable parameters describing the surface quality, according to ISO 4287 and ISO 4288, but unlike our research, we did not evaluate individual parameters but the similarity of the noise of the individual parameters [1, 5, 6].

From the measured profile of the surface, the shape and subsequently the waviness was first removed, resulting in a roughness profile according to the 4287 standards, which was drawn in sections Y separately (North-South) and especially in the X-axis (East-West). Based on standard 25178, the computer plotted a 3D profile of the surface. Finally, roughness parameter values were generated according to the 4287 standards (see Fig. 1) [1, 7, 8]. He proposed formatting by introducing other styles or by changing this template.

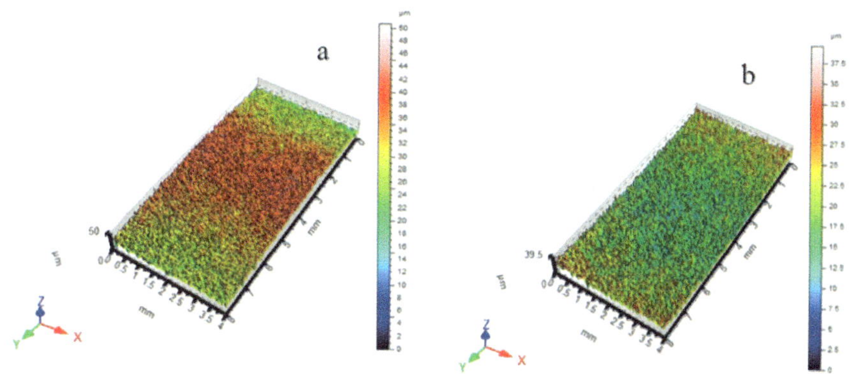

Fig. 1. (a) Original surface, (b) replica surface.

2   Materials and Methods

A classic approach to assessing compliance is to use hypothesis testing, individual parameters independent of each other. This, however, leads to the following problem, for example, I do not ignore the match of the Rp and Rv parameter between the form and the product, but I will reject the match between Parameters Rz, which is the sum of Rp and Rv.

The same problem is even when using 3D parameters, such as SP, Sv and Sz.

However, scientific practice requires a much more complex approach, such as solving the question of the consistency between the set of Ra, Rz, Rp and Rv parameters between the original and the replica.

2.1   General is the Procedure for Comparison of Surfaces

After creating the surface and its replicas with the above-described technologies, it is necessary to assess the harmony between the original surface and its replica. Here, however, we come across a fundamental problem. Using the necessary procedure described in (here to include a monograph reference), hypotheses hypothesis will need to be used, namely F tests and double-sided t-tests at pre-selected confidence levels [1, 6].

This, however, will only allow us to state that in a comparison pair of selected parameters, do not reject the consistency of their scattered or average values at a given confidential level.

However, if we consider that the individual parameters describing the surface quality together correlate, we will be more interested in the possible agreement (percentage) between the surface generated by the given technology. This agreement will focus not on the compliance of individual parameters, but the congruence of the clusters of individual parameters. To this approach, it is necessary to use the Factor Analysis methodology to find covariance's of characteristic parameters, and then attempt to calculate as few factors as possible, but which explain the most excellent possible variability of mentioned parameters. Selection of suitable parameters describing the amplitude properties of the resulting surface and its replicas [1, 7].

Both surfaces were scanned under repeatability conditions and described with five amplitude parameters, namely Ra, Rz, Rp, Rv and Rt (see Table 1). The question is whether the number of descriptive amplitude parameters is sufficient or whether it is possible to reduce their number while maintaining approximately the same carrier information on rated surfaces. To answer this question, the k-means Clustering method was first used, which decided to create two sets of parameters Ra, Rz, Rp, Rv and Rt for the original and replicated surface. As a basis for identifying individual clusters, the method of calculating distance based on Euclidean distance was chosen within the method [1, 8].

Table 1. Roughness parameters in accordance with ISO 4287 for original and replica.

Origin	Mean	St. Dev.	Med	Min	Max	Replica	Mean	St. Dev	Med.	Min	Max
Rp μm	4,35	0,64	4,40	2,70	6,09	Rp μm	3,93	0,79	3,91	2,6	6,44
Rv μm	4,44	0,74	4,44	2,82	6,17	Rv μm	4,05	0,73	4,13	2,52	5,62
Rz μm	8,79	1,29	8,88	6,10	12,1	Rz μm	7,97	1,45	7,96	5,12	11,4
Rt μm	11,7	1,81	11,8	7,31	6,10	Rt μm	10,5	2,07	10,3	6,64	16,4
Ra μm	0,87	0,29	1,87	1,21	2,53	Ra μm	1,67	0,27	1,73	1,01	2,25

2.2 Evaluation of the Degree of Similarity Based on "Abbott Firestone Curve"

An essential step in assessing the similarity of surfaces in replication is the computation of the material fraction See Fig. 2.

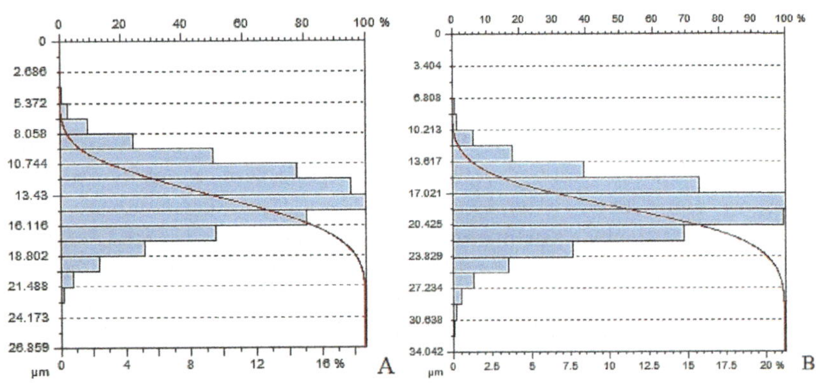

Fig. 2. Abbott firestone curve for (A) Original and (B) Replica.

Demonstrated Abbott Firestone curve, original surface and repaired demonstrate the similarity of both surfaces. Both curves were created as arithmetic averages of cumulated sums from individual scanned slices, see Fig. 2. From their shapes, it is possible to specify that in the case of the replicated surface there is a slight vertical displacement relative to the original surface, which is due to the unacceptability of the Siloflex© impression material. From the histogram shape, it can be stated that the replicated surface exhibits a slight obliquity again versus the original surface, which can again be explained by the imprinting properties of the mass.

However, it is noteworthy to note that these curves are so-called cumulative sums of zeta ordinal. Depending on their shape, it is possible to predict not only the character of the sensed surface but also the condition that is embedded in ISO 4288, the condition of the normality of the parameters of the described and measured surfaces. To emphasise these properties on the article of the described surfaces, the Abbot-Firestone curves were supplemented with histograms characterising the distribution of the zeta ordinal. By simple optical assessment, it is possible to characterise that the measured zeta ordinates have approximately Normal distributions with parameters μ and σ with which are unknown and determined using estimates of arithmetic means and standard deviations of individual parameters in the article according to ISO 4287 [1, 8].

All parameters, i.e. Ra, Rz, Rp, Rt and Rv, were divided into two basic clusters, where the number 1 denotes the original surface parameter cluster, on the other hand, the parameter number 2 for the surface replicated.

As can be seen from the above clustering graph (see Fig. 3), it is possible to state the closeness of individual parameters within own clusters number 1 and number 2 [1, 7].

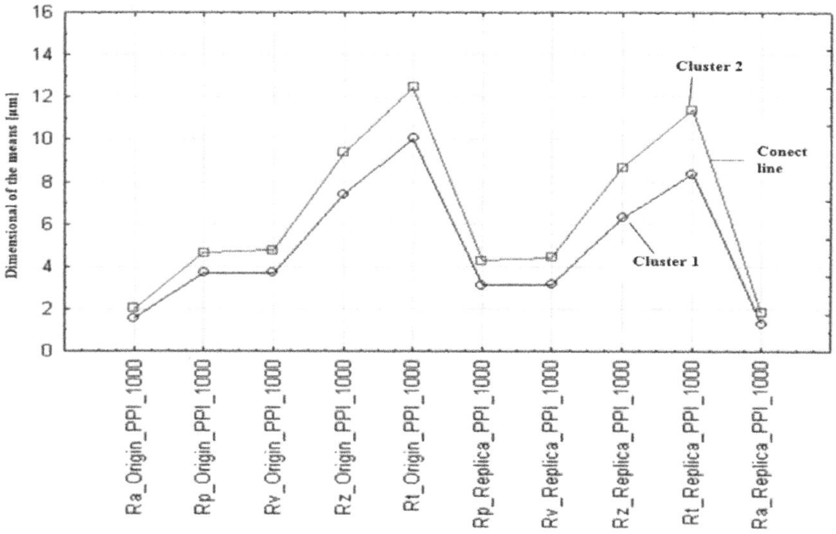

Fig. 3. The plot of means for each cluster.

In a more detailed study of the diagram of means for individual clusters, it can be stated that the maximum differences occur in parameters that do not meet the super-normality condition, which is currently very problematic when evaluating surface roughness parameters, both according to ISO 4287 and ISO 25 178. On the contrary, parameters of the Ra type exhibit considerable closeness of individual clusters see Fig. 3 because they are based on the principle of calculating estimates of arithmetic averages. It is therefore quite clear from this that the modern way of judging surfaces and their replicas cannot be based on the mere estimation of a parameter, as is often customary in practice, but on sets of parameters comprised of amplitude, frequency, or hybrid parameters. The resulting quality of the respective match between the original and the replica will be determined, as will be described, not only on the selected parameters but on and on their clusters. The necessity of this approach can be seen already in Fig. 3.

2.3   Analyse the Main Components

Analysis of significant components was chosen as the primary element of exploratory analysis. The reason was that Melon quotes are not necessary assumptions about truth-likeness selection. However, all of the amplitude above parameters were tested for data normality by two diametrically different methods at a confidence level of $1-\alpha = 0.95$ and hence the error rate of 0.05.

The first was the Anderson-Darling test, the second followed by a combined skew and acuity test.

For quality evaluation of the analysis of the main components, it was necessary to perform the Scree plot analysis of Fig. 6; the results can then be traced in Figs. 4 and 5.

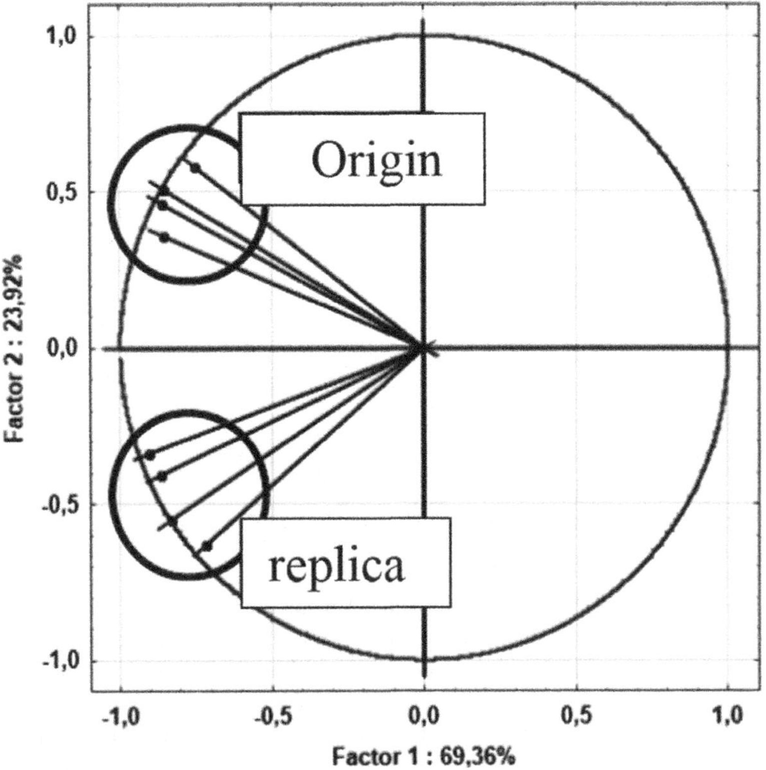

Fig. 4. Projection of the variables on the factor-plane.

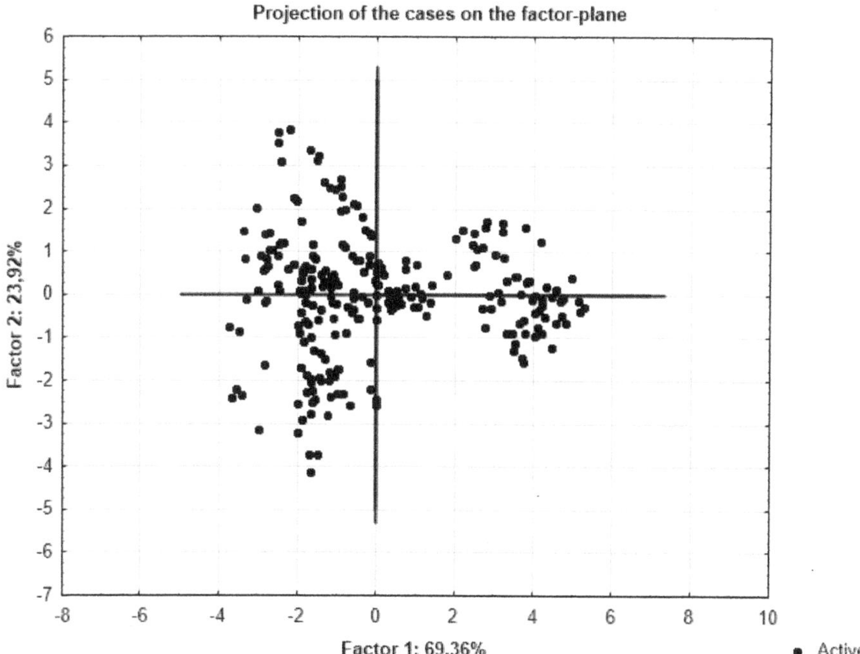

Fig. 5. Projection of the cases on the factor-plane routine diagram for optimising factor count.

Authors Meloun [8] is used to determine the number of statistically significant components. This number is characterised by a so-called breakpoint, before which the statistically significant main components are present, in the presented article, the amplitude factors of the original surface and its replicas.

On the contrary, the factors behind the fault point are statistically insignificant.

In the present article, as expected, the statistically significant factors 2, described below, are explained explaining a total of 90% of the dispersions of the factors. As can be seen from Fig. 6 for determining the statistical significance of individual factors, Kaiser's number one is chosen and the authors are recommended, which is marked with a horizontal line in the Scree fence and numerically gets one.

Altogether, out of ten factors, according to Fig. 6, Scree, the plot of the eighth diagram characterises the statistical insignificance.

From this relatively high percentage expression and further from the character of Fig. 4 it can be stated that the choice of the method and the characteristic factors have been carried out correctly. Of course, it is necessary to assess the degree of similarity between individual factors having clusters of characters, which will be done in this section below in the cluster analysis section.

Based on this claim, the Principal Component Analysis methodology (PCA) was used to reduce the dimension, the number of the parameters with the least loss of information that these parameters carry.

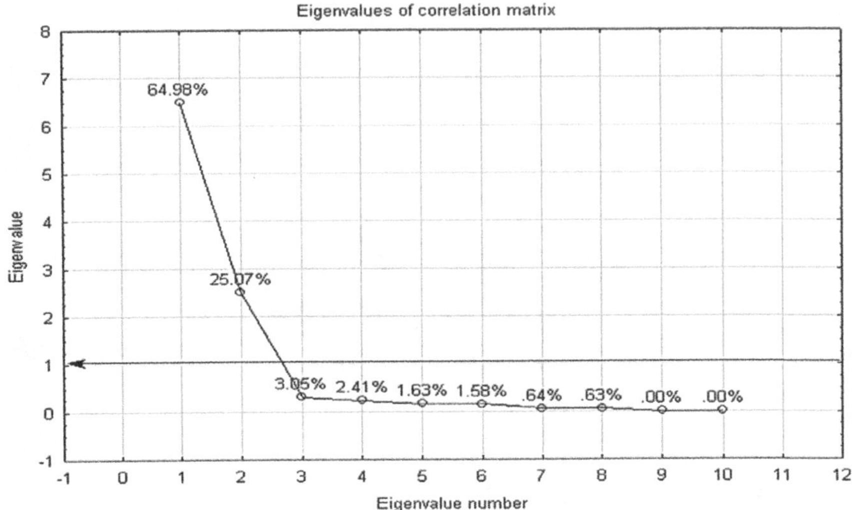

Fig. 6. Eigenvalues of the correlation matrix.

First, the projection of Factors first and second was made to the so-called Factor Plane.

Looking at the design graph (see Fig. 4), it can be stated that all amplitude parameters are describing both the original and the replicated surface created, like the previous method, two clusters that describe the surface quality of the surface.

It is also possible to state that the number of factors is sufficient because, like the planes in PCA results, factor first describes 69.36% scattering, the other 23.92%.

Thus, two factors describe more than 90% of the variance of their amplitude parameters. Based on these results, it is possible to adopt a hypothesis on the number of two factors. Furthermore, there is a correlation between the factors.

In the case of the projection of the measured results of the Ra, Rz, Rp and Rv parameters into the Factor Factor, for both types of surfaces, it is again apparent that there are two similarities between the surfaces [1, 8].

To numerically confirm the number of descriptive factors used, it is necessary to use the calculation method of the own numbers and then to draw up the Diagram of the footnote of own numbers - see Fig. 5.

We use the commonly used so-called Kaiser's first value as our critical value. After this calculation, we will again receive two primary factors describing more than 90% of the single surface scattering.

2.4   Performing the Cluster Analysis

The primary element for evaluating objects between the replica and the original described by the amplitude above parameters according to ISO 4287 is to determine the mutual similarity of the investigated objects - multidimensional objects as previously defined.

In this specific case, it is necessary to create two assemblies based on a correlation matrix that will determine the degree of similarity or disparity between the original and its replica. As a measure of similarity, we use the method of calculation according to Euler's principle and for the subsequent comparison of the Manhattan measure.

When comparing the results, it was stated that these results are not statistically different from each other and therefore, in connection with the correlation matrix, the Euclidian [1, 8].

As a way of clustering, the Hierarchical clustering was tested in general. For which the correlation coefficient was calculated, according to which a suitable hierarchical method was then selected. This method according to the coefficient as mentioned above was chosen as Wardova, which is also often recommended for tasks of the type described above [1, 8].

For a comparison of both surfaces in two clusters, the number of which has already been demonstrated above, it is possible to use a cloud analysis that finds proximity or distance of individual parameters within clusters.

For this procedure, Ward's method with Euclidean distance was used, which is highly suitable in this case, because it makes relatively smaller and more accurate clusters (Fig. 7).

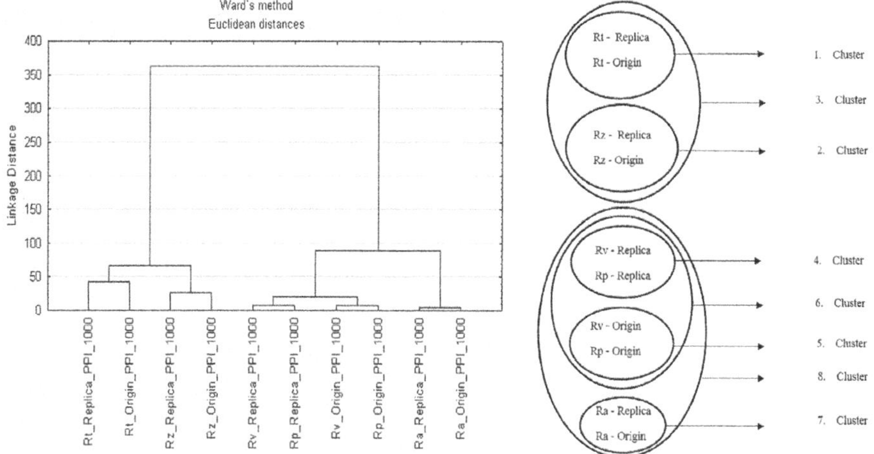

Fig. 7. Cluster analysis.

As can be seen from the above diagram, the degree of similarity expressed by the spacing of individual clusters within the individual parameters is relatively high.

3 Conclusion

The article describes replica production, its measurement and statistical evaluation. For this evaluation Factor Analysis is mainly used, Graph of the Own Numbers, followed by Cluster Analysis. Instead of using Rp, Rv, Ra, and Rz for the surface and its replica, it is statistically significant to use only two factors that represent the original eight parameters to 69%.

We do not need to judge one parameter against another (for example, Rz replica versus Rz form), using 2-fold analyses to produce the clusters as described by the factors. As can be seen from the graph "the Scree Plot", Kaizer's first criterion can be said that the variability of all parameters, both for the original and for the replica, can be expressed using only two factors that express 69% variability.

Therefore, we call factor first, a factor describing the properties of the original surface expressed by amplitude parameters and factor second, a factor describing replica surface properties expressed by amplitude parameters.

In conclusion, it is possible to state that the comparison of the original and replicated surfaces based on the methods described above, in particular, Clustering, PCS and Cluster analysis, is much more accurate and above all more representative than the classical method based on scattering and diameters tests.

In conclusion, it is possible to state that the article generally brings a new perspective on the quality of the replicated surfaces and the degree of their conformity with the original surfaces. This view is based on the principle of hierarchical clustering, which is not widely used in technical practice, despite its undisputed benefits. It is clear from the article that a purely fair assessment of the consistency using current ordinary t-tests often leads to misinterpretation.

References

1. Whitehouse DJ (2011) Handbook of surface and nanometrology, 2nd edn. CRC Press, Boca Raton. ISBN 978-1-4200-8201-2
2. Bilodeau MA, Brenner D (1999) Theory of multivariate statistics. Springer texts in statistics. Springer, New York
3. Bílek O (2016) Cutting tool performance in end milling of glass fibre-reinforced polymer composites. manufacturing technology. J Sci Res Prod 16(1): 12–16. Accessed 21 Mar 2017. ISSN 1-2132-489
4. Rolf-Dieter R, Thomas M (2001) Statistical analysis of extreme values: with applications to insurance, finance, hydrology and other fields, 2nd edn. Birkhäuser Verlag, Basel. ISBN 3-7643-6487-4
5. Zhang J, Feng C, Ma Y, Tang W, Wang S, Zhong X (2017) Non-destructive analysis of surface integrity in turning and grinding operations. Manuf Technol 17(3): 412–418
6. Hanzl P, Zetková I, Mach J (2017) Optimization of the pressure porous sample and its manufacturability by selective laser melting. Manuf Technol 17(1):34–38
7. Hnatkova E, Sanetrnik D, Pata V, Hausnerova B, Dvorak Z (2016) Mold surface analysis after injection molding of highly filled polymeric compounds. Manuf Technol 16(1):86–90
8. Meloun M, Militký MA, Forina M (1992) Chemometrics for analytical chemistry, 1st edn. Ellis Horwood, New York. ISBN 01-312-6376-5

Application of Acoustic Emission Signals Pattern Recognition for a Firearm Identification

Leszek Chałko, Paweł Maciąg, and Mirosław Rucki[✉]

Faculty of Mechanical Engineering, Kazimierz Pulaski University of Technology
and Humanities in Radom, Radom, Poland
m.rucki@uthrad.pl

Abstract. In the paper, important security issue of firearm identification is addressed. State of art on the acoustic pattern recognition is widely presented emphasizing that no research is reported being conducted on the firearm identification based on acoustic signal emitted during reloading. It was demonstrated that reloading sound pattern is dependent on manufacturing technologies, dimensions, materials and other features, and thus can serve as a recognizable signature of a firearm type. In particular, the firearms manufactured out of a metal piece with milling technologies, as well as molded guns expose higher stability of the main components, and their emitted acoustic signal is characterized by wider spectrum and smaller energy than that of the guns manufactured by stamping technology out of the flat sheet metal. The respective measurement system is presented and results of the measurements are discussed. This novel system is based on two microphones and parallel processing of two registered acoustic signals, which proved to be highly effective. The results confirmed the individual characteristics of the acoustic "signature" of different types of firearms and possibility to apply the registered reloading sound in the security recognition systems.

Keywords: Firearm technology · Sound pattern · Acoustic emission · Security

1 Introduction

When a human ear is exposed to impact noise, at a sound intensity of 120 dB or higher, there is a risk for acoustic trauma, changes to the hearing threshold, or noise-induced-hearing-loss [1]. Therefore numerous studies are performed to evaluate hearing protection systems used by police officers exposed during activities at the shooting range [2]. However, while the shotguns and artillery report is considered to be just a noise, it can be in fact very informative. Acoustic signal of the firearm discharge can be detected and classified [3]. For example, Defense Technical Information Center published a study of the spectral content of small arms fire at varying distances [4]. In the study, two components of the small-arms noise were distinguished: the muzzle blast and the sonic boom. In another approach aimed to a pattern recognition, shock wave, muzzle blast and reflections were distinguished, and gunshot signatures were divided into multiple

M. Diering et al. (Eds.): *Advances in Manufacturing II* - Volume 5, LNME, pp. 11–24, 2019.
https://doi.org/10.1007/978-3-030-18682-1_2

classes [5]. To analyze sound pressure levels generated by firearms with and without suppression devices, certified measuring devices are produced [6].

Another problem solved with acoustic signal analysis is the localization of a shooter. The respective systems use the muzzle blast and acoustic shock waves to compute the shooter's location [7]. An analysis based on Acoustic Gun Shot Detector for small fire arms was reported, with utilization of the Acoustical Characteristics of small firearms based on the recorded data and test results of field trials [8]. Even though the systems designed for gunshot localization are developed since the early 1990's, they still need to be improved. In particular, three specific problems related to the processing of gunshot acoustic signatures should be addressed: direction-of-arrival estimation, noise cancellation, and issues related to multipath propagation [9].

Recently, gunshot detection became an issue. It found widespread applications from systems aimed to protect elephants in Africa [10] to the ones installed in an urban neighborhood plagued by high rates of violent crimes [11]. Perhaps the most chilling challenge were the multiple school shootings occurring in February and May 2018 in USA, which motivated to install systems for instant detect of gunshots and identification of weapon in the schools [12].

However, a bullet leaves the gun with initial velocity ca. 1000 m/s, while sound velocity in the air of 20 °C is ca. 340 m/s. Obviously, when the gunshot sound is registered, it is too late. The below presented investigations addresses the issue of weapon identification based on acoustic emission, but at the moment of weapon reloading, which takes place usually several seconds before shooting. Proper identification of weapon reload sound may help in gunshot prevention, which seems to make it an important direction of researches.

In the project, the main objective of the researches is to work out the methodology of the acoustic emission measurement and its analysis aimed to the immediate identification of the gun type. In the paper, after state of art is shortly described in Sect. 2, the experimental conditions and apparatus are presented in Sect. 3 together with main objectives of the research and explanation of chosen methodology. In the Sect. 4, results of experimental measurement are presented and discussed, and in Sect. 5 some conclusions with possible future plans are proposed.

2 Acoustic Pattern Analysis State-of-Art

Human beings are able to recognize sounds and to identify their sources naturally. However, the challenge of acoustic signal recognition and discrimination between emissions from different acoustic emission (AE) is far from simple. The pattern-recognition methods can be based on computerized extracting features of AE waveforms from an unknown source, and then comparing these features with those predicted for possible sources using theoretically and experimentally determined models of sources and calibration studies [13]. A new approach was reported, based on an exhaustive screening taking into account all combinations of signal features extracted from the recorded acoustic emission signals. For each possible combination of signal features an investigation of the classification performance of the k-means algorithm was evaluated and calculated utilizing the Davies–Bouldin and Tou indices, Rousseeuw's silhouette validation method and Hubert's Gamma statistics. As a second step the numerical ranking was performed using the cluster validation methods results. This methodology was proposed as an automated evaluation of the number of natural clusters and their partitions without previous knowledge about the cluster structure of acoustic emission signals [14].

In the most recent solutions, after feature extraction and selection, further classification is based on artificial neural network (ANN) [15]. In order to recognize leaking of natural gas, the dominant frequencies of the AE leak signals were validated and the feature sets {Peak, Mean, Peak Frequency, Kurtosis} and {Mean, Peak Frequency} were applied. Other proposed leakage detection scheme using an acoustic emission sensor was based on kernel principal component analysis and the support vector machine (SVM) classifier for the leakage level recognition [16]. Different approach proposed to estimate the leakage rate of a valve in a natural gas pipeline via factor and cluster analysis of acoustic emission signals [17]. Three types of clustering algorithm—fuzzy C means, k-means and k-medoids—were used to classify leakage rates.

One of the most obvious applications of the acoustic signal recognition is the monitoring of friction couples. It was reported that during friction and wear experiments performed with a pin-on-disk-type sliding-friction tester with an electric current flowing between the specimens, an analysis of the frequency of the resulting AE signal showed that each phenomenon produced a distinct AE frequency spectrum [18]. Other study proposed application of acoustic emission analysis for in situ monitoring of sliding surfaces until their failure due to scuffing mechanism [19]. Here, three regimes of events have been specified; steady-state, pre-scuffing, and scuffing according to the friction behavior. During the analysis, acoustic signals for each regime have been decomposed with wavelet packets, with sub-band energies as an analysis features.

Vibro-acoustic noise emitted during various machine operations can be highly informative. Relationship between rock properties and the vibro-acoustic signal characteristics generated by drilling was investigated, and a special wideband acoustic sensor method was applied to distinguish the specific time–frequency characteristics generated by drilling in different types of rock [20]. In addition, time domain and time–frequency analyses were applied to analyze the differences between signal characteristics generated by different kinds of rock. It was found that the acoustic sensor approach presented better signal-to-noise ratio than the vibration sensor approach.

Many works are devoted to acoustic signal used for supervision of machining processes. For example, the improved efficiency of machine state identification from AE data was reached when features extracted in both time and frequency domains were combined and then reduced with the linear discriminant analysis [21]. A high-frequency acoustic emission signal with further acquired data was used to develop characteristic factors to predict product quality and to detect tool defects [22]. In one research, acoustic signals were captured for the entire grinding cycle until the abrasive grains of the girding wheel become dull [23]. Various features of the acoustic emission signatures were extracted from the time-domain and correlated with the surface roughness. Good condition and dull condition of the grinding wheel is predicted using machine-learning techniques such as decision tree, artificial neural network, and support vector machine. Another group reported researches on Acoustic Emission Monitoring of Robot Assisted Polishing [24]. Here, a pre-processing phase was introduced to remove the bias generated by the AE raw signals. The signals were subjected to two feature extraction procedures: (1) a conventional one based on statistical analysis, and (2) an advanced one based on wavelet packet transform (WPT). The statistical analysis and WPT pattern feature vectors were utilized as inputs to neural network for decision making data processing.

The application of acoustic emission analysis to the damage diagnostics took place in the middle of 20th century when Josef Kaiser detected audible sounds produced by deformation under tensile test on metallic specimen, which is known nowadays as "the Kaiser effect" [25]. Next, Schofield and Tatro improved the instrumentation and clarified the source of AE during plastic deformation. In the decade of 1960s, AE has been applied as a nondestructive testing (NDT) method in the aerospace industry, and in early 1970s research started on fiber-reinforced composites testing. Recently, an AE technique monitoring in plain and steel fiber reinforced concrete specimens was analyzed using unsupervised kernel fuzzy c-means pattern recognition and the principal component method to categorize various damage stages [26]. Results were reported on AE signatures of synthetic mooring ropes subjected to sinusoidal tension-tension loading in a controlled environment [27], a possible correlation between the dominant failure in glass fibre-reinforced plastics and their according acoustic emissions was reportedly evaluated [28], the Identification of Active Damage Processes acoustic emission method was reported to be applied to continuous monitoring of active destructive processes in reinforced concrete bridges [29].

Another large group of researches is related to the technical issues of speech recognition, so natural for humans. The fundamental concepts and component technologies for automatic speech recognition include several types of acoustic models—Gaussian mixture models (GMM), hidden Markov models (HMM), and deep neural networks (DNN), plus their major variants [30]. Feature statistics normalization in the cepstral domain is one of the most performing approaches for robust automatic speech and speaker recognition in noisy acoustic scenarios, using suitable linear or nonlinear transformations, including histogram equalization, in order to match the noisy speech statistics to the clean speech one [31]. Fuzzy elastic matching machine (FEMM) was proposed as a sequential pattern recognition tool, that helped to solve three basic problems, including evaluation, assignment, and training problems [32]. A study that investigated the use of non-conventional body-conductive acoustic sensors in human-human speech communication and automatic speech recognition was reported, where the body-conductive sensors were directly attached to the speaker and received the uttered speech through the skin and bones. The results proved higher robustness against environmental noise [33]. Noise issue was addressed in the context of tests used by military and civilian researchers, audiologists, and hearing technicians to assess performance of an individual in recognizing speech in background noise [34]. On the other hand, acoustic signal analysis of human speech was utilized for automatic evaluation of voice and speech disorders [35] and laryngeal pathology assessment [36]. A novel methodology was proposed to characterize voice diseases by using nonlinear dynamics, considering different complexity measures, mainly based on the analysis of the time delay embedded space [37].

Interesting study was reported on a musical-staff-inspired signal processing method for standard description expressions for discrete signals and characteristics of acoustic emission signals. In the proposed method, various AE signals with complex environments were mapped into the normalized musical space, and 4 new indexes were proposed to comprehensively describe the signal. Contour, amplitude, and signal changing rate, were considered key features and quantitatively expressed in a normalized musical space. It was emphasized that the processed information required small storage space and maintained high fidelity [38]. Even though we do not usually associate gunshots or gun reload sounds with a music, this approach may be helpful in the analysis of the respective acoustic signals.

3 Experimental Conditions and Apparatus

The measurement of acoustic characteristics is highly dependent on the physical conditions of the experiment. In particular, the field measurements are significantly different from the ones performed in the closed rooms, where reflected sounds play important role and have their impact on measurement results. Closed room measurements require appropriate definition of the acoustic conditions in order to choose proper method and apparatus. Important factors should be taken into consideration, such as physical dimensions of the analyzed object, dimensions of the room where measurement is performed, noise, air movement inside the room, and eventually humidity and temperature. In some cases measurement takes place in an acoustic anechoic chamber, where no reverberant sounds can be registered, or in the rooms with no sound absorption where acoustic waves bounce off the walls producing loud echoes. In the latter case, the walls are non-parallel, so all the sound energy is dispersed in a co-called dispersed field. In real rooms, sounds are partially absorbed and partially reflected, which may have an impact on the acoustic measurement. The respective fields corresponding with the direct and reflected sounds in an open space and closed rooms are shown in the Figs. 1 and 2.

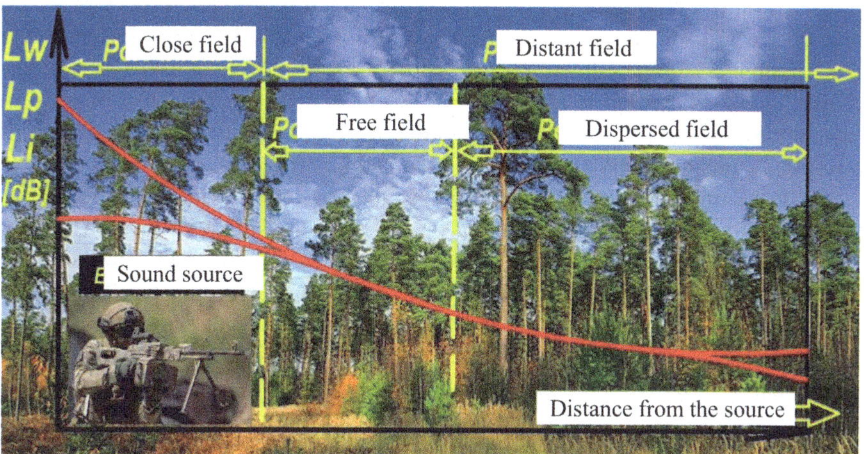

Fig. 1. Acoustic conditions in an open air.

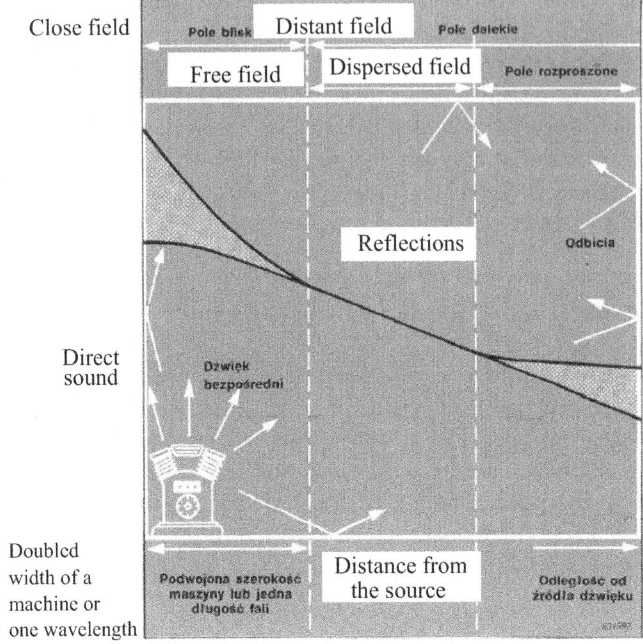

Fig. 2. Acoustic conditions in a closed room [39].

In the traditional measurement methods based on the sound pressure level or spectral analysis, it is not possible to identify properly the gun type from the emitted sound waves. In these devices the signal distributed in time is averaged, so that different signals of the same sound energy appear to be similar. Thus, the main idea was to combine two measurement systems: precise microphones with high class acoustic sensors and the oscilloscope able to register the signal in the real time. This way the voltage signal is related to the sound pressure, as follows: 20 dB lower pressure reduces voltage 10 times down to 5 mV, 40 dB lower reduces it 100 times down to 0.5 mV, etc. This relation is real-time type, so this signal can be analyzed with high frequency sampling performed by oscilloscope. Characteristics of the oscillogram corresponds unequivocally to the weapon, so can be treated as a weapon signature.

When the microphone is placed too close to the acoustic source, the sound level measurement result is highly dependent on the position. This phenomenon is particularly significant when the distance from the source is smaller than the longest emitted wavelength. This area is denoted in Figs. 1 and 2 as a "Close field." Thus, sound pressure level (SPL) measurement defines changes of the acoustic pressures in the free or dispersed field. SPL is the local pressure deviation caused by a sound wave related to the ambient atmospheric pressure, average or equilibrium. Sound pressure L_p [dB] is defined as follows:

$$L_p = 10 \cdot \log \frac{p}{p_0}, \tag{1}$$

where p and p_0 are sound pressure and reference sound pressure [Pa], respectively. The value of $p_0 = 2 \times 10^{-5}$ Pa.

When the acoustic measurement takes place in the close field, instead of sound pressure, sound intensity level (SIL) denoted L_I and sound power level (SWL) denoted L_W. The respective equations describe these values:

$$L_I = 10 \cdot \log \frac{I}{I_0}, \tag{2}$$

$$L_W = 10 \cdot \log \frac{W}{W_0}, \tag{3}$$

where I and I_0 are sound intensity in the measuring point [W/m^2] and reference sound intensity $I_0 = 1$ [pW/m^2], respectively. Similarly, W and W_0 are sound power of the examined source [W] and reference sound power $W_0 = 1$ [pW], respectively. Both SIL and SWL values are logarithmic ones related to human perception addressed by Weber–Fechner law postulating that the external stimulus is scaled into a logarithmic internal representation of sensation [40].

There are very few devices able to measure these energy parameters of the sound in real-time conditions. The Brüel & Kjaer 2260 analyzer with a sensing microphone system 3595 was chosen for the experimental research. Apart from standard software for one-microphone measurements, device had BZ7205 application that enabled to measure the sound power level with the sound intensity method. The system had actual calibration certificate No. 3290/2017 from the accredited laboratory HAIK sp. z o.o. The experiments were performed with reloading of the guns and with gunshots. Figure 3 (left) illustrates the registration of the reloading sound of AKM assault rifle in the laboratory, and the gunshot registration in the shooting range (right).

Fig. 3. Registration of reloading sound (left) [41] and gunshot (right) in the closed rooms.

In the previous research, the acoustic emission was registered with the digital oscilloscope DSO-2902 from the analogous visual monitoring system Heanworld HD2M256 [41]. In the new investigations, it was decided to connect the oscilloscope directly to the analog outputs #1 and 2. This way, the "pure" signal was obtained without any additional processing. Thus, a quick real-time signal was recorded with no deformations typically made by the corrections aimed to suit the analyzed frequencies to the human ear perception. Microphones ½" type 4189 had sensitivity 50 mV/Pa. In decibels, acoustic signal of 94 dB caused generation 50 mV alternating voltage in the microphone output. Similarly, signal 74 db (i.e. 20 dB lower) generated 10 times smaller voltage (i.e. 5 mV), signal 54 dB (i.e. 40 dB lower) generated 100 times smaller voltage and so on. This way, two independent systems registered simultaneously the same signal. This approach will determine the further research directions aimed on the weapon identification based on the reload sound. The abovementioned configuration of the measurement systems corresponds with the appropriate methodology of acoustic field identification in order to prevent possible threat to the human life and health.

Complete determination of the sound power emitted during the reload was not able due to very short duration time. Thus, it was decided to register sound signal during 2 s on the surface 0.5×0.5 m, i.e. 0.25 m^2. The results were registered for further spectral analysis in a ranges of 1/3 octave. The distance between registering microphones was 12 mm, so the sound waves of frequency between 25 and 10,000 Hz were registered, which corresponded with the sounds distinguished by human ears. Additionally, sound pressure level was registered in two channels for further comparative analysis. All the measurement cycles were repeated three times in order to minimize measurement uncertainty. Before each measurement series, the device was calibrated using the sound calibrator type 4231 (B&K).

4 Results and Discussion

Figure 4 presents photo of the gun PR-15 [42] and patterns registered with DSO-2902 oscilloscope during its reload. It can be seen that in the main part of the registered acoustic signal that corresponds with the direct sound emitted by the device, its peaks sequence is highly repeatable and recognizable.

Acoustic vibrations registered in the oscillograms, correspond both with the sounds directly emitted by the weapon moving mechanisms and with their later "ringing," i.e. declining sound emission after the full movement cycle is finished.

The reloading automatic cycle is started with pulling a gun shutter or slide backward, and ends with its strike against the motion limiter. Then the shutter or slide will snap forward, cambering a round from the loaded magazine. The final closure of shutter usually marks the last high peak in the oscillogram, though in some guns the shutter may emit a series of damped echoing sounds afterwards. In the latter case, it is impossible to determine the final moment of reloading from the registered acoustic signal.

Spectral analysis of the PR-15 pistol reloading sound is shown in Fig. 5. The sound signature of the firearm is determined by its dimensions, constructional features (shape and thickness of the details), mechanism principles (sequence of contact interactions between elements inside the gun during reload) and material acoustic emission

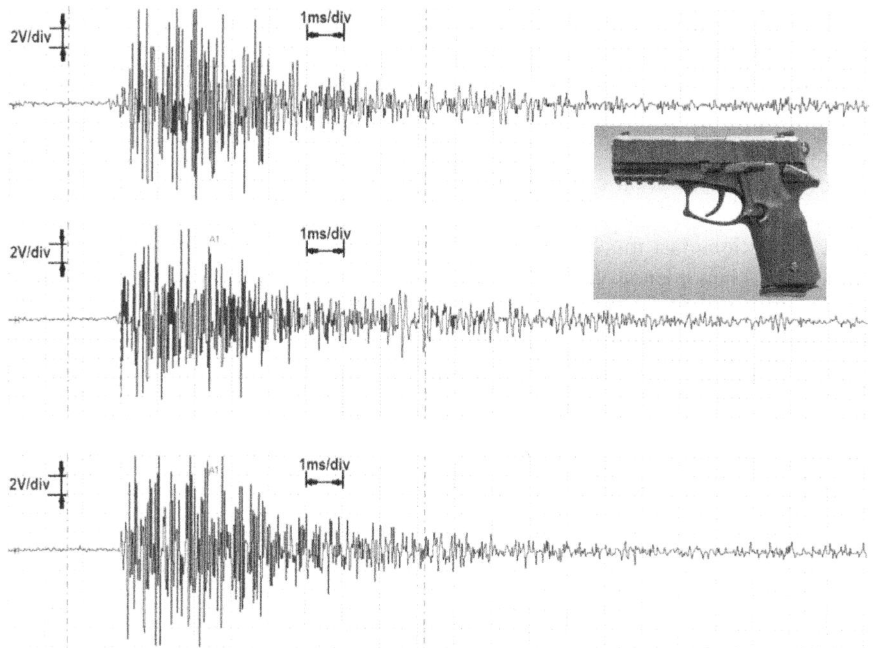

Fig. 4. Reload oscillograms for the gun PR-15 registered with DSO-2902 oscilloscope (photo from [42]).

characteristics. Futhermore, some technological issues may have impact on the acoustic emission. For example, the firearms manufactured out of a metal piece with milling technologies, as well as molded guns expose higher stability of the main components. These manufacturing technologies usually leave non-uniform thickness of the details' walls, so the emitted acoustic signal is characterized by wider spectrum and smaller energy than that of the guns manufactured by stamping technology out of the flat sheet metal.

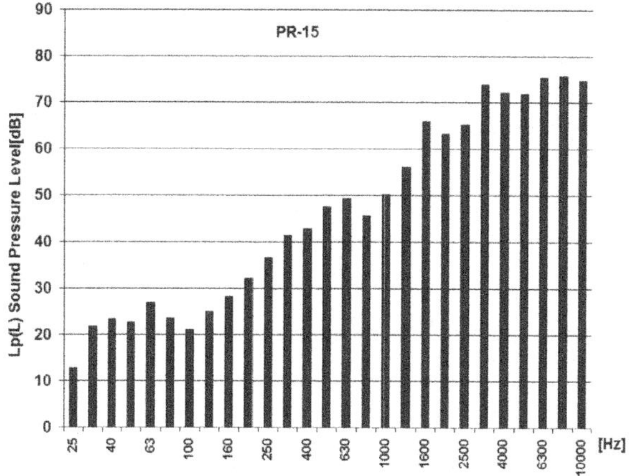

Fig. 5. Spectral analysis in the 1/3 octave band of sound pressure $L_p(L)$ for PR-15 reloading.

This observation is very important for further investigations and for the possible database of "gun signatures" necessary for weapon identification preventive systems. Irrespective on the distance from microphones, periodical oscillations remain the same for each gun type, the distance may only weaken signals and decrease the amplitudes of registered vibrations.

The same observation can be made concerning the gunshot. Example of the gunshot oscillogram for the same gun PR-15 is shown in Fig. 6. It should be noted that the sound oscillations last much longer than those after reloading, as it was presented in Fig. 4. This phenomenon can be explained referring to the experimental conditions, namely, the shooting tests were performed in the closed room (see Fig. 3). The reflections unavoidably were registered despite the numerous attempts to damp the echoes. As a result, the sound is still registered long after the gunshot took place.

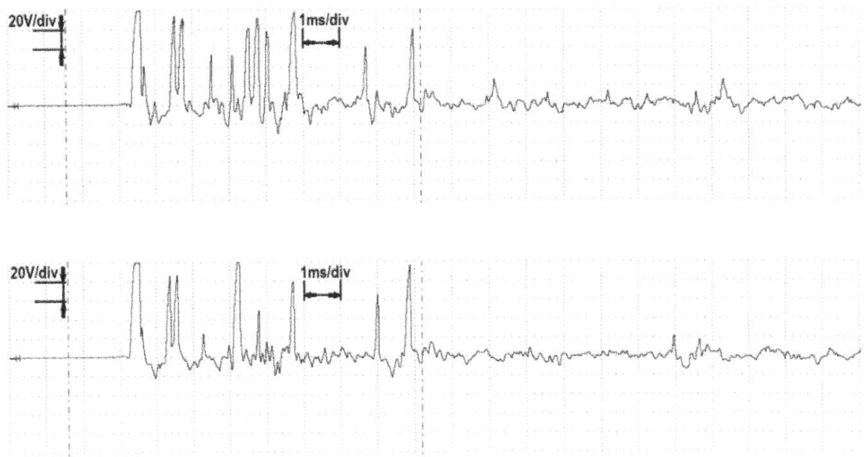

Fig. 6. Gunshot oscillograms for the gun PR-15.

In the oscillogram (Fig. 6), the attention must be paid to the first three peaks. The first one, very high and outstanding, corresponds with the very gunshot. The second two closely neighboring peaks resembling the letter M, are typical characteristics of automatic reload of the firearm. All the subsequent peaks that appear after ca. 3 ms later than the gunshot sound, are the echoes bouncing from the wall behind the sound source.

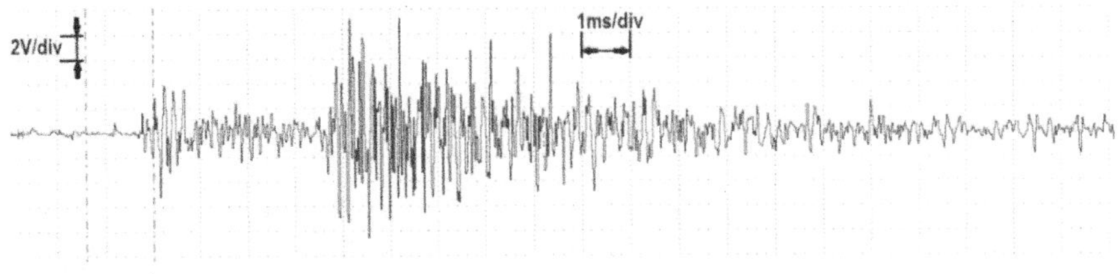

Fig. 7. MSBS GROT rifle reloading sound oscillogram.

In order to demonstrate a typical acoustic "signature" of reloading, Fig. 7 presents the oscillogram of reloading sound of other gun, MSBS GROT rifle. Its spectral analysis is seen in Fig. 8, where also photo of the gun is added.

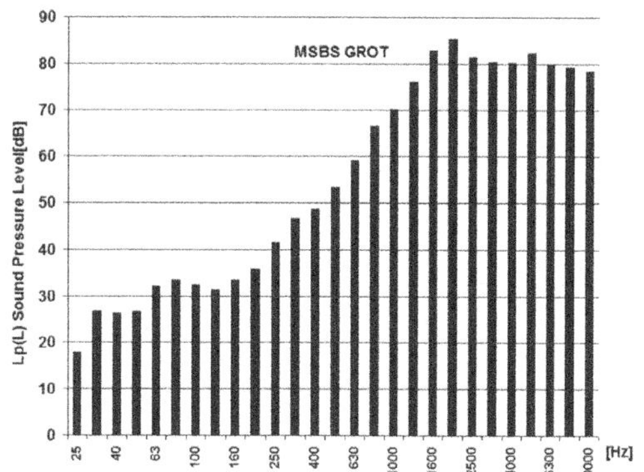

Fig. 8. Photo of the MSBS GROT rifle (left) [42] and spectral analysis in the 1/3 octave band of sound pressure $L_p(L)$ for its reloading (right).

Even without deepened analysis, obvious differences in oscillograms can be assigned to the larger dimensions of the rifle. The Table 1 contains the most important measured values, such as peak values of maximal sound pressure $L_{pk}(L)$ and equivalent continuous level of sound pressure $L_{eq}(L)$. Additionally, the peak value $L_{pk}(A)$ was added, obtained after typical correction with A-frequency-weighting.

Initial analysis of the obtained results demonstrated that the methods of acoustic emission measurement based on such parameters as Sound Pressure Level, Sound Power and Sound Intensity are not helpful in case of systems for identification of possible firearms usage. The main limitation is caused by the fact that the measurement results for abovementioned parameters are highly dependent on the distance of sound source from the registering device. However, the results of the measurements confirmed the recognizability of the acoustic "signature" of different types of firearms. This way, in case of criminal or terrorist usage of a gun, registered sound may help investigators to focus on the specific weapon type.

Table 1. Sound field parameters measured during tests.

Measured parameter => Type of firearm ⇓	$L_{eq}(L)$ (Equivalent continuous level of sound pressure)	$L_{pk}(L)$ Maximum pressure (Peak value)	$L_{pk}(A)$ Maximum pressure (Peak value)
PR-15 - reload	56.7 dB	66.1 dB	68.1 dB
PR-15 - shot	102.7 dB	132.5 dB	132.0 dB
MSBS GROT - reload	63.0 dB	74.6 dB	71.1 dB
MSBS GROT - shot	112.4 dB	133.8 dB	133.8 dB

Based on numerous measurement repetitions, it can be stated that the new correctly produced weapon of various types produced repeatable recognizable sound pattern which differs significantly from the one emitted by the defective or worn gun of the same type. This observation is very important for the early diagnostics of defects on any stage of exploitation from the very first upload to the very last shot.

In this respect, further researches are directed both to laboratory investigations and field applications. In the scientific analysis, it is especially important to determine the temperature effect on the registered sound signal. In the field application, where the gunfire should be recognized from a single sound record, it is important to work out the most effective recognition algorithm.

5   Conclusions

Presented initial analysis of the application of acoustic emission analysis for preventive identification of possible threat based on the reloading sound led to the following conclusions:

(1) technological features of a firearm, such as dimensions, constructional features (shape and thickness of the details), mechanism principles (sequence of contact interactions between elements inside the gun during reload) and material acoustic emission characteristics, determine an unique acoustic spectrum which can be identified as an individual "signature" of each type of gun both during reload and during the gunshot,
(2) novel measurement system based on two microphones and parallel processing of two registered acoustic signals proved to be highly effective,
(3) proposed novel analysis of the "acoustic signature" of the reloading sound is very helpful in the preventive identification of possible gunfire some time before it happens, instead of gunshot recognition after the tragedy had taken place,
(4) there are some limitations of the application of measurement systems available on the market to the firearm identification, that can be mostly overcame through appropriate choice of the measurement method and system configuration.

Acknowledgments. The research works had been performed due to financial support by the Ministry of Science and Higher Education in Poland, research funds No. 3270/22/P on the subject "Sound emission of the machines and devices." Authors express their gratitude to the Fabryka Broni "Łucznik" – Radom sp. z o.o. for enabling the measurements.

References

1. Guida HL, Diniz T, Kinoshita SK (2011) Acoustic and psychoacoustic analysis of the noise produced by the police force firearms. Braz J Otorhinolaryngol 77(2):163–170
2. Guida HL, Taxini CL, de Oliveira Gonçalves CG, Valenti VE (2014) Evaluation of hearing protection used by police officers in the shooting range. Braz J Otorhinolaryngol 80(6): 515–521

3. Luzi L, Gonzalez E, Bruillard P et al (2016) Acoustic firearm discharge detection and classification in an enclosed environment. J Acoust Soc Am 139(5):2723–2731
4. Peterson S, Schomer P (1994) Acoustic Analysis of Small Arms Fire, USACERL Technical Report EC-94/06. http://www.dtic.mil/dtic/tr/fulltext/u2/a278306.pdf
5. Libal U, Spyra K (2014) Wavelet based shock wave and muzzle blast classification for different supersonic projectiles. Expert Syst Appl 41(11):5097–5104
6. https://www.pcb.com/ContentStore/mktg/LD_Brochures/LD_LxTQPR_Firearm_Test_lowres.pdf. Accessed 29 Oct 2018
7. Akman Ç, Sönmez T, Özuğur Ö, Başlı AB, Leblebicioğlu MK (2018) Sensor fusion, sensitivity analysis and calibration in shooter localization systems. Sens Actuators, A 271:66–75
8. Pathrose N, Nair KR, Murali R, Rajesh KR, Mathew N, Vishnu S (2016) Analysis of acoustic signatures of small firearms for gun shot localization. In: Proceedings of 2016 IEEE annual india conference (INDICON), Bangalore, India, 16–18 Dec. 2016
9. Ramos A (2015) On acoustic gunshot localization systems. In: Proceedings of the 20th international conference of society for design and process science SDPS-2015, Fort Worth, TX, USA, pp 558–565, November 2015
10. Hrabina M, Sigmund M (2015) Acoustical detection of gunshots. In: Proceedings of the 25th international conference radioelektronika, Pardubice, Czech Republik, pp 150–153, 21–22 April 2015
11. Choi K, Librett M, Collins TJ (2014) An empirical evaluation: gunshot detection system and its effectiveness on police practices. Police Pract Res 15(1):48–61
12. R&D Conference Staff, Sensor Can Instantly Detect Gunshots, ID Weapons During School Shootings. https://www.rd100conference.com/news/item/130/sensor-can-instantly-detect-gunshots-id-weapons-during-school-shootings/
13. Scala CM, Coyle RA (1983) Pattern recognition and acoustic emission. NDT Int 16(6):339–343
14. Sause MGR, Gribov A, Unwin AR, Horn S (2012) Pattern recognition approach to identify natural clusters of acoustic emission signals. Pattern Recognit Lett 33(1):17–23
15. Li S, Song Y, Zhou G (2018) Leak detection of water distribution pipeline subject to failure of socket joint based on acoustic emission and pattern recognition. Measurement 115:39–44
16. Li Z, Zhang H, Tan D, Chen X, Lei H (2017) A novel acoustic emission detection module for leakage recognition in a gas pipeline valve. Process Saf Environ Prot 105:32–40
17. Zhu S, Li Z, Zhang S, Liang L, Zhang H (2018) Natural gas pipeline valve leakage rate estimation via factor and cluster analysis of acoustic emissions. Measurement 125:48–55
18. Hase A, Mishina H (2018) Identification and evaluation of wear phenomena under electric current by using an acoustic emission technique. Tribol Int 127:372–378
19. Saeidi F, Shevchik SA, Wasmer K (2016) Automatic detection of scuffing using acoustic emission. Tribol Int 94:112–117
20. Qin M, Wang K, Pan K, Sun T, Liu Z (2018) Analysis of signal characteristics from rock drilling based on vibration and acoustic sensor approaches. Appl Acoust 140:275–282
21. Liu J, Hu Y, Wu B, Wang Y (2018) An improved fault diagnosis approach for FDM process with acoustic emission. J Manuf Process 35:570–579
22. Albers A, Stürmlinger T, Wantzen K (2017) Prediction of the product quality of turned parts by real-time acoustic emission indicators. Procedia CIRP 63:348–353
23. Arun A, Rameshkumar K, Unnikrishnan D, Sumesh A (2018) Tool condition monitoring of cylindrical grinding process using acoustic emission sensor. Mater Today Proc 5(5):11888–11899, Part 2
24. Segreto T, Karam S, Teti R, Ramsing J (2015) Feature extraction and pattern recognition in acoustic emission monitoring of robot assisted polishing. Procedia CIRP 28:22–27

25. Dahmene F, Yaacoubi S, El-Mountassir M (2015) Acoustic emission of composites structures: story, success, and challenges. Phys Procedia 70:599–603
26. Behnia A, Chai HK, GhasemiGol M, Sepehrinezhad A, Mousa AA (2018) Advanced damage detection technique by integration of unsupervised clustering into acoustic emission. Engineering Fracture Mechanics, corrected proof, 5 July 2018, In press. https://doi.org/10.1016/j.engfracmech.2018.07.005
27. Bashir I, Walsh J, Thies PR et al (2017) Underwater acoustic emission monitoring – experimental investigations and acoustic signature recognition of synthetic mooring ropes. Appl Acoust 121:95–103
28. Bohmann T, Schlamp M, Ehrlich I (2018) Acoustic emission of material damages in glass fibre-reinforced plastics. Compos Part B: Eng 155:444–451
29. Goszczyńska B, Świt G, Trąmpczyński W (2016) Application of the IADP acoustic emission method to automatic control of traffic on reinforced concrete bridges to ensure their safe operation. Arch Civ Mech Eng 16(4):867–875
30. Li J, Deng L, Haeb-Umbach R, Gong Y (2016) Robust automatic speech recognition. Elsevier, Waltham
31. Squartini S, Principi E, Rotili R, Piazza F (2012) Environmental robust speech and speaker recognition through multi-channel histogram equalization. Neurocomputing 78(1):111–120
32. Shahmoradi S, Shouraki SB (2018) Evaluation of a novel fuzzy sequential pattern recognition tool (fuzzy elastic matching machine) and its applications in speech and handwriting recognition. Appl Soft Comput 62:315–327
33. Heracleous P, Even J, Sugaya F, Hashimoto M, Yoneyama A (2018) Exploiting alternative acoustic sensors for improved noise robustness in speech communication. Pattern Recognit Lett 112:191–197
34. Le Prell CG, Clavier OH (2017) Effects of noise on speech recognition: challenges for communication by service members. Hear Res 349:76–89
35. Maier A, Haderlein T, Eysholdt U et al (2009) PEAKS – a system for the automatic evaluation of voice and speech disorders. Speech Commun 51(5):425–437
36. Hemmerling D, Skalski A, Gajda J (2016) Voice data mining for laryngeal pathology assessment. Comput Biol Med 69:270–276
37. Travieso CM, Alonso JB, Orozco-Arroyave JR (2017) Detection of different voice diseases based on the nonlinear characterization of speech signals. Expert Syst Appl 82:184–195
38. Zheng W, Wu Ch (2016) An information processing method for acoustic emission signal inspired from musical staff. Mech Syst Sig Process 66–67:388–398
39. (1988) Measurement of sounds. Bruel&Kjaer, Naerum. (in Polish)
40. Dehaene S (2003) The neural basis of the Weber-Fechner law: a logarithmic mental number line. TRENDS Cognit Sci 7(4):145–147
41. Chałko L, Maciąg P (2018) Firearms identification based on acoustic signals. Sci Lett Rzeszow Univ Technol 35(298):261–273 (in Polish)
42. Fabryka Broni "Łucznik" – Radom sp. z o.o. www.fabrykabroni.pl. Accessed 29 Oct 2018

Automated System for Workpiece Leveling on a Machine Tool

Marcin Pelic, Tomasz Bartkowiak$^{(\boxtimes)}$, and Andrzej Gessner

Poznan University of Technology, Poznań, Poland
tomasz.bartkowiak@put.poznan.pl

Abstract. The paper presents a concept, a mathematical model and a control algorithm of an automated system for workpiece leveling. In the proposed solution, a workpiece is placed on a set of electrically-powered variable-length supports. The location of each support is not known before the leveling. A current orientation of workpiece is measured by a precise biaxial inclinometer. The leveling is performed in three stages which involve subsequent extensions of supports and measuring the resulting orientation. This system allows determining the location of each support and the required displacements of each support that are necessary for leveling the workpiece. The system is controlled by an industrial PLC. A single power source and motion controller as well as a distributed demultiplexer, which activates each support, are used to reduce the cost. The supports were manufactured and tested for accuracy and repeatability according to ISO 230-2:2014. The proposed solution can be used for layout and for leveling workpieces on a machine tool table.

Keywords: Layout · Machining · Levelling · Automation

Nomenclature

$\alpha_n,\ \beta_n$	angles between gravity vector and normal vector of the inclinometer to XZ and YZ plane in its local coordinate system at n-th stage of leveling
$a,\ b$	lengths of support extensions during leveling
$\mathbf{G}$	gravity vector
g	gravitational acceleration constant
$k_0,\ k_1,\ k_2$	normalizing coefficients
$\mathbf{N}$	normal vector of the plane
$P_0,\ P_1,\ P_2$	supports and support contact points

1 Introduction

Manufacturing body components of machine tools usually involves casting and subsequent machining of the cast iron workpieces. The evaluation of cast geometrical accuracy is performed manually during layout in which most important features and dimensions are checked and lines are marked using scriber or other layout tools. This process has not changed significantly since the beginning of the history of machine tools production. It is laborious and prone to human errors [1]. The lack of a substantial

© Springer Nature Switzerland AG 2019
M. Diering et al. (Eds.): *Advances in Manufacturing II* - Volume 5, LNME, pp. 25–36, 2019.
https://doi.org/10.1007/978-3-030-18682-1_3

progress in this field can be explained by the insufficient development of tools to automate it. The potential solution to that problem is a 3D scanning and then the comparison of the measured point cloud with their CAD representations. These include photogrammetry, laser or structural light scanning. Those techniques have been widely used in the industry [2–4] for process control and computer-aided inspection. Despite the undoubted advantages offered by those systems, including the comprehensive comparison between castings and the model [1, 5], there are still barriers that can be hard to overcome by small- and medium-size production enterprises, namely, the price of these optical measuring devices might not be compensated by the potential profits. In addition, not all potential customers need comprehensive and complex solutions for their products.

The accuracy of the layout process largely depends on the accuracy of the cast leveling on the layout table. This activity may be complicated due to the weight (which can range from single kilograms to several tons), the complex shapes of the casts themselves and its stiffness including contact stiffness. What is more, the location of the supports for layout might not be identical as during volume or precise milling. This may cause deformations and displacement of workpiece geometry and scratched lines due to the gravity.

The proper workpiece orientation can also provide the possibility to minimize the volume to be machined [5]. There exist universal fixtures that allow workpiece orientation around one or two axes and they are usually used in case of job production. In batch or mass production fixing systems, dedicated to a single type of parts, are common solutions. A fixing orientation or leveling is often adjusted manually by an operator, based on his practice and experience. If a higher precision is required, scratched mark-out lines are used as a reference.

Once a workpiece is fixed on machine tool table, mutual relations (translation and rotation) between the workpiece and CNC machine tool coordinate systems must be determined. In this operation, usually, a probe mounted on the spindle or other, integrated with machine tool, measurement system is used. The coordinate systems are aligned either by changing the orientation of workpiece, what includes leveling, or rotating machine tool axis (e.g. table or spindle). Then the origin of the coordinate system is translated to the desired location.

Most of the solutions that can be found in the literature require a rigid platform on which a workpiece has to be fixed or treats workpiece as a platform [6–8]. This platform can be oriented using parallel mechanisms [9, 10]. There are endeavors to include workpiece rigidness for very precise leveling [11, 12].

Proper leveling of the workpiece is crucial both in layout and subsequent machining process. In this paper, we present the concept of a system for an automatic leveling of workpieces on a table or in the machine working space. It involves a set of active supports which lengths are adjustable by stepper motors which powers screw/nut mechanisms. The paper describes the mathematical model which is included in the control algorithm. The supports were manufactured and tested for positioning accuracy and repeatability. The presented concept is a subject of patent application [13].

2 Concept

An automated leveling of the workpiece requires a set of actuators which are placed under the leveling object and can be extended or shortened what change its orientation. It is assumed that the workpiece during this process is placed on three supports as it is a minimum number of constraints to reduce the mobility and to change orientation about two axes. In the simplest case, the support marked P0 is a fixed support, while the remaining two P1 and P2 can change their length to some extent. However, there are cases in which the height of the P0 support should be adjusted. Therefore, in the schematic diagram, it was marked as a variable-length support (Fig. 1). The workpiece is to be leveled taking into account one base surface. This surface is usually a reference for other dimensions in layout and is the first feature to be machined. The quality of surface, especially for cast iron parts, might not high, what might be evident in high surface roughness and various defects (blowholes, sand holes, elephant skin, porosity) and mechanical damages [14]. We propose to use the inclinometer which is encapsulated inside a casing. The casing size should be adjusted to the character of surface which is a reference for leveling (Fig. 2). The inclinometer is placed on the workpiece surface which is a base for further marking-out or machining and the workpiece is to be levelled based on that surface.

To perform the leveling, it requires a biaxial precision inclinometer to monitor the deviation of the normal vector from the gravity vector. That sensor is connected to a controller that automatically receives and processes data from the tilt sensor and controls the height of each individual support. A single power source and motion controller as well as a distributed demultiplexer, which activates each support, are used to reduce the cost. A schematic diagram of the flow of signals in the device is shown in Fig. 3. Each support consists of a stepper motor which is powered by drive controller if

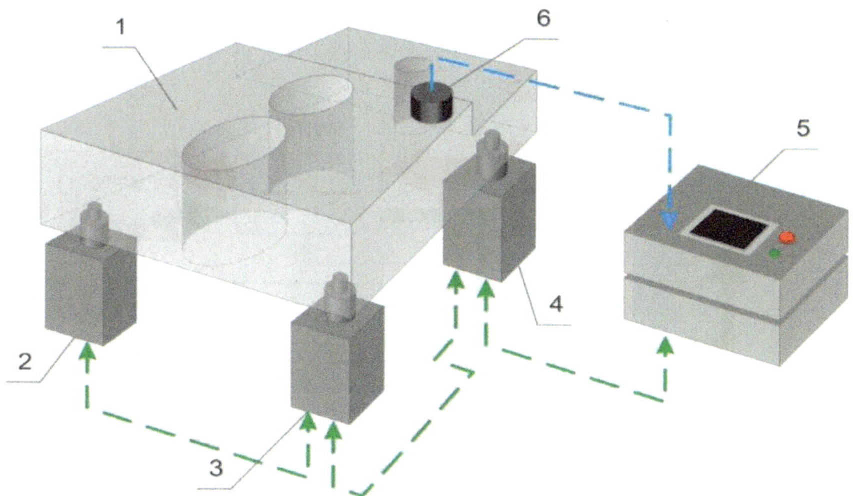

Fig. 1. Schematic diagram of an automated workpiece leveling system: 1 – WORKPIECE, 2 – P0 support, 3 – P1 support, 4 – P2 support, 5 – controller, 6 – inclinometer.

switch is set on. The power of the output shaft is transmitted through the nut/screw mechanism what allows linear displacement of the support shaft. The designed device is intended for leveling, another fixing system is required to remove all degrees of freedom prior to the machining.

Fig. 2. View of the inclinometer sensor inside a casing which is placed on the workpiece surface.

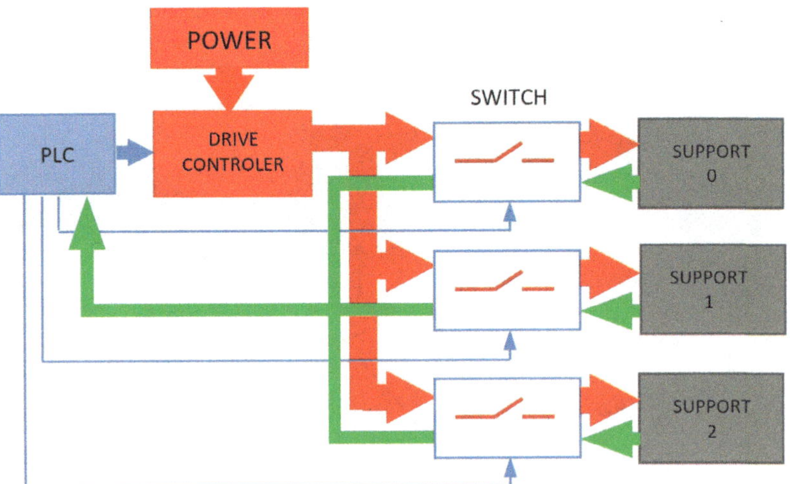

Fig. 3. Schematic diagram of the signal flow between PLC, drive controller and supports.

3   Control Algorithm

The use of the switches not only lowers the cost of the whole system, but it also allows for scaling. Applying the appropriate addressing system for the supports allows practically unlimited development. This might be essential for larger components or more geometrically complex components which require more than three supports. However, in this study, the leveling takes place on three supports, so our main focus was on that number only. Sample configuration of the system for cast part that is placed on three supports is depicted in Fig. 4.

In the first one, controlled supports are set to zero positions and inclinometer orientation is registered. In the designed system, the initial location of all supports is unknown and can be calculated as described in Sect. 4. The idea of system operation is based on the comparison of the angles α and β which describe deviations between the normal vector $\mathbf{N}$, which is related to the inclinometer base surface, and the gravitational vector $\mathbf{G}$ in the inclinometer coordinate system. This is performed in three stages. Then support P_1 is extended by length a and orientation is also registered. Finally, the support P_1 returns to its initial position and support P_2 is displaced by length b what results in different measured orientation. Based on the results of the measurements, the coordinates of the support P_1 and P_2 in respect to P_0 are calculated. What is more, displacement values z_{k1} (for P_1) and z_{k2} (for P_2) for which the normal vector of the inclinometer will be parallel to the gravitational vector are also derived. This is based on the assumption that the system has no information about the position of the supports and the orientation of the inclinometer relative to them, what simplifies the use of the device. The description of the mathematical model that allows the calculation of support location and lengths for leveling are described in next chapter.

Fig. 4. Sample configuration of the leveling systems for cast iron workpiece with three active supports and controller.

The system service is limited to the stable positioning of the castings on the supports and the proper mounting of the inclinometer. We assume that workpiece does not slide during the leveling. All three stages are shown graphically in Fig. 5. The control algorithm can be presented in form of a diagram what is depicted in Fig. 6.

In case any of calculated displacement z_{k1} or z_{k2} take negative values, all three supports (including P_0) have to be change their lengths by offset that is equal to lowest value of pair z_{k1} and z_{k2}.

4 Mathematical Model

Coordinates of contact points between support and workpiece can be formulated for each support: $P_0(x_0, y_0, z_0)$, $P_1(x_1, y_1, z_1)$ and $P_2(x_2, y_2, z_2)$. We assume that point P_0 is the center of coordinate system, what means that: $x_0 = 0$, $y_0 = 0$ and $z_0 = 0$. Normal vector of the plane created from contact points can be obtained from cross product:

$$\mathbf{N} = (P_2 - P_0) \times (P_1 - P_0). \tag{1}$$

The constituents of normal vector for initial stage, where all supports are zeroed are described the following formulas:

$$N_{x0} = k_0(-y_2 z_1 + y_1 z_2), \tag{2}$$

$$N_{y0} = k_0(x_2 z_1 - x_1 z_2), \tag{3}$$

$$N_{z0} = k_0(-x_2 y_1 - x_1 y_2). \tag{4}$$

Similar relations can be derived for next stages:

$$N_{x1} = k_1(-y_2(z_1 + a) + y_1 z_2). \tag{5}$$

$$N_{y1} = k_1(x_2(z_1 + a) + x_1 z_2). \tag{6}$$

$$N_{z1} = k_1(-x_2 y_1 - x_1 y_2). \tag{7}$$

$$N_{x2} = k_2(-y_2 z_1 + y_1(z_2 + b)). \tag{8}$$

$$N_{y2} = k_2(x_2 z_1 + x_1(z_2 + b)). \tag{9}$$

$$N_{z2} = k_2(-x_2 y_1 - x_1 y_2). \tag{10}$$

where: k_0, k_1, k_2 are the normalizing coefficient for inclinometer measurements, N_{xn}, N_{yn}, N_{zn} are constituents of normal vector $\mathbf{N}_n$ for measured surface at n-th stage; x_1, y_1, z_1, x_2, y_2, z_2 constitute coordinates of contact point between workpiece and support P_1

and P_2 accordingly and a, b are the extensions of P_1 and P_2. Components of gravity vector in the inclinometer coordinate systems during each stage of leveling can be described by:

$$g_{xn} = -g \cdot \sin(\alpha_n) \cdot \cos(\beta_n).$$ (11)

$$g_{yn} = -g \cdot \cos(\alpha_n) \cdot \sin(\beta_n).$$ (12)

$$g_{zn} = -g \cdot \cos(\alpha_n) \cdot \cos(\beta_n).$$ (13)

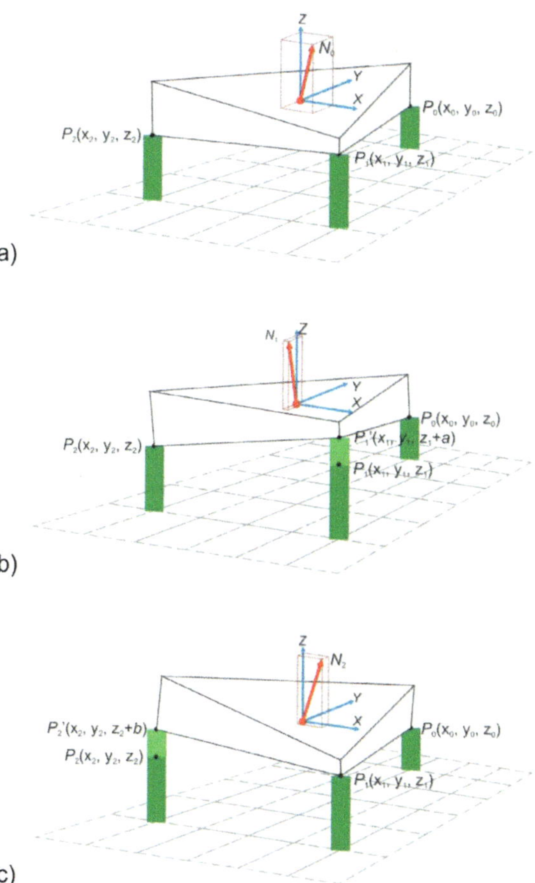

Fig. 5. Stages of leveling: (a) registration of orientation at zero position, (b) extension of P1 support by length a and registration of orientation, (c) zeroing P1 support, extension of P2 support by length b and registration of orientation.

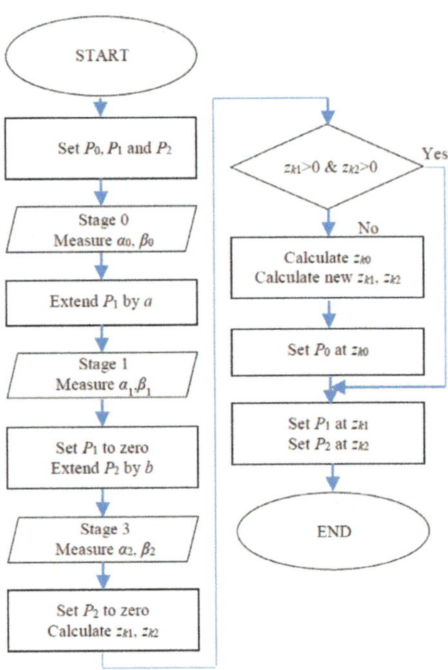

Fig. 6. Control algorithm for the leveling system.

where: g – gravitational acceleration constant; g_{xn}, g_{yn}, g_{zn} – constituents of gravity vector described in inclinometer coordinate system for n-th stage, α_n and β_n – angles between gravity vector and normal vector of the inclinometer to XZ and YZ plane in its local coordinate system at n-th stage of leveling.

In order to determine coordinates of P_1 and P_2, it requires to solve the following set of equation where x_1, y_1, z_1, x_2, y_2, z_2, k_1, k_2 and k_3 are the unknowns:

$$\begin{cases} N_{x0} = g_{x0} \\ N_{y0} = g_{y0} \\ N_{z0} = g_{z0} \\ N_{x1} = g_{x1} \\ N_{y1} = g_{y1} \\ N_{z1} = g_{z1} \\ N_{x2} = g_{x2} \\ N_{y2} = g_{y2} \\ N_{z2} = g_{z2} \end{cases} \tag{14}$$

The required extensions z_{k1} and z_{k2} for supports P_1 and P_2 for which normal of vector of inclinometer becomes parallel to gravity vector can expressed by:

$$z_{k1} = -z_1.$$

(15)

$$z_{k2} = -z_2.$$

(16)

5 Accuracy and Repeatability Testing

Due to the work condition, in which the load force is always directed axially downwards, the positioning accuracy test was carried out unidirectionally in accordance with ISO 230-2:2014 (E) "Test code for machine tools Part 2: Determination of accuracy and repeatability of positioning of numerically controlled axes". The study concerned three supports which are 235 mm high and they can carry maximal 3 kN load. The linear motion was achieved by trapezoidal screw TR16 with a pitch of 2 mm. Torque generated by stepper motor is directly transmitted to the nut which is a part of rotor. The device is not equipped with any measuring system, but only with the limit switch terminating the power at the reference position. The measuring stroke was assumed between the minimum and maximum extension of the pin in relation to its zero position deviated from its lower plane by 235 mm. During the test (to reduce the impact of backlash) the devices were loaded with a force of 1.6 kN.

The test was carried out with a Renishaw XL-80 laser interferometer equipped with an XC-80 temperature compensator. The interferometer worked in the vertical measurement system shown in Fig. 7.

During the test, the support was placed under the load actuator for its initial controlled loading. Seven measurement positions were selected for which displacement measurement was carried out from one (bottom-up) approach direction. For each point, the measurement was repeated five times. The force of 1.6 kN was applied by a pneumatic actuator $\varnothing 100 \times 100$ mm, in which the pressure was controlled by means of a proportional pressure valve connected to the laboratory power supply. Between the actuator and the device there was a force sensor KMB82-K-10kN-0000-D, which was connected to the MD150T panel display to control its value. The stepper motor of the support was connected to the control system. The image of the test stand is shown in Fig. 8.

The tests were carried out for three devices and the extension of their support pins in the range of 0–30 [mm]. The maximum travel of axes is 35 [mm]. This difference results from the software end positions implemented in the control system. This software feature moves the range of work of individual devices so that their reference position is at the set height, what is supposed to eliminate geometrical errors in their assembly. Positioning accuracy tests were carried out unidirectionally for 7 selected points. The methodology and test plan are described in ISO 230-2 standard.

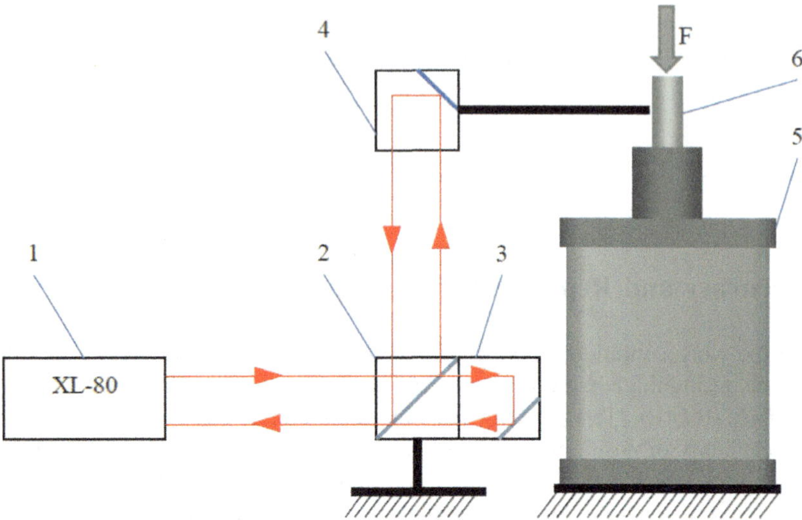

Fig. 7. Schematic diagram of measuring system for positioning accuracy and repeatability of a suport: 1 – light source and detectors, 2 – beam splitter, 3 – reflector, 4 – moveable reflector, 5 – support, 6 – pin.

The test results were aggregated in the form of unidirectional repeatability of positioning $R\uparrow$, unidirectional systematic positional deviation of an axis $E\uparrow$ and unidirectional accuracy of positioning of an axis $A\uparrow$ and are presented in Table 1. More details on the calculation of those parameters is available in ISO 230-2 standard.

Table 1. Measured parameters according to ISO 230-2 standard.

	ID0001	ID0002	ID0003
$R\uparrow$ [µm]	13.5	22.3	13.0
$E\uparrow$ [µm]	69.8	87.3	48.1
$A\uparrow$ [µm]	81.6	98.1	55.1

The lowest unidirectional accuracy of positioning of an axis is observed for ID0002 device and is equal to 81.6 µm, the highest for ID0003 - equal to 55.1 µm. It should be remembered that the drive in these instruments is a trapezoid screw and no compensating algorithm has been implemented. The variability of statistical error is important from the point of view of the device usage. In all three cases, the absolute value of this error increases with the extension value. This means that it will be most advantageous to use these devices at small extensions, and in the future to implement a compensating algorithm.

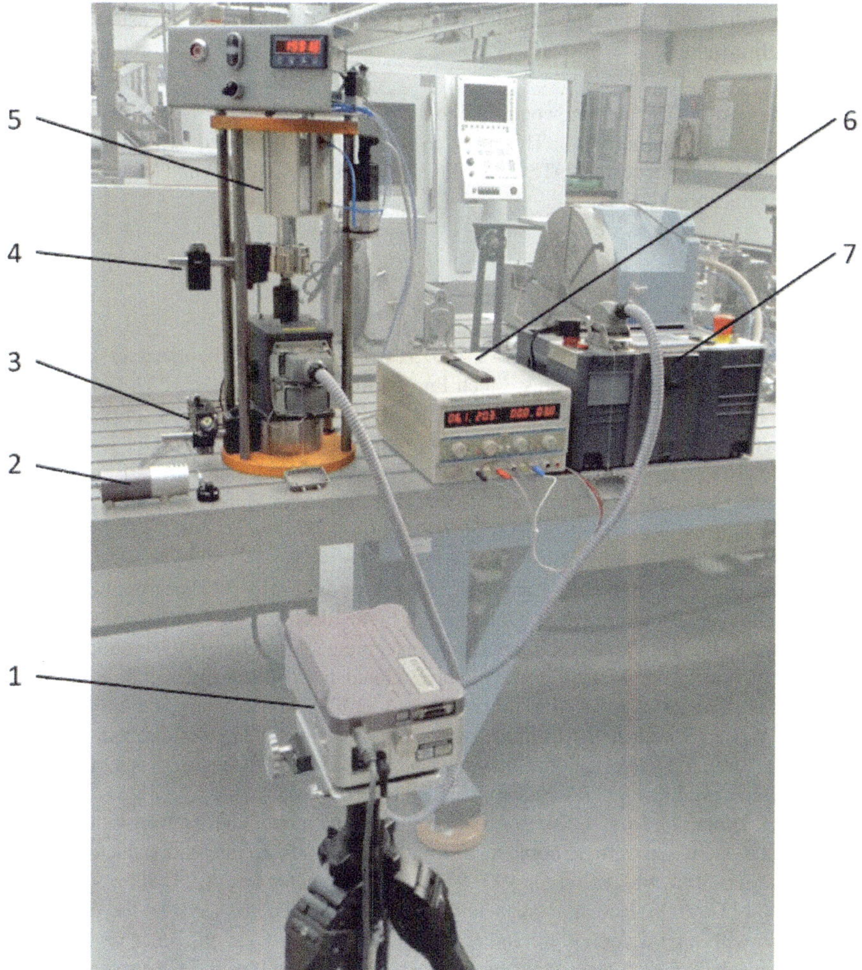

Fig. 8. Test stand for measuring accuracy and repeatability of supports under load: 1 – XL-80 laser interferometer, 2 – XC-80 compensator, 3 – beam splitter and static reflector, 4 – moveable reflector, 5 – pneumatic actuator with force meter, 6 – laboratory DC power supply, 7 – support control system.

6 Summary

In this work, we presented the concept of the automated workpiece leveling system. The main advantage is that it requires precise support positioning before the leveling. The number of supports can be easily increased thanks to modular control system.

We believe that this application will improve the efficiency of layout process and, more generally, workpiece leveling on the machine tool and on layout tables. This, in consequence, should contribute to higher precision, time and cost reduction. The system has been successfully implemented in DMG Mori manufacturing center. There is still an ongoing research into the repeatability of support positioning and measurement accuracy of inclinometer.

Acknowledgments. The authors would like to thank Waldemar Adam from DMG Mori for his thoughtful comments on our work. This research was partially supported by National Centre for Research and Development (INNOTECH K3/IN3/15/226458/NCBR/14) and Ministry of Science and Higher Education (DSPB).

References

1. Gessner A, Staniek R, Bartkowiak T (2015) Computer-aided alignment of castings and machining optimization. Proc Inst Mech Eng Part C: J Mech Eng Sci 229(3):485–492
2. Hsieh TH, Jywe WY, Huang HL, Chen SL (2011) Development of a laser-based measurement system for evaluation of the scraping workpiece quality. Opt Lasers Eng 49 (8):1045–1053
3. Malesa M, Malowany K, Tomczak U, Siwek B, Kujawińska M, Siemińska-Lewandowsk M (2013) Application of 3D digital image correlation in maintenance and process control in industry. Comput Ind 64(9):1301–1315
4. Ryu WJ, Kang YJ, Baik SH, Kang SJ (2008) A study on the 3-D measurement by using digital projection moiré method. Optik - Int J Light Electron Opt 119(10):453–458
5. Gessner A, Staniek R (2013) Optimizing machining of machine tool casting bodies by means of optical scanning. In: ASME international mechanical engineering congress and exposition, vol 2A. Advanced Manufacturing, V02AT02A061
6. Zhang Y, Yao S, Ren Z (2010) Study on leveling mechanism and control method of a rigid platform. In: 2010 international conference on measuring technology and mechatronics automation, pp 600–603. IEEE Computer Society, Changsha
7. Zhang J, Huang D, Lu C (2007) Research on dynamic model and control strategy of auto-leveling system for vehicle-borne platform. In: 2007 IEEE international conference on mechatronics and automation, pp 973–977. IEEE Computer Society, Harbin
8. Yang H, Li G (2006) Research on an automatically leveling control system for vehicle-borne platform with high accuracy. In: 2nd IEEE/ASME international conference on mechatronic and embedded systems and applications, pp 1–5. IEEE Intelligent Transportation Systems Society, Beijing
9. Chen DH, Xiao F, Zhu DC, Gu QH (2011) Analyse of the automatic leveling system of parallel support mechanism based on screw theory. Adv Mater Res 211–212:1077–1081
10. Myszkowski A, Bartkowiak T, Gessner A (2015) Kinematics of a novel type positioning table for cast alignment on machine tool. In: ASME international design engineering technical conferences and computers and information in engineering conference, volume 5C: 39th mechanisms and robotics conference, V05CT08A032
11. Liu Y, Fang S, Otsubo H, Sumida T (2013) Simulation and research on the automatic leveling of a precision stage. Comput Aided Des 45(3):717–722
12. Fang S, Liu Y, Otsubo H et al (2012) An automatic leveling method for the stage of precision machining center. Int J Adv Manuf Technol 61:303–309
13. Patent: P.228721 System for automated levelling of casts ["Układ do automatycznego poziomowania odlewu"]
14. Kambayashi H, Kurokawa Y, Ota H, Hoshiyama Y, Miyake H (2007) Evaluation with surface analysis equipment, of casting defects in cast iron articles (review). Mater Sci Forum, 539–543. (PART 2), pp 1110–1115

Technological Assurance of Machining Accuracy of Crankshaft

Alexeẏ Kotliar, Yevheniia Basova[✉], Maryna Ivanova[✉],
Magomediemin Gasanov, and Ivan Sazhniev

National Technical University "Kharkiv Polytechnic Institute",
Kharkiv, Ukraine
e.v.basova.khpi@gmail.com, ivanovamarynal@gmail.com

Abstract. The typical technological processes of manufacturing crankshafts are considered. The main directions for technological assurance of the accuracy and quality of machining these parts are given. For a compensation of an influence of cutting force on a quality and an accuracy of manufacturing crankshafts, the design of the following steady rest was proposed. The studies of a dependence of the total cutting force for grinding wheels with different grit from ultimate strength of the material, width of main journal and infeed speed were made. The design calculations of the spring were performed and the value of the pressing force developed by spring was obtained, which is capable to level the influence of the cutting force on the deformation processes during crankshaft machining. The elastic deformations which occur when grinding the crankshaft main journal with and without the proposed steady rest were estimated by simulation modeling with finite elements method. The values of pressing forces, which are necessary to compensate the influence of the total cutting force on the shape accuracy of shaft, were obtained.

Keywords: Crankshaft · Main journals · Following steady rest · Variable rigidity · Cutting forces

1 Introduction

Modern mechanical engineering is characterized by the intensification of technological processes for manufacturing of critical parts of machines, to the quality of which the high requirements are made.

It is known that crankshaft constructively and technologically is complex part. Moreover it are considered ones of the most critical and stressful part of internal combustion engine, as it affects on reliability of the assembly and of its structure in whole. The operating conditions of the crankshafts and associated with them parts require precise dimensions and the correct relative positions of the individual elements.

From the analysis of the manufacturing techniques of such parts, it was established that crankshafts are made from carbon, chromium-manganese, chromium-nickel-molybdenum steels, and also from special high-strength cast irons. Workpieces for medium-sized steel crankshaft are made by forging in closed dies on hammers or presses. It is worth noting that due to the high requirements to mechanical strength of

© Springer Nature Switzerland AG 2019
M. Diering et al. (Eds.): *Advances in Manufacturing II* - Volume 5, LNME, pp. 37–51, 2019.
https://doi.org/10.1007/978-3-030-18682-1_4

the crankshaft, a fibers arrangement of the workpiece material has a great importance in order to avoid their cutting during subsequent machining. In addition, the complexity of the geometrical and constructional forms of the crankshaft, its lack of rigidity, high requirements to an accuracy of machining surfaces cause special demands of choice of workpiece locating methods, fixturing methods and processing methods, as well as the sequence and combination of operations and the choice of metal-cutting equipment, machining attachments and tooling. In this case, the main bases of the crankshaft, as a rule, are the bearing surfaces of the main journals. In some cases, surfaces of the center holes are chosen as technological bases. In some operations, when processing the crankshaft in the centers due to its relatively low stiffness, the outer surfaces of the pre-treated necks are used as additional technological bases. Main journals, which are later used as technological bases for machining of crankpins and other surfaces, can be operated on conventional lathes. However, since the crankshaft is not enough rigid part and it tends to bend and twist under the action of cutting forces during processing, especially when one-sidedly drive lathes, specialized machine tools are used to process the main journals of the multithrow crankshaft. The peculiarity of such equipment is in the presence of a central or two-way drive, whose task is to reduce bending and twisting moments. To simplify the designs of such machine tools or utilize standard equipment, various kinds of devices and steady rests are used to eliminate significant bends and twisting of crankshafts. Their task is to eliminate the elastic deformations caused by cutting forces, taking into account the low and variable rigidity of the part.

From the analysis of existing structures such devices, it was found that the task of developing following steady rest for efficiently machining parts with variable rigidity, which depends on its angle of rotation, is promising.

2 Literature Review

From the analysis of recent researches it has been established that much attention is paid to the issues of ensuring the qualitative characteristics of the crankshaft and its further reliable utilizing [1–10].

For example, the authors of work [2] have presented a study by static simulation of a crankshaft and a single cylinder 4-stroke diesel engine. The analysis was accomplished for finding critical location in crankshaft.

Authors of research [3] have conducted a dynamic simulation on a crankshaft for a single cylinder four stroke camless Engine. Finite element analysis was performed in research to obtain the variation of stress magnitude at critical locations. The analysis was done for different engine speeds and as a result critical engine speed and critical region on the crankshaft were obtained.

But it is worth mentioning that works [1–5] were based on the fact that the crankshaft has ideal operating characteristics.

Author of paper [6], taking into account the qualitative characteristics of the crankshaft, have presented the results of the analysis of how sea waves affect the angular speed of a propulsion. In the article, the method of assessing the state of the ship's engine was considered. Implementation of the method is possible, including with the appropriate qualitative characteristics of the crankshaft.

In article [7] describes the details of the method designed to determine parameters of vehicle's internal combustion engine with compression ignition during road tests. The method requires simultaneous measurements for the crankshaft rotation frequency, fuel pressure in the injector, pressure in the combustion chamber, air pressure and temperature in the intake system [7]. It is not possible to obtain an adequate mathematical model using this method if the crankshaft does not meet the criteria for quality and accuracy.

In the article [8] authors have considered the crankshaft (taking into account the fact that the part corresponds to high quality and accuracy characteristics) and presented the model of the crankshaft taking into account the coupling of bending and torsional vibrations. This allowed the authors present two approaches to modeling torsional and bending vibrations in crank systems and indicate the application area of the proposed methodology.

In addition, it is worth to notice that an assurance of production productivity of crankshafts is also of interest [11, 12]. This requires a search for new technological solutions to optimize the manufacturing process of such parts [13–15].

Authors of paper [16] have attempted to focus on to study the various methods preferred for designing and optimizing the crankshaft for safer working under various boundary conditions by various researchers.

However, it should be remembered that the technology of machining the crankshaft is a complex task, the solution of which requires taking into account of its design complexity and low rigidity, the presence of significant and not uniform deformation under the action of cutting forces and, at the same time, the need to ensure the required accuracy.

Development and application of various auxiliary devices for complex parts manufacturing allow obtaining the required quality characteristics of the products [17–20].

The flexible fixture characterized by high level of flexibility was proposed for CNC multiaxis machining operation [21]. Experimental research of it is presented in paper [22], which provides sufficient accessibility and allows to perform multiaxis machining of parts.

However, current works aimed at optimizing the technological process of manufacturing a crankshaft on grinding operations by the development of auxiliary following compensating devices could weren't found.

3 Research Methodology

3.1 Mathematical Substantiation of the Mechanical Component of the Design of the Following Steady Rest

A decision to develop an intermediate following support for grinding especially the middle part of the crankshaft was made. The purpose of the support is to be able to compensate for the lack of rigidity along the length of the shaft and uneven rotation angle.

It should be noted that the development of a following steady rest for the efficient processing of crankshafts requires an analysis of technological conditions when the detail is been manufacturing. From the point of view of the analysis of the technological process of crankshafts manufacturing, it is known that grinding operations of main and crankpin journals are the most responsible operations for manufacturing such parts. When performing these operations under the action of cutting forces, elastic deformations of the elements of the technological system and mainly workpiece are occurred, which lead to the appearance of form errors of the processed crankshaft's journals. It follows from this that the most important parameters to ensure the accuracy and quality of crankshafts are the machining conditions and cutting forces those influence on the part during the manufacturing process.

Estimation of the value of the tangential component of the cutting force P_z for cylindrical plunge grinding is possible by the formula (1). This formula is based on experimental studies of mentioned process, performed by professor Stepanov M.S. and Khodakov L.V. in the machine tool laboratory of the Kharkiv machine-tool building plant [23].

$$P_z = 2,254 \frac{\sigma_t^{0,342} \cdot H^{0,258} \cdot V_p^{0,945}}{z^{0.051} \cdot S_p^{0.073} \cdot S_{axial}^{0.073} \cdot t_{axial}^{0.026}} \cdot B, [N] \tag{1}$$

where σ_t – the ultimate strength of the workpiece material at high temperatures (600 °C), kgf/mm^2; H – sonic index of grinding wheel; z – grit of grinding wheel; V_p – infeed speed, mm/min; S_p – peripheral speed of workpiece rotation, m/min; S_{axial} – longitudinal speed of grinding wheel dressing, mm/min; t_{axial} – dressing depth, mm; B – grinding width, mm.

In the work [23] the researchers noted that the radial component P_y of the cutting force when peripheral speed of workpiece rotation is in a range 30 … 70 m/min is in 2.2 times bigger than P_z.

Then the total cutting force can be determined by the formula (2):

$$P = \sqrt{P_z^2 + P_y^2} = \sqrt{P_z^2 + (2.2 \cdot P_z)^2} \approx 2.417 \cdot P_z, [N] \tag{2}$$

Presumably this total cutting force can cause an elastic deflection of a part and further loss of quality.

3.2 Mathematical Substantiation of the Elements of the Following Steady Rest for Compensation of an Elastic Deflection

For a compensation of an influence of cutting force on a quality and an accuracy of manufacturing crankshafts, the design of the following steady rest was proposed [24]. One of the main and distinguishing features of the proposed device is a profile cam (Fig. 1). The geometrical profile of the cam is determined from a polar diagram of the deformation of cross-section of a main journal of crankshaft, depending on the applied total cutting force when grinding [24] (Fig. 2).

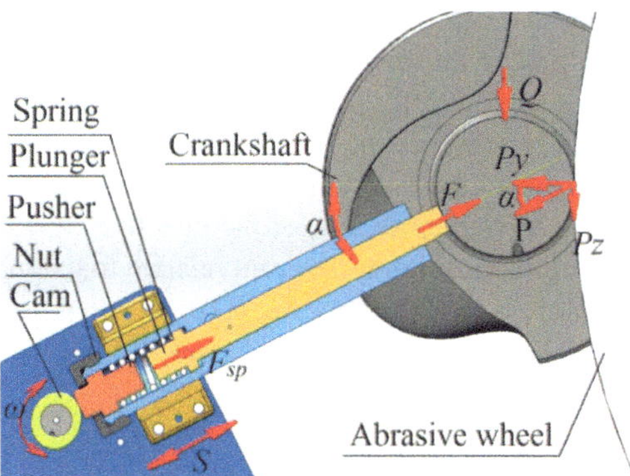

Fig. 1. Scheme of the work of the following steady rest when grinding main journal of crankshaft: P_y, P_z – cutting force components; P - total cutting force; Q - crankshaft weight; F - balancing force; S - plunger movement; F_{sp} – initial spring strain force; ω - angular speed of the cam rotation; α – angle between the closing cutting force P and the horizontal plane.

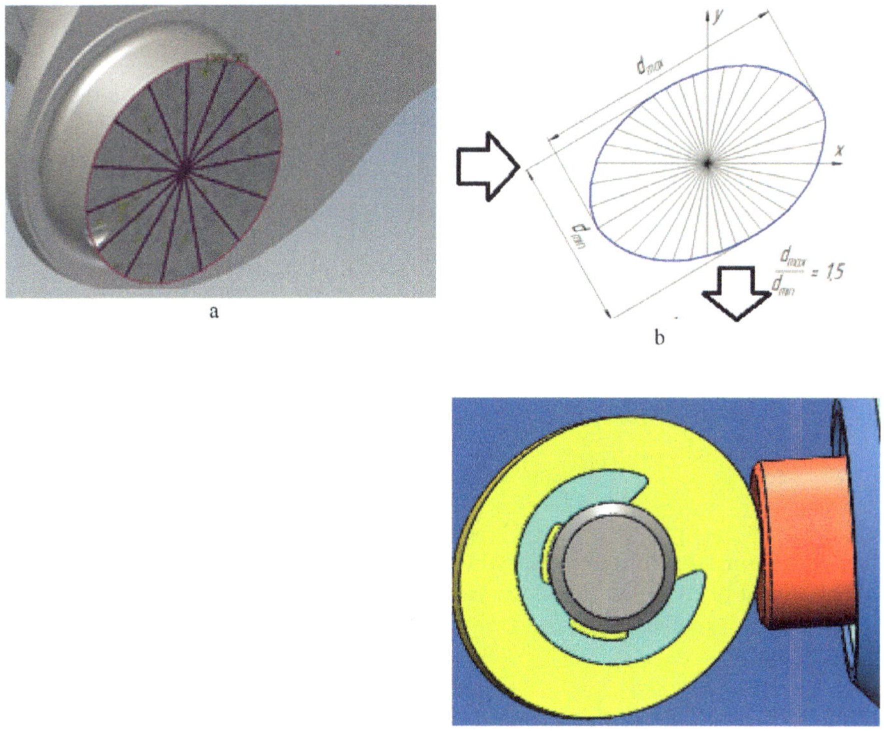

Fig. 2. A principle of the profile cam plotting: a – main journal cross-section; b – a polar diagram of the deformation of a cross-section of a crankshaft main journal; c – a profile cam included in the following steady rest construction.

The cam is actuated by electric motor and rotates moving the pusher, which moves the plunger with a carbide plate through the spring (See Fig. 1). The plunger applies a compensation force F to the crankshaft journal, thereby reducing its elastic deflection. A return of the plunger to its original position, as well as the constant contact of the pusher with the cam is provided by a spring. Spring stiffness and initial pressing force are adjusted by nut. An essential condition for the effective work of the steady rest is the determination of the initial position of the cam relative to the main journal and their subsequent synchronous rotation with the same rotation speed.

The angle between the total cutting force P and the horizontal plane α can be determined from the formula (3)

$$\alpha = \arctan\frac{P_z}{P_y} \tag{3}$$

An important element of the device design to ensure its adequate work is a spring. Therefore, an important stage in the design of the following steady rest is determination of the geometric dimensions and stiffness of the spring.

The pressing force of spring when deformation is maximal can be determined from the formula (4):

$$F_3 = \frac{F_2}{(1-\delta)} \tag{4}$$

where F_2 – pressing force of spring when working deformation; δ – relative inertia clearance.

Then the critical speed V_c of the spring movement is determined by the formula (5)

$$V_c = \frac{\tau_3 \cdot \left(1 - \frac{F_2}{F_3}\right)}{\sqrt{2 \cdot G \cdot \rho \cdot 10^{-3}}} \tag{5}$$

where τ_3 – maximum tangential stress of spring; G – shear modulus; ρ – dynamic density of material.

The design of the steady rest can be considered reliable if condition of safe operation during $1 \cdot 10^4$ h is fulfill (6):

$$\frac{V_{max}}{V_c} < 1 \tag{6}$$

where V_{max} – maximum moving speed of the moveable spring end.

Fulfillment of condition (6) allows calculating the spring stiffness (7), determining the number of its coils (8) and the average diameter (9):

$$c = \frac{F_2 - F_1}{h} \tag{7}$$

$$n_1 = \frac{c_1}{c} + n_2 \tag{8}$$

$$D = D_1 + 2d \tag{9}$$

where h – a spring stroke; F_1 – a pre-strain force, n_2 – the number of bearing coils that equals 1,5, c_1 – a stiffness of one coil, d – a wire diameter, D_1 – a spring inside diameter.

The design parameters of the spring can be optimized based on the results of computer simulation by the finite element method.

4 Results

To check the efficiency of the proposed steady rest construction the crankshaft grinding process was simulate with next parameters: a main journal diameter d_{mj} = 53 mm; rotation rate of crankshaft n_{cr} = 90 rev/min; infeed speed V_P = 0.9 mm/min; width of main journal B = 29.5 mm; wheel speed V_w = 50 m/sec. According to formula (2) total cutting force 367 N.

The studies of a dependence of the total cutting force for grinding wheels with different grit (1 – z = 10 μm; 2 – z = 16 μm; 3 – z = 25 μm; 4 – z = 40 μm) at temperature T = 600 °C from ultimate strength of the material (Fig. 3a), width of main journal (Fig. 3b) and infeed speed (Fig. 3c) were made.

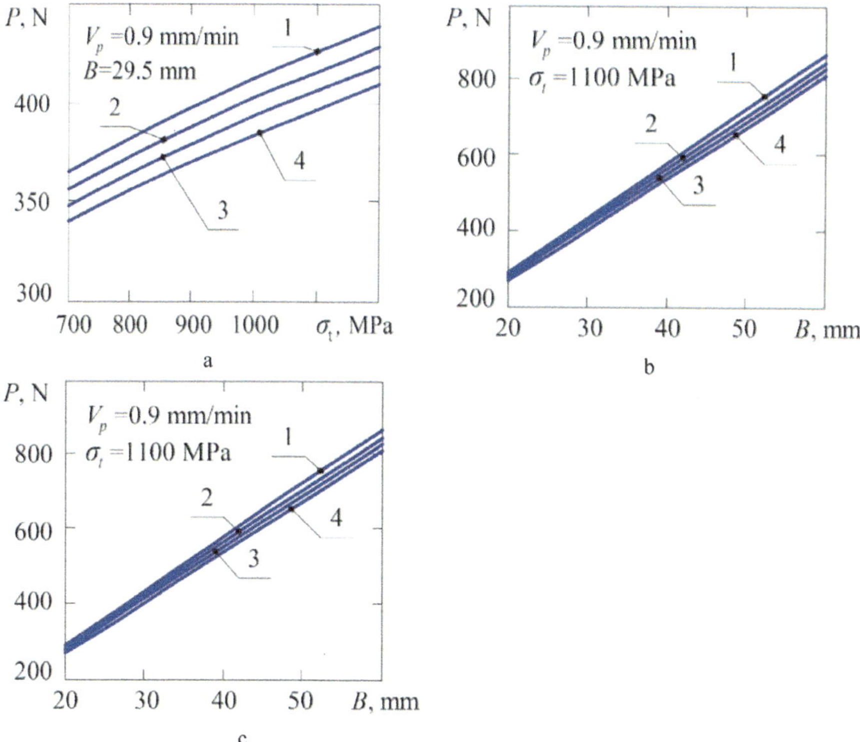

Fig. 3. A dependence of the total cutting force from: a – ultimate strength of the material; b – width of main journal; c – infeed speed.

From the Fig. 3 it can be seen that value of the total cutting force increases in direct proportion to an increase in an infeed speed and a width of the main journal. A rising of a ultimate strength of a material also results in an increase of a cutting force, however an increasing of the grit of grinding wheel the cutting force decreases insignificantly.

Knowing the values of the total cutting force occuring during infeed grinding, the design calculations of the spring (Table 1) were performed and the value of the pressing force developed by spring was obtained, which is capable to level the influence of the cutting force on the deformation processes during crankshaft machining.

To estimate the elastic deformations which occur when grinding the crankshaft main journal with and without the proposed steady rest a simulation modeling was performed. This made it possible to establish the maximum displacements of the processed journal in accordance with its circular diagram along the X, Y axes and at the expected point of contact with the steady rest (Figs. 4 and 5), and to determine the maximum displacements on the crankshaft as a whole (Tables 2 and 3).

As can be seen from Fig. 5 the polar diagram of deformations of the crankshaft main journal has eccentricity along the axes X and Y (See Fig. 5a, b), but at the point of the supposed contact of the main journal and steady rest the polar diagram of deformations has the desired shape (See Fig. 6). In addition, the simulation results showed that the smallest elastic deformations of the crankshaft are observed when the angle between the directions of the total cutting force and the pressing force of the steady rest is equal 156°.

Knowing the values of crankshaft main journal displacements and geometrical parameters of the spring, the pressing forces, which are necessary to compensate the influence of the total cutting force on the shape accuracy of shaft, were obtained (Table 3).

Table 1. The geometrical parameters of the working spring of the following steady rest.

Name of parameter	Description	Value
Pressing force of spring when maximum deformation, N	F_3	680
Wire diameter, mm	d	5
Stiffness of one coil, N/mm	C_1	310
Spring stiffness, N/mm	C	40
Number of coils	n_1	10
Spring inside diameter, mm	D_1	32
Maximum deformation, mm	S_3	17
Length in free condition, mm	l_0	67

An example of the results of modeling the loading by forces (total force and pressure force) of the crankshaft main journal by the finite elements method is shown on the Fig. 7. In addition, the figure clearly shows the points of contact between the steady rest and the grinding wheel with the workpiece. In addition, the points of contact between the steady rest and the grinding wheel with the workpiece are clearly indicated in the figure. This, in turn, allows us to estimate the distribution of elastic deformations

Table 2. Investigation of deformation processes when grinding of crankshaft main journal is without using following steady rest.

Angle of crankshaft rotation, [°]	Maximum displacement of the main journal along the axis X, [μm]	Maximum displacement of the main journal along the axis Y, [μm]	Maximum displacement of the main journal at an expected point of contact, [μm]	Maximum displacement of the crankshaft, [μm]
0	2,28	2,3	2,12	3
24	2,26	2,47	2,34	3,3
48	2,24	2,4	2,21	3,2
72	2,15	2,2	2,08	2,9
96	1,96	2,05	1,9	3,2
120	2,04	1,78	1,85	2,3
144	2,15	1,72	1,9	2,6
168	2,3	1,8	2,08	3
192	2,4	1,9	2,2	3,3
216	2,42	1,94	2,2	3,3
240	2,4	1,92	2,11	3,1
264	2,26	1,86	1,97	2,4
288	2,03	1,85	1,82	2,2
312	2,02	1,96	1,86	2,4
336	2,1	2,18	1,97	2,8

(displacements) of the crankshaft during machining. As the results of computer simulation showed, the maximum elastic deformations are observed when the cam is rotated on an angle of 192° and equals $0.69 \cdot 10^{-3}$ mm for the processed journal and $0.7 \cdot 10^{-3}$ mm – for crankshaft.

When comparing the data obtained from the simulation of machining with and without using of the following steady rest, it was established that its application makes it possible to reduce elastic deformations by an average of 87% (Figs. 8 and 9).

This allows saying about positive trend to reduce the negative effects of cutting force on crankshaft main journal by application the proposed following steady rest. However, it was found that theoretical calculations of the required spring stiffness have an error (Fig. 8). And to result in the value of main journal deformation to 0, it is necessary to perform an experimental optimization of the spring parameters.

Thus it can be mentioned that the proposed design of following steady rest is relevant and operative for improving the quality and accuracy of machining the crankshaft, as well as reducing the manufacture time due to the possibility to intensify of cutting conditions without the threat of increase of deformations and shape errors.

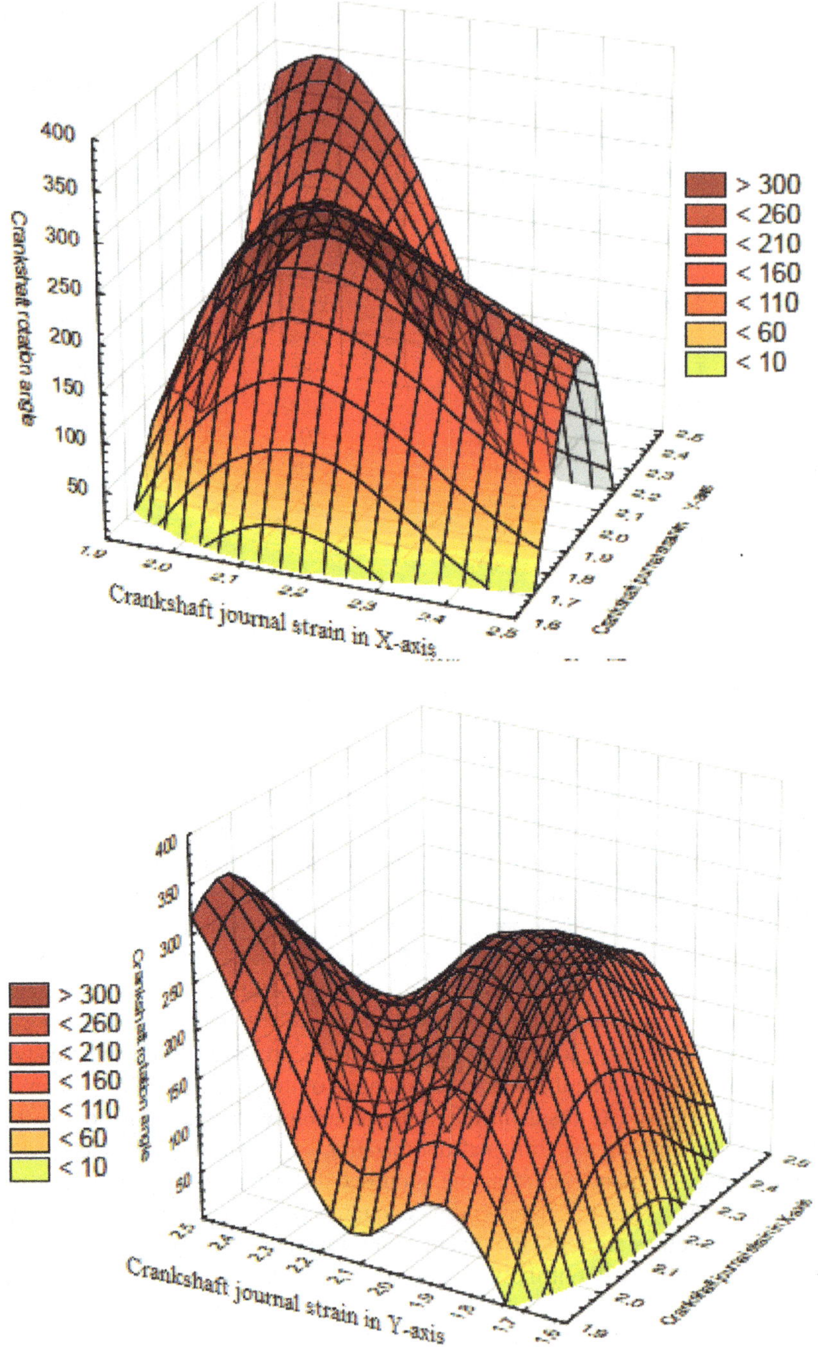

Fig. 4. The tension distribution of processed crankshaft main journal in X and Y axes.

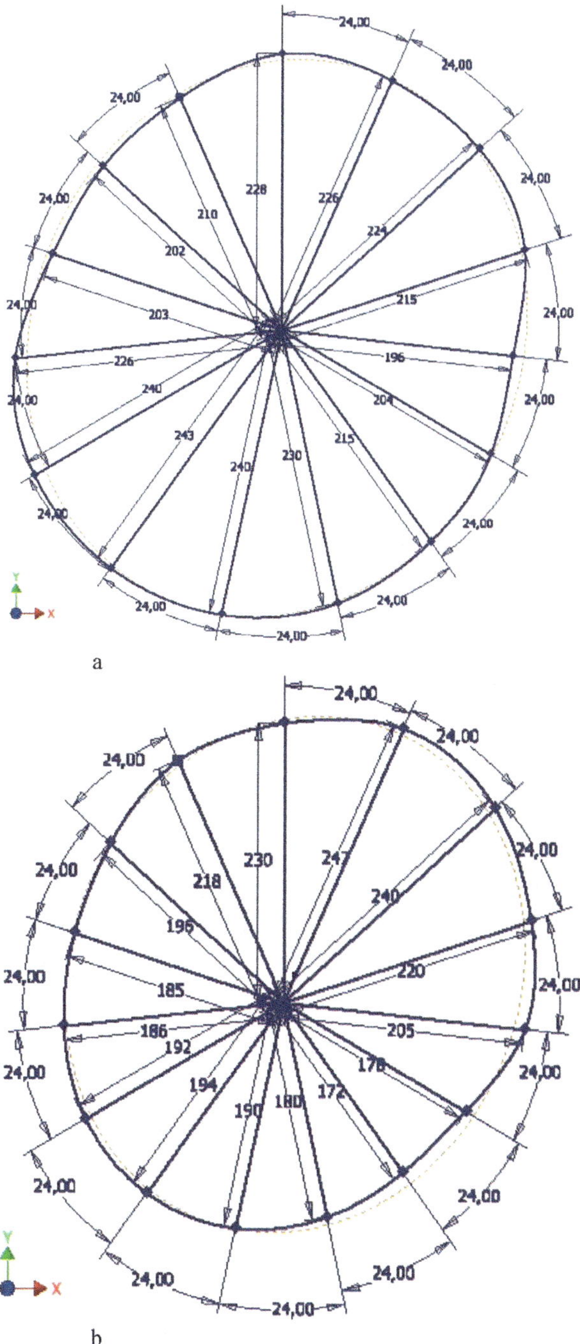

Fig. 5. The polar diagrams of elastic deformations of the crankshaft main journal which are got from simulation results: a – displacements in X-direction; b – displacements in Y-direction.

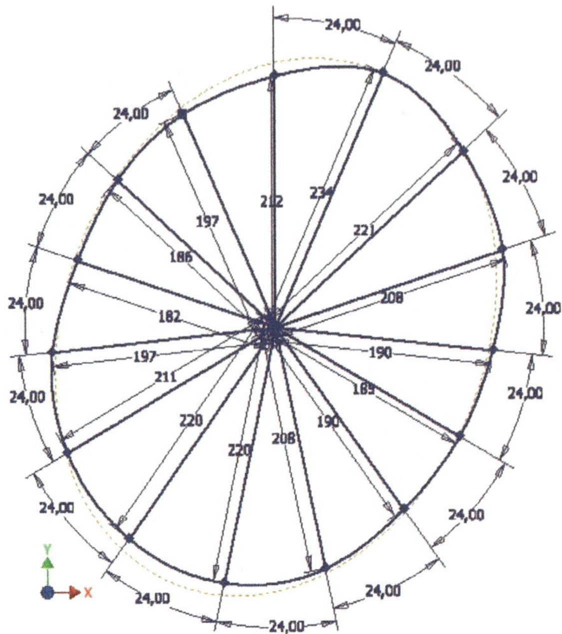

Fig. 6. The polar diagram of elastic deformations of the crankshaft main journal which are got from simulation results. Displacements at the contact point of the main journal and steady rest.

Table 3. Investigation of deformation processes when grinding of crankshaft main journal is with using following steady rest.

Angle of crankshaft rotation, [°]	Spring pressing force, [N]	Maximum displacement of the main journal along the axis X, [μm]	Maximum displacement of the main journal along the axis Y, [μm]	Maximum displacement of the main journal at the point of contact, [μm]	Maximum displacement of the crankshaft, [μm]
0	445	0,15	0,45	0,52	0,52
24	471	0,2	0,45	0,68	0,68
48	445	0,12	0,3	0,54	0,54
72	419	0,12	0,17	0,37	0,45
96	393	0,21	0,09	0,21	0,5
120	393	0,21	0,09	0,29	0,54
144	419	0,12	0,17	0,38	0,54
168	445	0,16	0,27	0,53	0,55
192	471	0,3	0,38	0,69	0,7
216	445	0,2	0,24	0,53	0,54
240	419	0,16	0,10	0,37	0,38
264	393	0,24	0,05	0,21	0,47
288	367	0,3	0,15	0,15	0,55
312	393	0,23	0,07	0,24	0,53
336	419	0,1	0,13	0,39	0,5

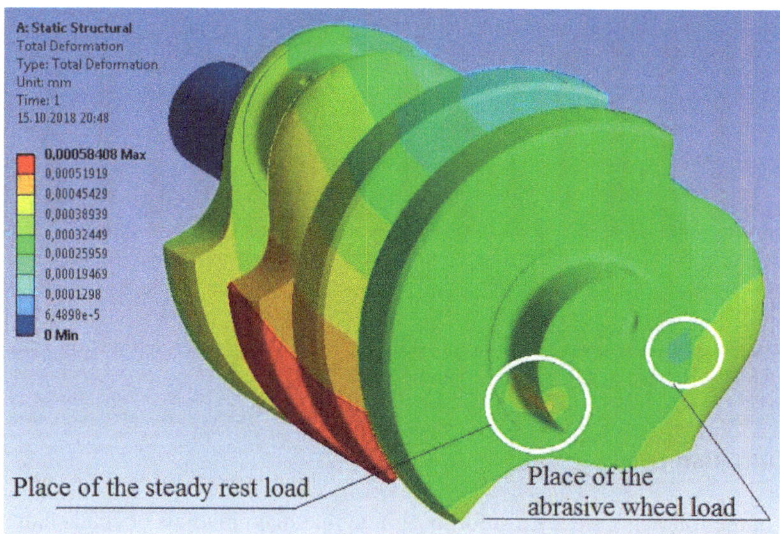

Fig. 7. Investigation of the crankshaft deformations under the influence of the forces from the steady rest and the grinding wheel

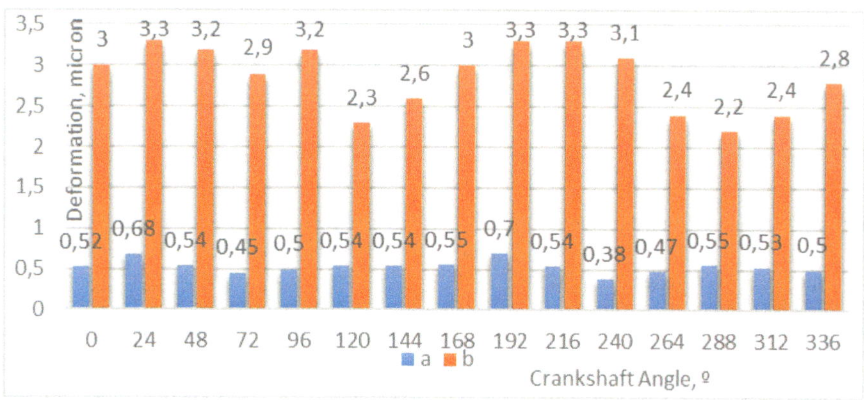

Fig. 8. Comparative analysis of the maximum deformations of the crankshaft depends on an angle of cam rotation: a – with steady rest, b – without steady rest

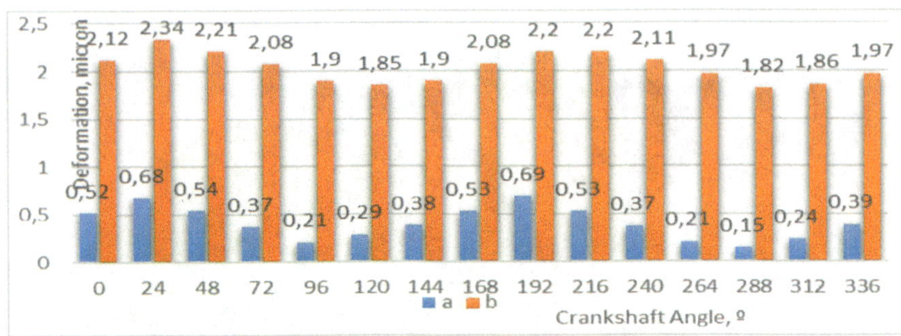

Fig. 9. Comparative analysis of the deformations of the main journal depends on an angle of cam rotation: a – with steady rest, b – without steady rest

5 Conclusions

Using of the following steady rest for machining the main journals of crankshaft allows compensating the influence of cutting forces and minimizing the deformation component of the grinding process. This will permit to stabilize the thickness of cut by ensuring the constancy of the infeed cycle and will result in a reduction of roughness and an increase of shape accuracy of the machined main journal.

The using of the steady rest may increase the crankshaft machining productivity due to the intensification of cutting conditions without the threat of appearance of deformations and shape errors of the processed journals.

References

1. Jia Y, Tian H, Chen C, Wang L (2017) Predicting the availability of production lines by combining simulation and surrogate model. Adv Prod Eng Manage 12(3):285–295. https://doi.org/10.14743/apem2017.3.259
2. Guru Bramhananda Reddy M, Rajendar Prasad C, Uday Kiran Reddy M, Naveen Kumar L (2017) Design & analysis of crankshaft by forged steel & composite material. Int J Eng Trends Technol (IJETT) 46(1). https://doi.org/10.14445/22315381/IJETT-V46P207
3. Gill KJS, Rosha P, Chander S, Bharaj RS (2014) Durability analysis of lightweight crankshafts design using geometrically restricted finite element simulation techniques for camless engines. In: International conference of advance research and innovation (ICARI-2014), pp 56–68
4. Yadav VD, Jamdar VM, Jadhav GS, Gunavant PS, Mohite PS (2017) Reverse engineering of crank shaft. In: International advanced research journal in science, engineering and technology. national conference on design, manufacturing, energy & thermal engineering (NCDMETE-2017), pp 20–22. https://doi.org/10.17148/IARJSET/NCDMETE.2017.06
5. Senthil kumar M, Ragunathan S, Suresh M, Srinivashan VR (2014) A study of crankpin failure in I.C. engine. Mech Eng Int J (MEIJ) 1(3):1–11
6. Dereszewski M (2014) Validation of failure detection by crankshaft angular speed analysis under sea waving conditions. Diagnostyka 15(3):59–63
7. Grytsyuk O, Vrublevskyi O (2018) Investigations of diesel engine in the road test. Diagnostyka 19(2):89–94. http://dx.doi.org/10.29354/diag/10.29354/diag/90279

8. Burdzik B, Pankiewicz J, Wądołowski M (2016) Analysis of bending-torsional vibrations for diagnostics. Diagnostyka 17(4):79–84

9. Pavlenko I, Simonovskiy V, Ivanov V et al (2019) Implementation of artificial neural network for identification of bearing stiffness characteristics according to the research of critical frequencies of the rotor. In: Ivanov V, et al. (eds.) Advances in design, simulation and manufacturing. DSMIE-2018. Lecture Notes in Mechanical Engineering, pp 325–335. https://doi.org/10.1007/978-3-319-93587-4_34

10. Pavlenko IV, Simonovskiy VI, Demianenko MM (2017) Dynamic analysis of centrifugal machines rotors supported on ball bearings by combined application of 3D and beam finite element models. In: IOP conference series: materials science and engineering, 233 (1):012053. https://doi.org/10.1088/1757-899X/233/1/012053

11. Pawar PP, Dalvi SD, Rane S, Divakaran CB (2015) Evaluation of crankshaft manufacturing methods - an overview of material removal and additive processes. Int Res J Eng Technol (IRJET) 02(04):118–122

12. Rekha RS, Nallusamy S, Vijayakumar R, Saravanan S (2015) Modelling and analysis of crank shaft with metal matrix composites. Int J Appl Eng Res 10(62):133–137

13. Trojanowska J, Kolinski A, Galusik D et al (2018) A methodology of improvement of manufacturing productivity through increasing operational efficiency of the production process. In: Hamrol A, Ciszak O, Legutko S, Jurczyk M (eds) Advances in manufacturing. Lecture Notes in Mechanical Engineering, pp 23–32. Springer, Cham

14. Rewers P, Trojanowska J, Diakun J et al (2018) A study of priority rules for a levelled production plan. In: Hamrol A, Ciszak O, Legutko S, Jurczyk M (eds) Advances in manufacturing. Lecture Notes in Mechanical Engineering, pp 111–120. Springer, Cham

15. Araújo AF, Varela MLR, Gomes MS (2018) Development of an intelligent and automated system for lean industrial production adding maximum productivity and efficiency in the production process. In: Hamrol A, Ciszak O, Legutko S, Jurczyk M (eds) Advances in manufacturing. Lecture Notes in Mechanical Engineering, pp 131–140. Springer, Cham

16. Mahendrakar S, Kulkarni PR (2016) A review on design and vibration analysis of a crank shaft by FEA and experimental approach. Int J Sci Dev Res (IJSDR) 1(11):146–149

17. Bi ZM, Zhang WJ (2001) Flexible fixture design and automation: review, issues and future directions. Int J Prod Res 39:2867–2894. https://doi.org/10.1080/00207540110054579

18. Karpus VE, Ivanov VA (2012) Locating accuracy of shafts in V-blocks. Russ Eng Res 32(2):144–150. https://doi.org/10.3103/S1068798X1202013X

19. Gothwal S, Raj T (2017) Different aspects in design and development of flexible fixtures: review and future directions. Int J Serv Oper Manage 26(3):386–410. https://doi.org/10.1504/IJSOM.2017.081944

20. Krol O, Sokolov V (2018) Development of models and research into tooling for machining centers. East-Eur J Enterp Technol 3, 1(93):12–22. https://doi.org/10.15587/1729-4061.2018.131778

21. Karpus V, Ivanov V, Dehtiarov I, Zajac J, Kurochkina V (2019) Technological assurance of complex parts manufacturing. In: Ivanov V et al (eds) Advances in design, simulation and manufacturing. DSMIE 2018. Lecture Notes in Mechanical Engineering, pp 51–61. Springer, Cham. https://doi.org/10.1007/978-3-319-93587-4_6

22. Ivanov V, Dehtiarov I, Denysenko Y et al (2018) Experimental diagnostic research of fixture. Diagnostyka 19(3):3–9. https://doi.org/10.29354/diag/92293

23. Evtuhov VG et al (2009) Modelirovanie protsessa kruglogo vreznogo shlifovaniya. Bulletin SumDU 2:124–129. [Russian]

24. Kotliar A, Gasanov M, Basova Y, Panamariova O, Gubskyi S (2019) Ensuring the reliability and performance criterias of crankshafts. Diagnostyka 20(1):23–32. https://doi.org/10.29354/diag/99605

Nonparametric Assessment of Surface Shaping by Hybrid Manufacturing Technology

Sara Dudzińska[✉], Daniel Grochała, Emilia Bachtiak-Radka,
and Stefan Berczyński

West Pomeranian University of Technology, Szczecin, Poland
sara.dudzinska@zut.edu.pl

Abstract. Development of surface metrology tries to respond to the needs of the industrial environment and to keep up with increasingly sophisticated manufacturing methods and products. However, a multitude of surface texture parameters, as well as the research equipment available on the market, often complicates the estimation and comparison of surfaces, e.g. between different branches. The authors present a method of nonparametric assessment of milled and then burnished surfaces which is alternative instrument to parametric assessment of surface topography. This method gives directly answers, whether the treatment was conducted in line with expectations. This instrument can also be helpful in a very wide spectrum of technological cases, where the cloud of points is collected by stylus and optical systems. The measurements were conducted with an AltiSurf A520 multisensor instrument, manufactured by Altimet according to own developed methodology.

Keywords: Surface topography · Hybrid manufacturing · Surface metrology · Ball burnishing · Nonparametric assessment

1 Introduction

In recent years much attention in world literature has been devoted not only to the conscious shaping of the surface [1–8], but also to the problems associated with the measurement of the created, often very specific, surfaces [9–12]. Thanks to the development of technology and desire to reduce production costs while preserving high quality of manufactured parts, it is more common to integrate different technological operation in one. This is called hybrid manufacturing [13].

Hybridization of machining operation is sometimes the only alternative when it comes to producing a desirable element in high quality and lower price. Karunakaran et al. [14] and Zhu et al. [13] emphasize that hybrid manufacturing allows to increase advantages of each operation and minimize its drawbacks at the same time. Thanks to the use of such technologies, it is possible to manufacture elements that are very complicated and difficult to make by conventional methods.

The development of this type of technology and dedicated machines requires a unique involvement not only from technologists and process engineers, but also from metrologists [15, 16]. In the case of the production of surfaces with very specific properties the limitation to conventional methods of 2D surface analysis and parametric

© Springer Nature Switzerland AG 2019
M. Diering et al. (Eds.): *Advances in Manufacturing II* - Volume 5, LNME, pp. 52–62, 2019.
https://doi.org/10.1007/978-3-030-18682-1_5

assessment of the 3D surface texture can lead to wrong conclusions. MacAulay et al. [17] points out that even surface texture parameters are not suitable for characterizing an individual surface feature. In addition, it should be kept in mind that the surface processing by hybrid treatment is the result of many tasks and in this case the procedures for measurement should be properly understood and prepared. In recent years, many authors have been addressing this issue. Tosello et al. [10] emphasize that it is necessary to create standardized measurement methods, as well as simplify methods that can be used in industry. This is particularly important in the case of plastics processing. Obtaining the proper structure of the plastic parts is related to the appropriate surface structure of the injection molds. In case when it is necessary to manufacture very smooth and reflective plastic parts [9], the molds should be prepared in an appropriate manner. However this process is long and expensive. That is why, researchers are conducting experiments to reduce production costs [1, 11, 18–22] as well as achieving high accuracy of shape and dimension [23, 24]. One of the method resulting in high surface smoothness is a hybrid treatment which involves milling and burnishing [2, 5, 18, 20, 25]. Such surfaces should be analyzed in respect of morphology, which allows nonparametric surface texture assessment [26, 27].

For several years, scientists have pointed out that the analysis of 2D surface topography is not sufficient to determine the functionality manufactured surfaces. Krolczyk et al. [27] points out directly that there is just no way to effectively characterize 3D surface in a 2D way. In addition, as the contemporary paradigm in modern production it is so important to achieve high-quality at lower production cost. Which is why scientists are trying to develop such methods of 3D surface description that will facilitate the controlling process of the surface functionality during production and wear prediction during exploitation. Alfyorova [26] points out that for this purpose it is significant to develop research on nonparametric evaluation of surface topography. She emphasizes that in contrast to parametric surface description, nonparametric gives the possibility of a more detailed assessment. Therefore, according to contemporary trends in the presented article, the authors made an attempt to nonparametric evaluation of milled and burnished surface to obtain the answer, whether the treatment was carried out with the appropriate parameters, and the surface changed morphologically.

1.1 Isotropic Surfaces

The surface isotropy means the same surface structure in all directions, which should be interpreted as a perfectly symmetrical structure in relation to all possible symmetry axes [28]. There are various possibilities to determine the degree of surface isotropy, but the most common is the autocorrelation function [28]. The anisotropic surfaces have an asymmetric, elongated and slender shape of this function (Fig. 1a), while isotropic surfaces have the function shape more oval, sometimes round and symmetric (Fig. 1b). Isotropy is given in percentage: from 0% to 100%. If an isotropic degree is less than 20%, it indicates to anisotropic surface. However, isotropic surfaces have an isotropic degree higher than 80%.

Of course it is also good to estimate topographic characteristic based on spatial parameters (Table 1), which gives information about texture strength and the uniformity of the texture in all directions [29].

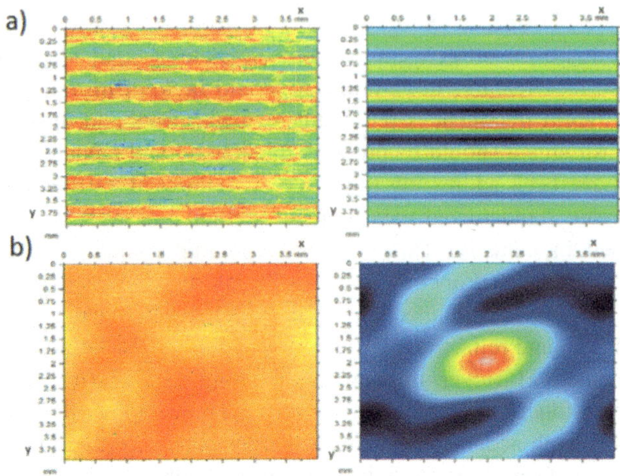

Fig. 1. Experimental values of Sq and Sa parameters after burnishing surface isotropy: (a) anisotropic milled surface with fwm = 0.3 mm - 5.81%, (b) isotropic - 65.4% surface after burnishing with interval feed of fwb = 0.02 mm [25].

Table 1. Description of spatial parameters (according ISO 25178) [28].

	Name of parameter	Parameter description
Sal	Auto-correlation length	"Horizontal distance of the autocorrelation function (tx, ty) which has the fastest decay to a specified value s, with 0 < s < 1"
Str	Texture-aspect ratio	"This is the ratio of the shortest decrease length at 0.2 from the autocorrelation, on the greatest length"
Std	Texture direction	"This parameter calculates the main angle for the texture of the surface, given by the maximum of the polar spectrum. This parameter has a meaning if Str is lower than 0.5"

Surface isotropy also gives very important information about tribological properties and subsequent material wear during exploitation. Matuszewski et al. [30] and Bustillo et al. [31] showed that the material wear during exploitation depends on the geometric surface structure. They also prove, that the isotropic surfaces wear out less much.

2   Materials and Methods

2.1   Sample Preparation

On the grounds of previous tests related to correlation burnishing force F_b and surface isotropy parameters, which were examined in Surface Metrology Laboratory on West Pomeranian University of Technology, another series of experiments was designed and conducted to examine the nonparametric assessment of isotropy. The $100 \times 100 \times 20$ mm 42CrMo4 steel samples, thermally improved to 35 ± 2HRC was used. The chemical composition of the 42CrMo4 steel is presented in the Table 2.

Table 2. The chemical composition of the 42CrMo4 steel.

42CrMo4 chemical composition	C	Mn	Si	P	S	Cr	Ni	Mo	W	V	Cu
	0,38–0,45	0,17–0,37	0,17–0,37	max. 0,035	max. 0,035	0,9–1,2	max. 0,3	0,15–0,25	max 0,2	max 0,05	max 0,25

Milling was conducted on DMG DMU 60MONOBLOCK machining center. The technical specification of this machining center is presented in the Table 3.

Table 3. The technical specification of DMG DMU 60MONOBLOCK [32].

DMG DMU 60MONOBLOCK technical data	Work area	Main drive motor spindle	Fixed table
	X/Y/Z-axis: 630/560/560 mm Max. rapid traverse: 30 m/min Max. feed rate: 30 000 mm/min Max. acceleration X/Y/Z: 6/7/4 m/s^2	Output: 15/10 kW Max. spindle speed: 12 000 rpm	Dimensions: 1000 × 600 mm Max. load: 700 kg

A WNT R1000G.42.6.M16.IK torus (Fig. 2 and Table 4) with six inserts - diameter of $d_\mathrm{p} = 10$ mm (RD.X1003 MOT – WTN1205) was used.

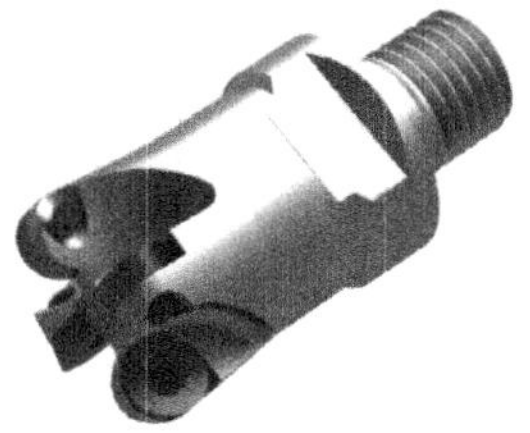

Fig. 2. WNT R1000G.42.6.M16.IK torus [33].

Sample surface was shaped with milling depth of $a_p = 0.5$. Machining velocity while milling was $v_c = 100$ m/min, cutting tip feed $f_z = 0.1$ mm, and transverse feed $f_{wm} = 0.1$ mm and $f_{wm} = 0.7$ mm.

A one-ball hydrostatic tool with bellows actuator, with a ZrO2 tip - diameter of $d_b = 10$ mm. Burnishing velocity v_b was 8 m/min and transverse feed was $f_{wb} = 0.12$ mm. Burnishing was conducted with force F_b in the range of 200÷1400 N, with increase step of 200 N.

Table 4. The specification of WNT R1000G.42.6.M16.IK torus [33].

Description	d_1 [mm]	A [mm]	a [mm]	d_2 [mm]	d_3 [mm]
R1000G.42.6.M16.IK	42	43	2.8	17	29

The scheme of samples preparation is presented in the Fig. 3.

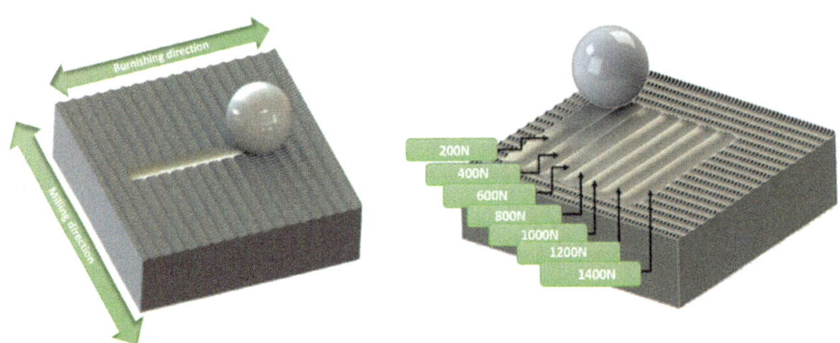

Fig. 3. Preparation of samples for experiments.

2.2   Surface Measurement

Surface measurements were undertaken in Surface Metrology Laboratory on West Pomeranian University of Technology on the AltiSurf A520 multisensor, manufactured by Altimet fitted with a CL1 chromatic confocal sensor. The description of the instrument is given in Table 5. Measurements were conducted on 2,5 × 2,5 mm areas.

AltiMap PREMIUM 6.2 was used to analyze data and determine surface topography. Surface topography analysis was based on a developed methodology, which involved:

- determining threshold level to delete unreliable surface point data,
- conducting surface leveling (with mean area),
- determining some 3D SG parameters in compliance with ISO 25178.

Table 5. Measuring instrument.

Measurement method	Instrument/software	Parameters
Optical	AltiSurf A520/CL1 sensor/AltiMap PREMIUM 6.2	Range: 130 µm Vertical resolution: 8 nm Scanning resolution along X axes: 0,47 µm and Y axes: 5 µm

3 Case Studies and Discussion

The values of selected amplitude and spatial parameters after milling are in Table 6. Burnishing data shows, that amplitude surface roughness parameters values obtained after milling were significantly reduced (Fig. 4), which means much smoother surface. In case of milled surface with a transverse feed of $f_{wm} = 0.1$, burnishing reduced those parameters up to 40,5%. The burnishing of milled surfaces with the transverse feed value $f_{wm} = 0.1$ was more effective than with $f_{wm} = 0.7$ (in this case the reduction of amplitude surface parameters was about 10%). The most noticeable reduction of S_q and S_a parameters was in the case of milling with $f_{wm} = 0.1$ and burnishing with $F_b = 600$ N (Fig. 4).

Table 6. Surface roughness parameters after milling; transverse feed: $f_{wm} = 0.1$ mm; $f_{wm} = 0.7$ mm.

Name of parameter	Parameter description	Unit	$f_{wm} = 0{,}1$ mm	$f_{wm} = 0{,}7$ mm
Sq	Root mean square height	µm	0.474	4.81
Sa	Arithmetical mean height	µm	0.367	3.92
Sal	Autocorrelation length	mm	0.0248	0.215
Str	Texture-aspect ratio		0.0249	0.211

In both cases ($f_{wm} = 0.1$ mm and $f_{wm} = 0.7$ mm) the values of S_{al} parameters increased. That means more uniform structure. The biggest S_{al} values were present in samples milled with $f_{wm} = 0.1$ mm and burnished with $F_b = 800$ N. In all cases S_{tr}

parameters values increased insignificantly, what means that those set of technological parameters does not provide isotropic surface (Fig. 5).

Surface isotropy and autocorrelation function gave very interesting conclusions (Fig. 6). In case of milled surfaces with a transverse feed f_{wm} = 0.1 mm and burnished with burnishing force F_b = (200–1200 N) it is clear to observe that in the first steps of burnishing process (F_b = 200) traces left by the milling tool are dominant. However, when using higher value forces (F_b = 400–1200 N) burnishing marks start to dominate. Isotropy after milling was 2,48%. After burnishing with F_b = 800 N surface isotropy increased to 33,2%, which is still far from pure isotropy surface. In case of milled surfaces with a transverse feed f_{wm} = 0.7 mm and burnishing force of 600 N milling traces are still visible. Force values exceeding 1000 N provides burnishing marks. Using autocorrelation function in conjunction with parametric assessment provide better surface analysis which translates to better conclusions. This gives more efficient ways of technological process planning.

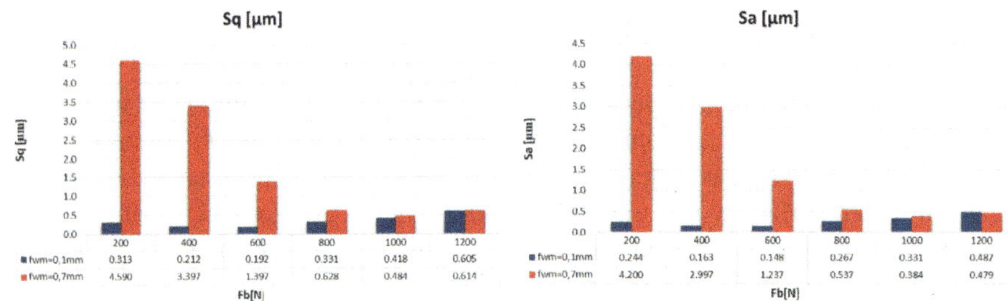

Fig. 4. Experimental values of Sq and Sa parameters after burnishing.

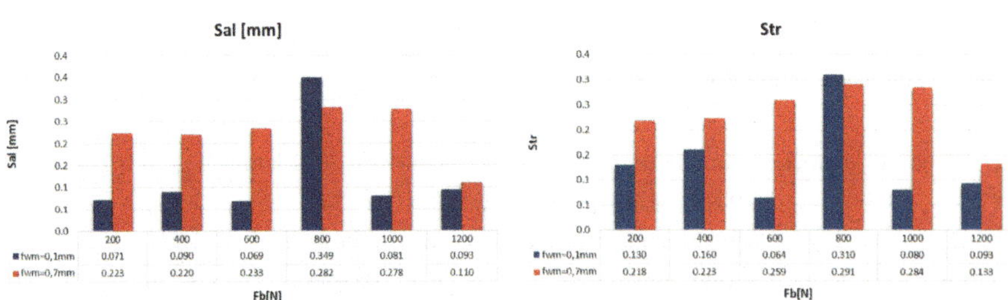

Fig. 5. Experimental values of Sal and Str parameters after burnishing.

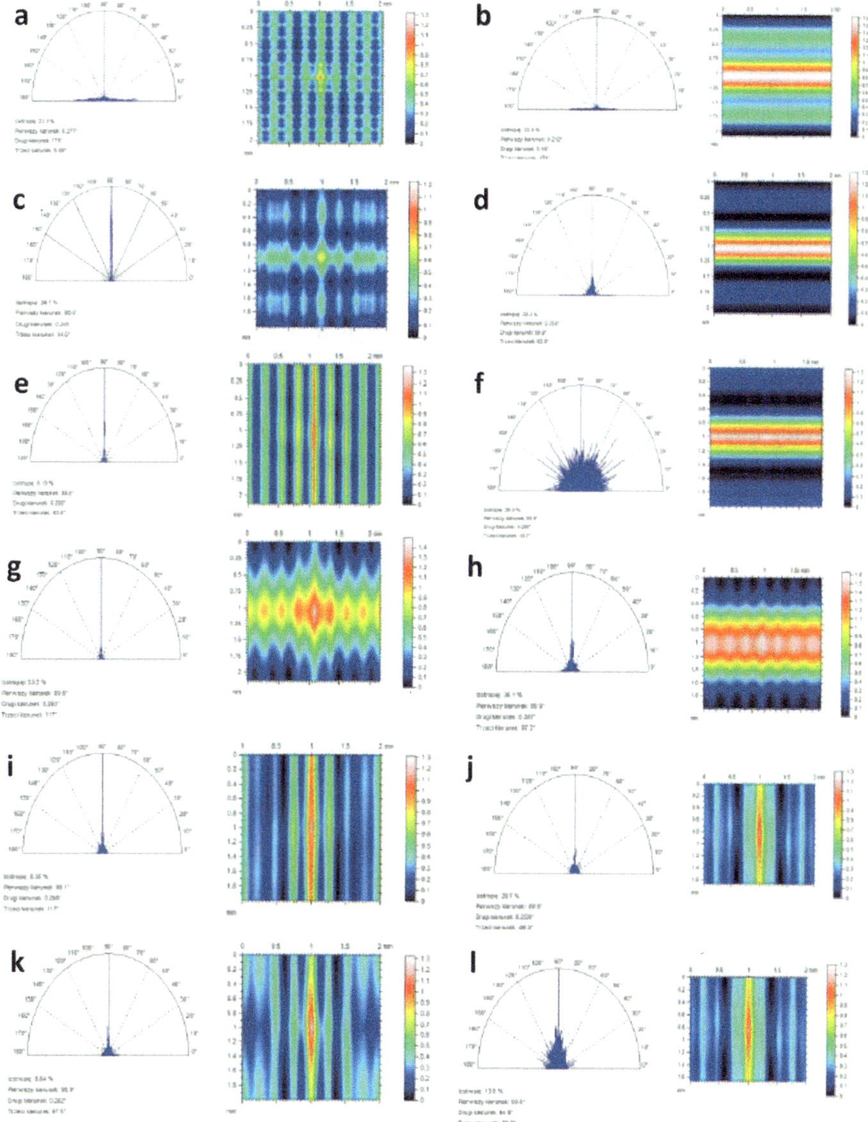

Fig. 6. Analysis of surface structure direction and autocorrelation function of examined milled and burnished samples *(a: f_{wm}= 0.1 mm, F_b= 200 N; b: f_{wm}= 0.7 mm, F_b= 200 N; c: f_{wm}= 0.1 mm, F_b= 400 N; d: f_{wm}= 0.7 mm, F_b= 400 N; e: f_{wm}= 0.1 mm, F_b= 600 N; f: f_{wm}= 0.7 mm, F_b= 600 N; g: f_{wm}= 0.1 mm, F_b= 800 N; h: f_{wm}= 0.7 mm, F_b= 800 N; i: f_{wm}= 0.1 mm, F_b= 1000 N; j: f_{wm}= 0.7 mm, F_b= 1000 N; k: f_{wm}= 0.1 mm, F_b= 1200 N; l: f_{wm}= 0.7 mm, F_b= 1200 N).*

4 Summary and Conclusions

Autocorrelation function combined with spatial surface parameters greatly improves efficiency assessment of machining procedures (exchange of surface texture from anisotropic to isotropic). It definitely facilitates the engineers work. In this way, the subjective evaluation of the created surface can be avoided (depending on the knowledge and practice acquired by the process engineer). More efficient selection of technological parameters is possible thanks to additional criteria, i.e. isotropy and the autocorrelation function. It is especially important in hybrid manufacturing, where different machining operations are combined. Despite the great advantages of using non-parametric tools to assess the geometric structure of the surface after hybrid machining, it is impossible to get the answer to the question whether the traces left by the tool in the previous operation were removed and the structure completely changed. Therefore, new methods and tools should be developed to facilitate correlation between surfaces obtained in subsequent machining operations.

References

1. Pessoles X, Tournier C (2009) Automatic polishing process of plastic injection molds on a 5-axis milling center. J Mater Process Technol 209(7):3665–3673
2. Yuan XL, Sun YW, Gao LS, Jiang SL (2016) Effect of roller burnishing process parameters on the surface roughness and microhardness for TA2 alloy. Int J Adv Manuf Technol 85(5–8):1373–1383
3. Maheshwari AS, Gawande RR (2016) Improving surface quality of AA 6351 by the stiff burnishing technique. Jordan J Mech Ind Eng 10(4):245–251
4. Denkena B, Grove T, Maiss O (2017) Surface texturing of rolling elements by hard ball-end milling and burnishing. Int J Adv Manuf Technol 93(9–12):3713–3721
5. El-Taweel TA, El-Axir MH (2009) Analysis and optimization of the ball burnishing process through the Taguchi technique. Int J Adv Manuf Technol 41(3–4):301–310
6. Garn R, Schubert A, Zeidler H (2011) Analysis of the effect of vibrations on the micro-EDM process at the workpiece surface. Prec Eng J Int Soc Prec Eng Nanotechnol 35(2):364–368
7. Dandekar CR, Shin YC, Barnes J (2010) Machinability improvement of titanium alloy (Ti-6Al-4V) via LAM and hybrid machining. Int J Mach Tools Manuf 50(2):174–182
8. Lauwers B (2011) Surface integrity in hybrid machining processes. In: 1st CIRP Conference on Surface Integrity (CSI), vol 19, p 11
9. Crisan N, Descartes S, Berthier Y, Cavoret J, Baud D, Montalbano F (2016) Tribological assessment of the interface injection mold/plastic part. Tribol Int 100:388–399
10. Tosello G, Haitjema H, Leach RK, Quagliotti D, Gasparin S, Hansen HN (2016) An international comparison of surface texture parameters quantification on polymer artefacts using optical instruments. CIRP Ann Manuf Technol 65(1):529–532
11. Chu WS, Kim CS, Lee HT, Choi JO, Park JI, Song JH, Jang KH, Ahn SH (2014) Hybrid manufacturing in micro/nano scale: a review. Int J Prec Eng Manuf Green Technol 1(1):75–92
12. Wang J, Pagani L, Leach RK, Zeng WH, Colosimo BM, Zhou LP (2017) Study of weighted fusion methods for the measurement of surface geometry. Prec Eng J Int Soc Prec Eng Nanotechnol 47:111–121
13. Zhu Z, Dhokia VG, Nassehi A, Newman ST (2013) A review of hybrid manufacturing processes - state of the art and future perspectives. Int J Comput Integr Manuf 26(7):596–615

14. Karunakaran KP, Suryakumar S, Pushpa V, Akula S (2010) Low cost integration of additive and subtractive processes for hybrid layered manufacturing. Robot Comput Integr Manuf 26 (5):490–499
15. Townsend A, Senin N, Blunt L, Leach RK, Taylor JS (2016) Surface texture metrology for metal additive manufacturing: a review. Prec Eng J Int Soc Prec Eng Nanotechnol 46:34–47
16. Triantaphyllou A, Giusca CL, Macaulay GD, Roerig F, Hoebel M, Leach RK, Tomita B, Milne KA (2015) Surface texture measurement for additive manufacturing. Surf Topogr Metrol Prop 3(2):8
17. MacAulay GD, Senin N, Giusca CL, Leach RK (2016) Study of manufacturing and measurement reproducibility on a laser textured structured surface. Measurement 94:942–948
18. Shiou FJ, Chuang CH (2010) Precision surface finish of the mold steel PDS5 using an innovative ball burnishing tool embedded with a load cell. Prec Eng J Int Soc Prec Eng Nanotechnol 34(1):76–84
19. Shiou FJ, Chen CCA, Li WT (2006) Automated surface finishing of plastic injection mold steel with spherical grinding and ball burnishing processes. Int J Adv Manuf Technol 28(1–2):61–66
20. Shiou FJ, Chen CH (2003) Determination of optimal ball-burnishing parameters for plastic injection moulding steel. Int J Adv Manuf Technol 21(3):177–185
21. Macedo PB, Alem D, Santos M, Lage M, Moreno A (2016) Hybrid manufacturing and remanufacturing lot-sizing problem with stochastic demand, return, and setup costs. Int J Adv Manuf Technol 82(5–8):1241–1257
22. Sagbas A (2011) Analysis and optimization of surface roughness in the ball burnishing process using response surface methodology and desirability function. Adv Eng Softw 42 (11):992–998
23. Kong LB, Cheung CF, Jiang XQ, Lee WB, To S, Blunt L, Scott P (2010) Characterization of surface generation of optical microstructures using a pattern and feature parametric analysis method. Prec Eng J Int Soc Prec Eng Nanotechnol 34(4):755–766
24. Colledani M, Tolio T, Fischer A, Iung B, Lanza G, Schmitt R, Vancza J (2014) Design and management of manufacturing systems for production quality. CIRP Ann Manuf Technol 63 (2):773–796
25. Bachtiak-Radka E, Dudzinska S, Grochala D, Berczynski S, Olszak W (2017) The influence of CNC milling and ball burnishing on shaping complex 3D surfaces. Surf Topogr Metrol Prop 5(1):7
26. Alfyorova EA (2018) Nonparametric estimation of deformation relief. Letters on Materials-Pis Ma O Materialakh 8(2):220–224
27. Krolczyk GM, Maruda RW, Krolczyk JB, Nieslony P, Wojciechowski S, Legutko S (2018) Parametric and nonparametric description of the surface topography in the dry and MQCL cutting conditions. Measurement 121:225–239
28. ISO 25178-2:2012 - Geometrical product specifications (GPS) - surface texture: areal – Part 2: terms, definitions and surface texture parameters
29. Jiang XJ, Whitehouse DJ (2012) Technological shifts in surface metrology. CIRP Ann Manuf Technol 61(2):815–836
30. Matuszewski M, Mikolajczyk T, Pimenov DY, Styp-Rekowski M (2017) Influence of structure isotropy of machined surface on the wear process. Int J Adv Manuf Technol 88(9–12):2477–2483
31. Bustillo A, Pimenov DY, Matuszewski M, Mikolajczyk T (2018) Using artificial intelligence models for the prediction of surface wear based on surface isotropy levels. Robot Comput Integr Manuf 53:215–227

32. DMU 60/80/100 monoBlock classic Series. https://hamofa.eu/wp-content/uploads/2017/11/DMG-DMU-80.pdf. Accessed 11 Jan 2019
33. WNT Catalogue. https://www.wnt.com/pl.html. Accessed 11 Jan 2019

Analysis of the Geometry and Surface of the Knife Blade After Milling with a Various Strategies

Jakub Czyżycki[✉], Paweł Twardowski, and Natalia Znojkiewicz

Institute of Mechanical Technology, Poznan University of Technology,
ul. Piotrowo 3, 60-965 Poznań, Poland
jakub.r.czyzycki@doctorate.put.poznan.pl

Abstract. The main objective of the paper was the analysis of knife blades and knife bevels made by milling and grinding. The first part of the work focuses on the typical knife construction, materials used for knives, and parameters describing production quality. The research determined the influence of various milling methods on the machining time, surface roughness of the obtained surface, topography of the surface and thickness of the remaining machining allowance for further processing. The research has shown that the morph milling strategy allows obtaining the most accurate surface and the shape of the knife bevel. The proper programming of the tool path had an important influence on the final shape of the knives. A hand-ground knife requiring a lot of experience despite the longest execution time has the smallest surface roughness and allowance for further processing of the assumed value.

Keywords: Milling · Grinding · Knife main bevel · Knives

1 Introduction

Each knife in order to fulfill its task, which is cut in an appropriate manner requires the correct blade geometry, the cross-section called main bevel determines how knife will perform in its job [1].

Currently, knife bevels are made mainly by grinding on grinding wheels or abrasive belts, or are forged. Due to the increasing availability of numerically controlled machine tools, milling returns to use. The construction of a knife on the example of a full-tang knife with two scales is shown in the Fig. 1.

The basic materials used for knife blades are:

- stainless steels,
- non-alloyed (carbon) steels,
- powder steels,
- titanium alloys,
- ceramics.

Due to the contact of knives with various environments, most knives are made of stainless steel such as 4H13, 420, 440, Sandvik 12C27, 8Cr13MoV, N690 and AEB-L.

© Springer Nature Switzerland AG 2019
M. Diering et al. (Eds.): *Advances in Manufacturing II* - Volume 5, LNME, pp. 63–81, 2019.
https://doi.org/10.1007/978-3-030-18682-1_6

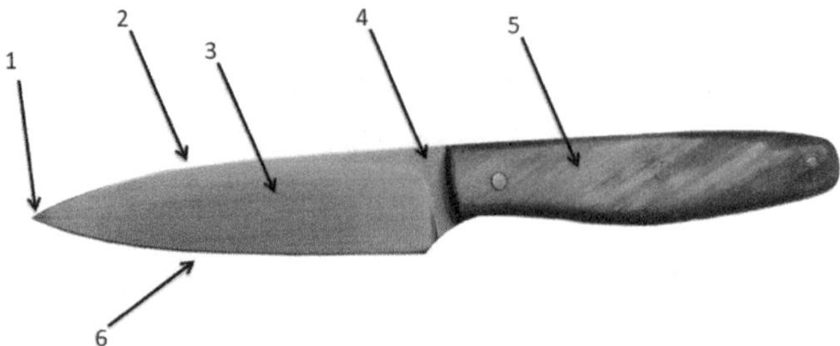

Fig. 1. Construction of a knife: 1 – tip, 2 – spine, 3 – knife main bevel, 4 – blade flat, 5 – handle, 6 – cutting edge.

An alternative used successfully in kitchen knives that allow much longer to maintain its sharpness are ceramic knives, they have very good corrosion resistance despite small impact strength [1–3].

Among the many parameters determining the usefulness of the knife, hardness, ductility, elasticity and impact strength are the most important.

Another important parameter is the choice of the type of knife bevel, that is its geometry. As the work presented [4], the smaller the angle of the blade, the lower the resistance to cutting food products. Of the many types of grinds (Fig. 2), four basic can be distinguished [5]. Flat grind is the most popular one and easy to make. Hollow grind is used in folding and hunting knives. Convex grind have great durability. Chisel grind is the easiest to make, but usually used in tools, and industrial blades.

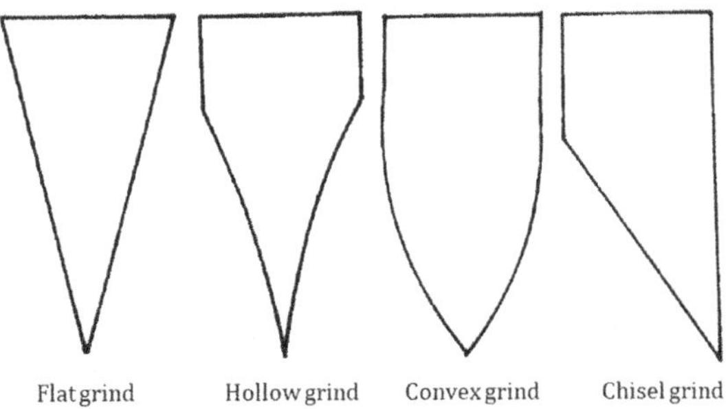

Fig. 2. General type of knife grinds [5].

Milling as one of the methods of making knives requires the use of a numerically controlled machine tool along with CAM software. The major problems occurring during precise milling on CNC machines concern the formation of machined surface texture. There are many studies which are being focused on modeling and experimental investigating of the formation of surface irregularities in milling processes [6, 7].

Ball-end mills are successfully used for the machining of 3D surfaces in injection molds, dies and parts of the aerospace industry. The aforementioned advantage of ball-end mills cutters is connected with their disadvantage, which is affected by the variable inclination angle of the machined surface causing the change of the active length of the cutting edge of the cutter. The result of this are temporary changes in the components of the total force which affects the dimensional and shape accuracy as well as the roughness of the surface [8, 9]. Wojciechowski et al. [10] showed in studies that the inclination of a ball nose end mill at an angle of 45° to the workpiece compared to the plain milling positively influences the improvement of Ra and Rz parameters of surface roughness over a wide range of applied rotational speeds.

In order to obtain a good quality of the surface and to extend the life of the tool, it is recommended to work with a ball-end mills inclined at an angle (β) of 15° to the workpiece (Fig. 3). The X, Y, Z coordinate system refers to the tool, and X', Y', Z' refers to a workpiece [11, 12].

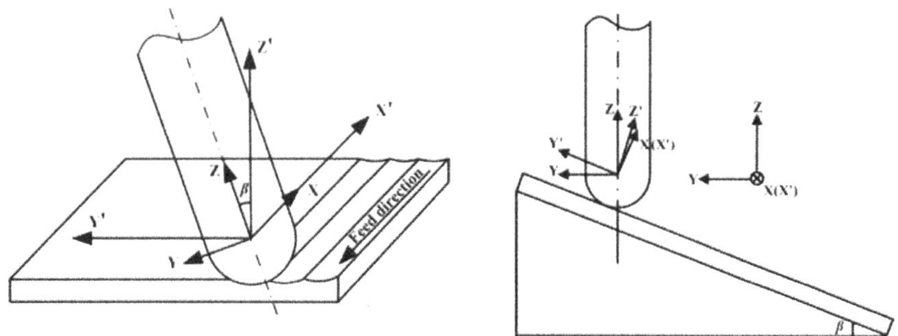

Fig. 3. The inclination of the ball-end mill relative to the workpiece [11].

The smaller the angle of inclination of the ball nose end mill the higher the vibration amplitude and the result is an increase in the surface roughness parameter Rz. This is confirmed by the tests carried out during micro milling using different angles of inclination of the tool [13].

Due to the necessity of using a tool inclined to the workpiece, a situation may arise in which it will be necessary to use a tool with a length bigger than with the plain milling routine. In Wojciechowski et al. work [14], research were carried out showing the differences between the length of the cutter and surface roughness. In the case of cutters with a longer length ($l = 85$ mm), the basic parameters of Sa and Sz roughness were over three times higher than in the case of a shorter cutter ($l = 35$). The reason for the decrease in surface roughness during milling with a longer tool was the increase in dynamic deflections of the tools due to milling forces.

In addition to the correct positioning of the tool relative to the material being processed, the selection of the right tool strategy is a very important aspect. Cam software have a wide selection of strategies for each machining. As shown by research [15, 16], among ten different strategies of the same surface, significant differences in tool deflection, machining time and surface roughness errors were obtained.

The factors mentioned above are a direct result of the choice of strategy, whereas there are indirect effects that are not so obvious. An example is the choice of surface treatment strategies for forging tools. The choice of an inadequate strategy can lead to less durability of the tool and deterioration of the surface roughness of the obtained effects. The research shows that the optimal strategy is to choose the tool paths along the material flow direction during the forging process. The difference in the durability of the tool can be as high as 30% [17].

Some studies pay attention to the fact that the choice of strategy depends on the amount of electricity consumed by the machine tool. The research [18] shows that with the increase of the tool feed speed, the cutting time and the amount of energy used decrease. This is related to the increased amount of material being machined during larger feeds, which in turn affects the lower energy consumption. As it can be see, choosing the right strategy affects many factors related not only to the workpiece.

The impact of the choice of strategy and tool is presented in the current article. The main purpose of this article is to compare milling with different methods to traditional grinding, the method most commonly used in making knife blades.

2 Experimental Details

2.1 Research and Objective Range

In the research, flat bars of 3 mm thick made of 4H13 martensitic stainless steel (1.4034) were used as the material to be milled. This steel is intended for the production of household objects, measuring instruments, surgical instruments, springs, and has also found its use in hunting knives. The material subjected to grinding was a flat bar made of NMWV cold tool steel (1.2510). This steel is used for sheet metal cutting tools, guides, shear blades and measuring instruments.

The chemical composition of 4H13 steel is shown in Table 1 [19].

Table 1. Chemical composition of 4H13 steel.

Steel type	Chemical composition (%)									
	C	Mn	Si	P	S	Cu	Cr	Ni	Mo	Other
4H13	0,36 0,45	max 0,80	max 0,80	max 0,040	max 0,030	max 0,030	12,00 14,00	max 0,60	–	–

The chemical composition of O1 steel is shown in Table 2 [20].

Table 2. Chemical composition of O1 steel.

Steel type	Chemical composition (%)									
	C	Mn	Si	P	S	Cu	Cr	Ni	Mo	Other
O1	0,90 1,00	1,00 1,30	0,15 0,40	max 0,030	max 0,030	–	0,50 0,70	–	–	–

FRAISA monolithic milling cutters made of fine-grained cemented carbide were used (Figs. 4, 5 and 6). Machining parameters for milling are shown in Table 3. The model of the knife made in Autodesk Inventor Professional being the object of research is shown in Fig. 7.

Cylindrical end mill with a diameter of $D = 10$ mm, number of teeth $z = 4$, tool rake angle $\gamma_0 = -10°$, tool helix angle $\lambda_s = 55°$ and working length is $L = 31$ mm.

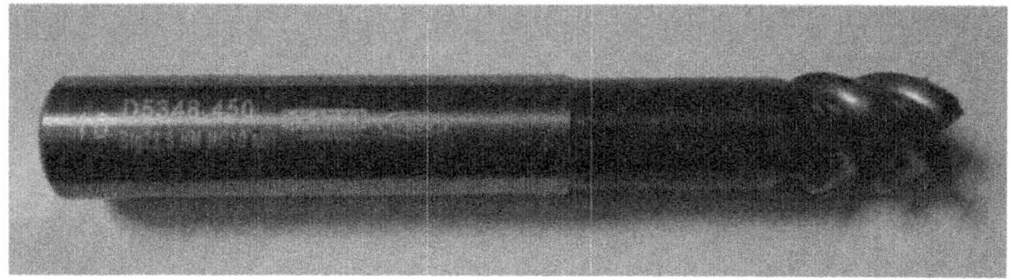

Fig. 4. Cylindrical end mill: FRAISA D5348 450.

Ball nose end mills with a diameter of $D = 10$ mm, and $D = 12$ mm, number of teeth $z = 2$, tool rake angle $\gamma_0 = -10°$, tool helix angle $\lambda_s = 30°$ and working length is $L = 31$ mm.

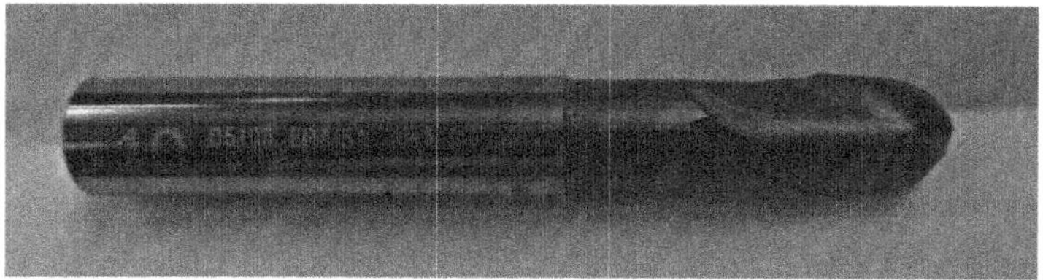

Fig. 5. Ball nose end mill: FRAISA D5100.

Corner radius end mill with a diameter of $D = 12$ mm, number of teeth $z = 4$, tool rake angle $\gamma_0 = -5°$, tool helix angle $\lambda_s = 30°$ and working length is $L = 37$ mm.

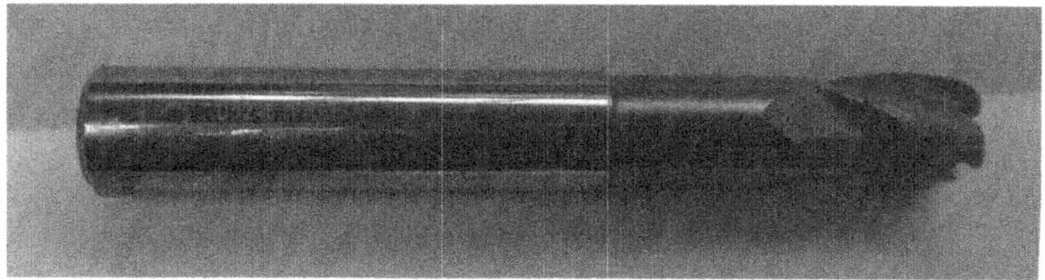

Fig. 6. Corner radius end mill: FRAISA U5246.

Table 3. Cutting parameters.

Cutting parameters	Cylindrical end mill	Ball nose end mill	Ball nose end mill	Corner radius end mill
Tool diameter D [mm]	10	10	12	12
Cutting speed v_c [m/min]	60	118	140	130
Feed per tooth f_z [mm]	0,045	0,130	0,130	0,130
Rotation speed n [rev/min]	1910	3750	3750	3750

Fig. 7. The default knife model.

In order to be able to make a knife using the milling method on a numerically controlled center, it was necessary to adjust its geometry to the needs of mounting the flat bars on the machine tool. Figure 8 shows the knife model adapted for machining.

The relieving holes were changed into four holes arranged in one line, two of which served as a base, and two as mounting. A fin was also added on the spine of the knife so as to be able to press it to the machining fixture as close as possible to the tip, eliminating vibrations of the flat bar during machining.

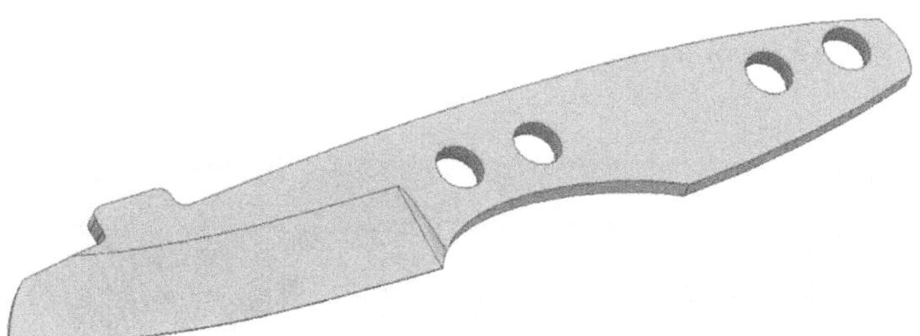

Fig. 8. Adapted model of the knife.

Figure 9 shows the knife model attached to the machining fixture. The center of the coordinate system is determined in the axis of the first locating pin at the height of the flat bar.

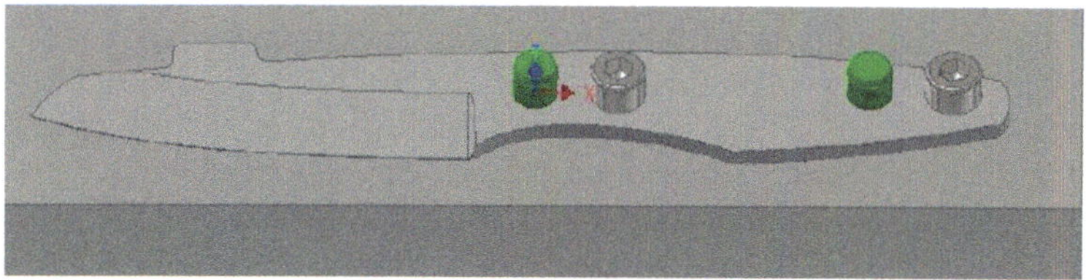

Fig. 9. Mounting knife to the machining fixture.

Every milling operation was made by DMC 70V machining center made by Deckel Maho. When milling the shape of the knife, a fixing plate was used in the front part of the flat bar (Fig. 10).

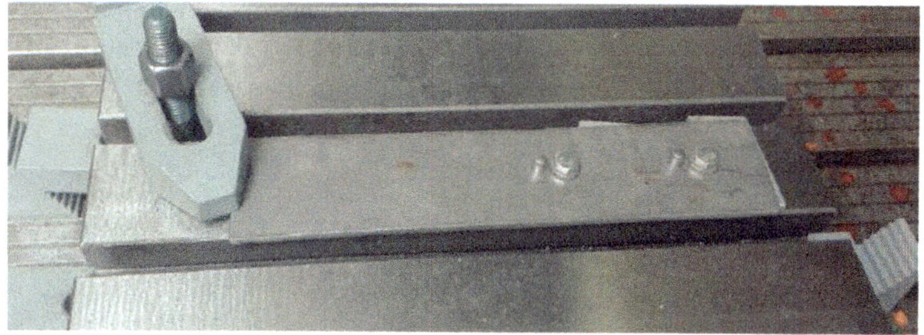

Fig. 10. Mounting the flat bar in the machining fixture.

The second fastening is shown in Fig. 11. The place of the pressure plate is moved to the place of the fin on the spine of the knife.

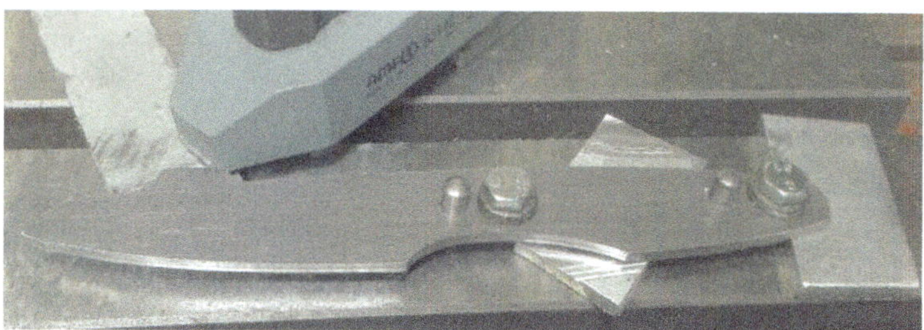

Fig. 11. Mounting the knife after cutting out the shape.

The next fix allowing for the milling on the other side of the knife required turning the knife to the other side and reassembling together with the pressure plate on the opposite side.

All machining programs were made in Autodesk Inventor Professional with an HSM overlay. The milling of the knife shape (Fig. 12) was made using a contour cut.

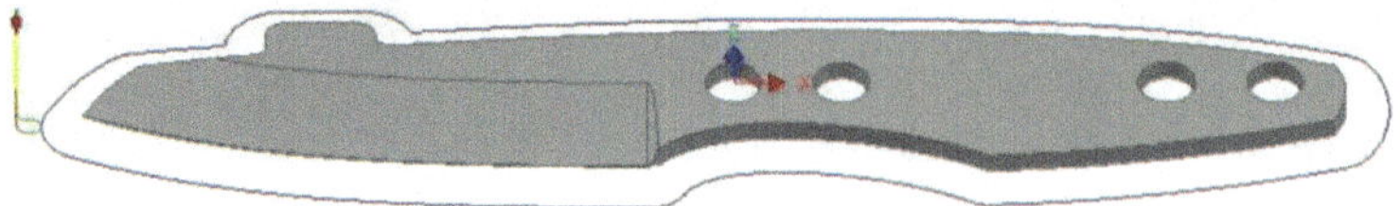

Fig. 12. Tool path for cutting the knife contour

The milling of the knife main bevels has been made using two different strategies:

- parallel - parallel paths in the XY plane, they follow the work surface at a constant distance Z. The path lines may run one way or an additional path may be imposed which intersects the previous one creating a grid in both directions. The parallel machining was carried out in two variants. The first (Fig. 13) assumes concurrent and counter-rotational milling and the second (Fig. 14) using only concurrent milling.

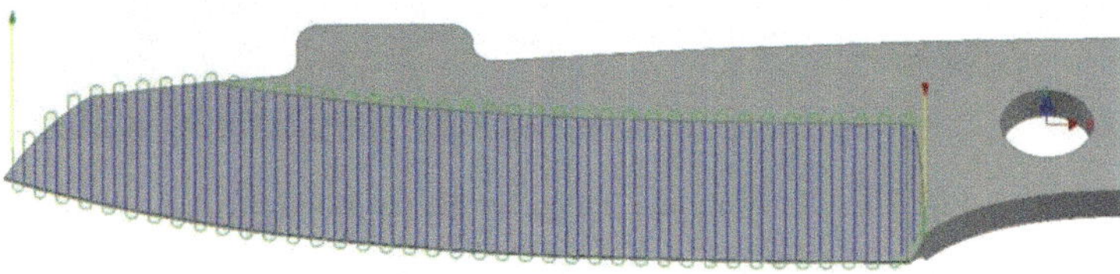

Fig. 13. Parallel milling first variant.

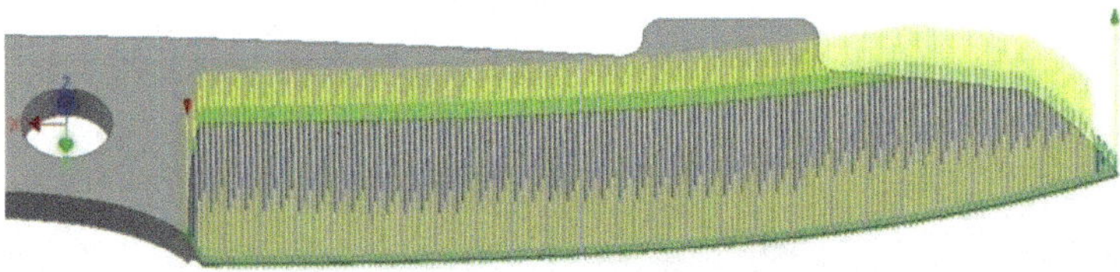

Fig. 14. Parallel milling second variant.

- morph - strategy assuming the processing of narrow spaces between two designated lines. The upper and lower line of the knife main bevel defines the processing area. The machining of the grinding with the Morph strategy is shown in Fig. 15.

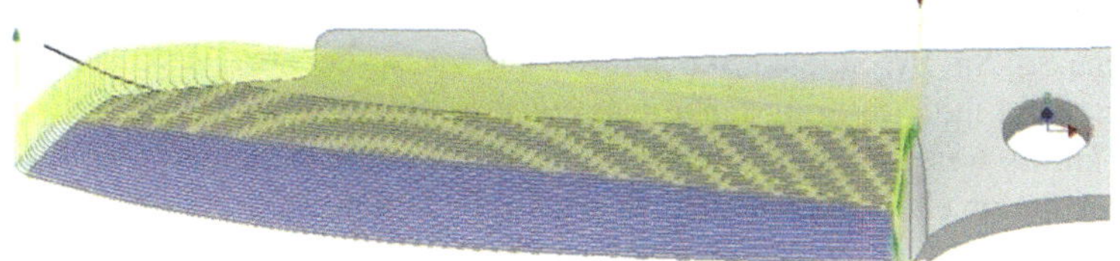

Fig. 15. Morph milling strategy.

2.2   Research Method

The following measurements were carried out in the field of research:

- the main time for milling main bevels,
- surface roughness of the surface,
- topography of the surface of the main bevels,
- the thickness of the material left on the cutting edge.

The roughness measurement of the prepared main bevels was based on four roughness parameters: *Ra, Rz, Rt* and *Rsm*. The measurements were made on a portable roughness measuring instrument W5 made by Hommel-Etamic with a resolution of 6 nm and a measuring range of 320 µm (Figure 16).

Fig. 16. Portable roughness measuring instrument W5 made by Hommel-Etamic.

The study of the surface topography of the main bevels was carried out on a Hommelwerke T8000 profilograph whose resolution is from 1 nm and the measurement range is 1200 µm. The place of measurement determined the end of the fin on the back of the knife viewed from the tip (Fig. 17).

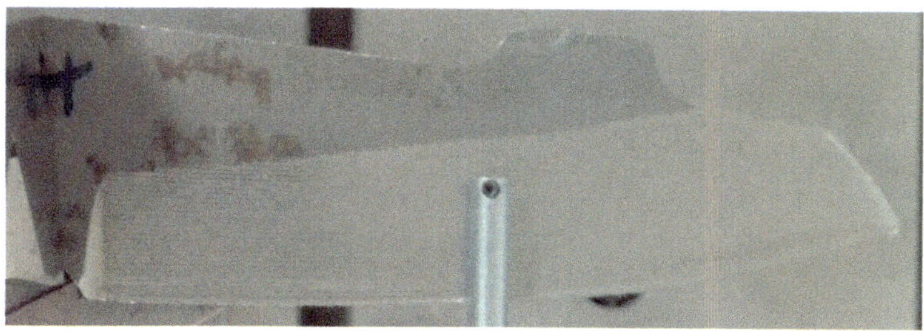

Fig. 17. Place of measurement of surface topography.

Based on the topography measurements, the following parameters were determined: Sa, Sq, Sp, Sv, St, Ssk, Sku and Sz. The measuring area was a field measuring 3×3 mm and the measuring range was equal to 160 μm.

The knives were checked in terms of the geometry of the knife in the place where the two main bevels come together, in which the cutting edge will be created in the further processing after hardening. The remaining allowance determines the subsequent thickness of the cutting edge. A characteristic element of each knife has been immortalized in photographs taken with the ZEISS microscope (12:1 magnification) (Fig. 18) with the AxioCam MRc module.

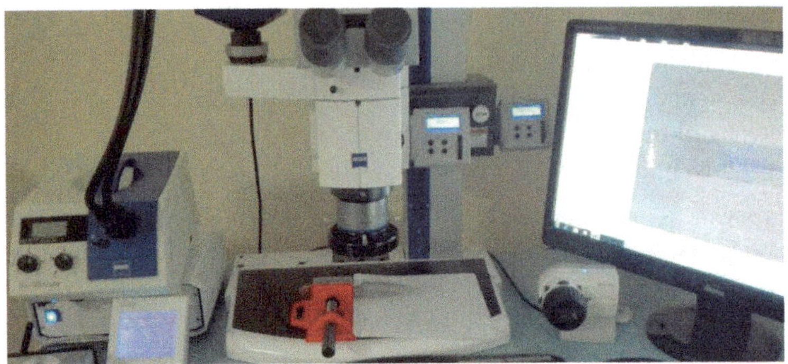

Fig. 18. ZEISS microscope with the AxioCam MRc module.

3   Results and Discussion

The result of the research were four knives made using CAM machining and one knife ground by hand. The following pictures show the next sides of knives with cut-outs. All milling methods except the first one were performed using coolant

- Parallel milling without coolant (parallel variant I) (Fig. 19) and with coolant (parallel variant II) (Fig. 20) - made using a ball nose end mill with diameter D = 12 mm. The distance between the toolpaths was 1 mm.

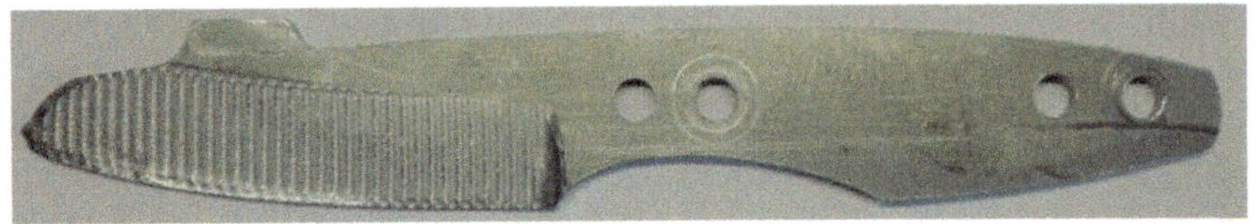

Fig. 19. Knife with main bevel made by parallel strategy variant I.

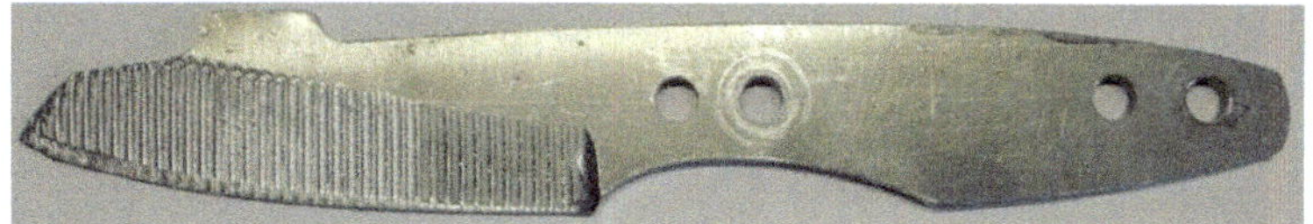

Fig. 20. Knife with main bevel made by parallel strategy variant II.

On the basis of the photographs, in the first variant it can be see the correct shape of the main bevel, while the tip has deformed. The second variant has a reduced height of main bevel in its initial part. The use of coolant affected the lack of discoloration on the blade of the knife.

- parallel milling with coolant – made using a ball nose end mill with diameter D = 10 mm (Figs. 21 and 22). The distance between the toolpaths was 0,5 mm. Two sides of knives were made with this strategy, the first one is named A and the second one is B.

Fig. 21. Knife with main bevel made by parallel strategy (A).

The use of a line spacing equal to 0.5 mm, a smaller diameter tool and a lack of counter-milling resulted in a visible improvement in the quality of the surface after processing visible to the unaided eye.

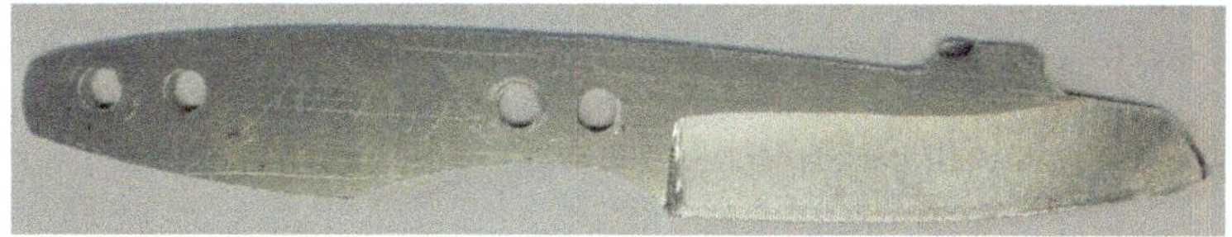

Fig. 22. Knife with main bevel made by parallel strategy (B).

Figure 23 shows a comparison of the main bevel made to the main bevel the knife model. Visible reduction of the grinding height on this side of the knife may be the result of the reduction of the cross-section of the knife after making the first main bevel, so that during cutting the other side, knife has been deformed.

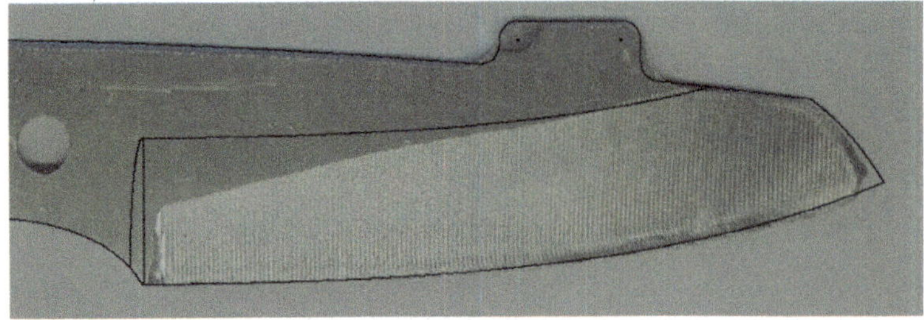

Fig. 23. Comparison of the main bevel made to the main bevel of the knife model.

- morph milling – made using ball nose end mill with diameter D = 10 mm (Figs. 24 and 25).

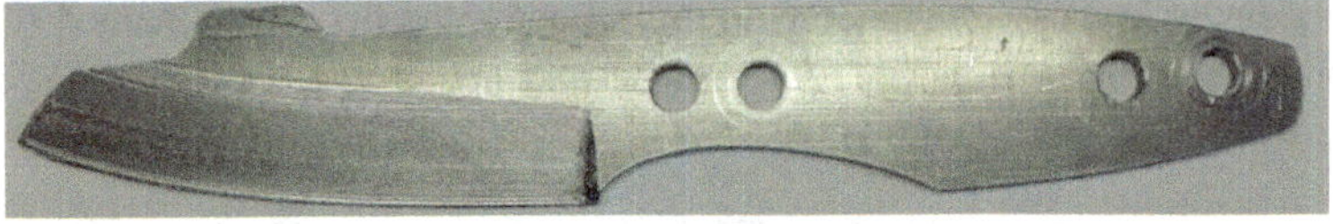

Fig. 24. Knife with main bevel made by morph strategy (ball nose end mill).

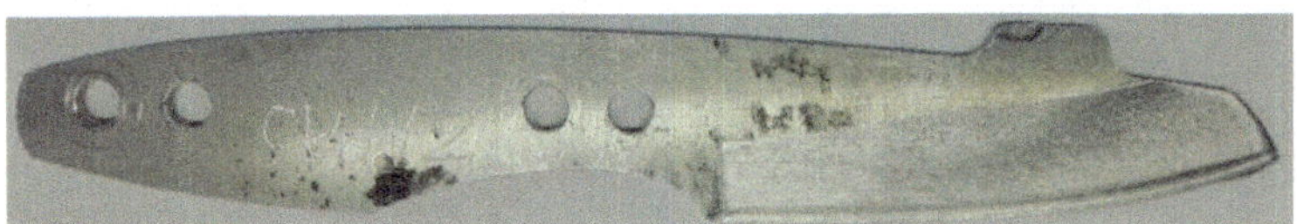

Fig. 25. Other side of the knife with main bevel made by morph strategy (ball nose end mill).

In contrast to previous strategies in the described case, it can be noticed only a slight dimensional distortion of the main bevel on the other side of the knife. This is probably due to a different force distribution during machining using this strategy. The use of a tool with a smaller diameter and compacted lines of the tool path, influenced similarly as in the case of parallel strategy A and B to obtain a smooth ground surface.

- morph milling – made using corner radius end mill with diameter D = 12 mm (Figs. 26 and 27).

Fig. 26. Knife with main bevel made by morph strategy (corner radius end mill).

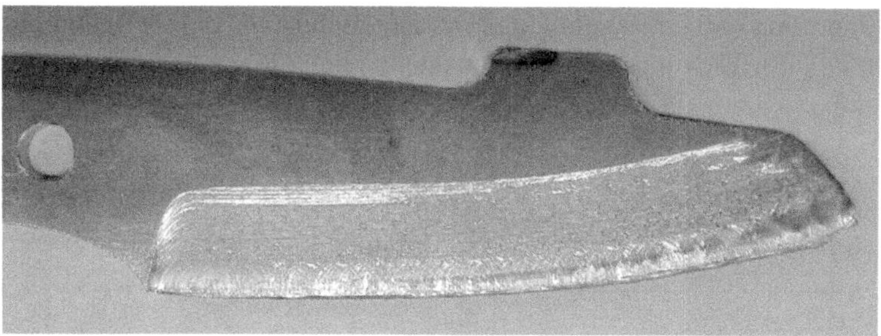

Fig. 27. Other side of the knife with main bevel made by morph strategy (corner radius end mill).

On the basis of the photos, one can notice a significant distortion of the shape of the main bevel, in particular on the side shown in Fig. 26. The tip as well as the area near the cutting edge is jagged and irregular. Significant depressions on the surface of the blade are visible. The reason for the deformed surface is probably too large diameter of the tool and too dense arrangement of the paths for this tool. In order to check whether the diameter of the tool affects this problem, the same strategy should be carried out using a torus mill with a much smaller diameter.

- grinding by hand (Figs. 28 and 29).

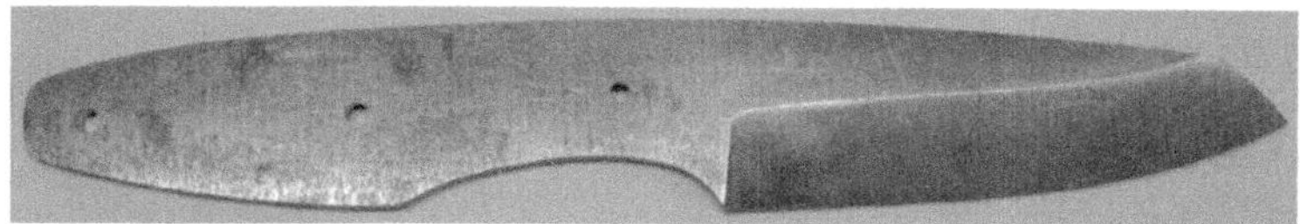

Fig. 28. Knife with main bevel made by grinding.

Fig. 29. Other side of the knife with main bevel made by grinding.

A handmade main bevel is symmetrical, starts in one place on both sides and coincides with the same point on the spine. The height of the main bevel is equal along the entire length. The surface of the blade is very smooth, with no visible distortions. No need to mount the knife in the mounting fixture for processing, allows to make holes allowing for later assembly of handle scales.

A comparison of the machining times of all grinding and bevels milling methods is shown in Fig. 30. The times shown in the diagram refer to the main bevels on one side of the knife.

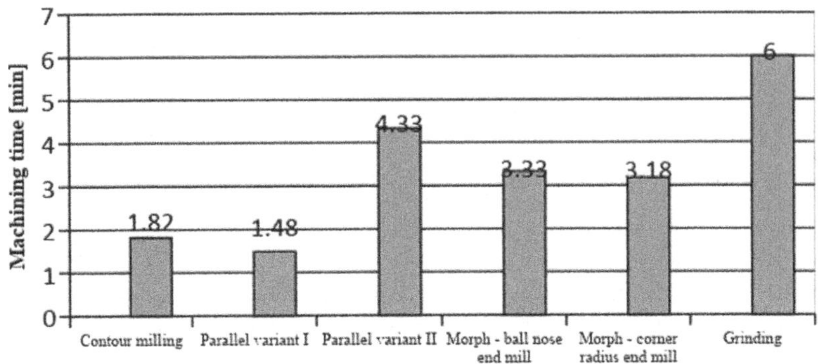

Fig. 30. The influence of the machining strategy on the machining time.

The first variant of Parallel milling obtained the shortest time due to the combination of concurrent milling with counter-rotary milling, so that the tool made a working movement all along its path. The use of only concurrent milling and the need to return the tool and the density of the rows in the path increased the time three times. Manual grinding, which requires a lot of experience, lasted for the longest time reaching 6 min.

In addition, a comparison of the actual machining time to the simulation in Inventor was made (Fig. 31).

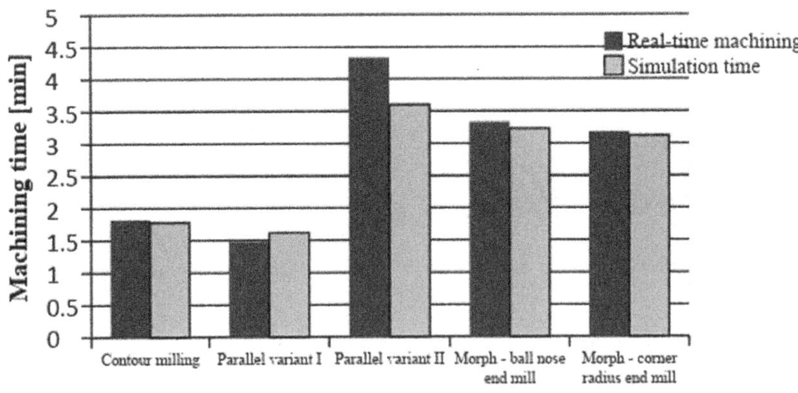

Fig. 31. Real-time machining comparison with machining simulation time.

Based on the graph, it can be see discrepancies in real time and simulation time. This is probably related to the problem of determining the actual feed speed of the tool due to the possibilities that the machine tool provides on a given section of the tool path.

Figures 32 and 33 present a summary of the measured roughness parameters of bevels made of knives. Main bevel made with the parallel variant I strategy with the use of coolant and other one using a corner radius end mill were not included in the list due to exceeding the measuring range of the device.

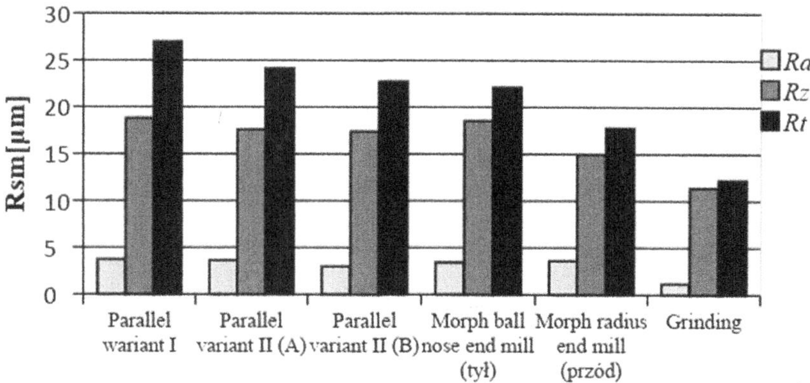

Fig. 32. List of roughness parameters of tested main bevels.

On the basis of the graph, it can be stated that the main bevel made by grinding has the best surface quality, while the strategy of the parallel variant I has obtained the worst surface quality among the measured ones. The elimination of counter-milling in the parallel variant II strategy allowed for an improvement in surface finish at the expense of longer machining time. The obtained roughness of the tested knives with the exception of the parallel strategy in the variant I is sufficient to avoid further manual processing, and replace it with, for example, sandblasting or glass blasting.

The average width of the profile elements shown in Fig. 33 illustrates the effect in parallel strategy on the distance between the toolpaths. Using the paths with 1 mm spacing and then changing this value to 0.5 mm, it can be see the effect of this change in the decrease in the roughness parameter *Rsm*.

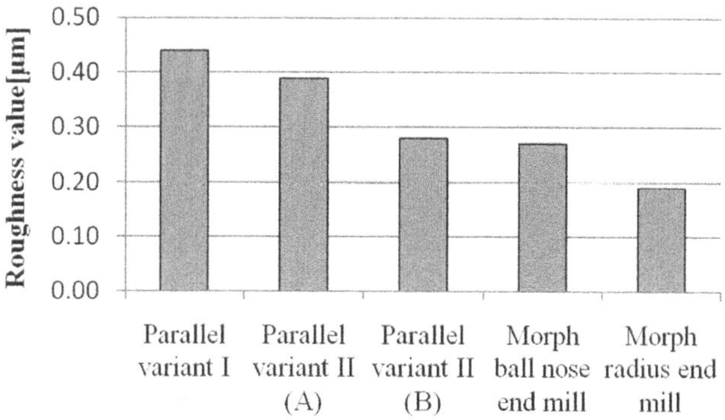

Fig. 33. Average width of the profile elements *Rsm*.

The obtained topographies differ in height, peak distribution and the characteristic representation of the tool path. Due to the exceeding of the measuring range for the main bevels made with the parallel variant II method and for morph milling using the corner radius end mill, they were not included in the following list.

The following list (Figs. 34, 35, 36 and 37) show the diversity of the surface obtained depending on the strategy used. On the first picture (Fig. 34), the effects of using a gap of 1 mm between paths are visible, although the St parameter for this blade has obtained the lowest value among milled knives. A knife made with the morph strategy (Fig. 35) is characterized by the largest overall height of the profile, as it can be noticed the largest elevation is in a single place, which indicates the possibility of the needle running over the impurity or a local burr on the surface.

The topography of the main bevel made by grinding has a typical surface after grinding with scratches with sharp inclines and depressions. The height of the profile in this case is the smallest of all tested blades.

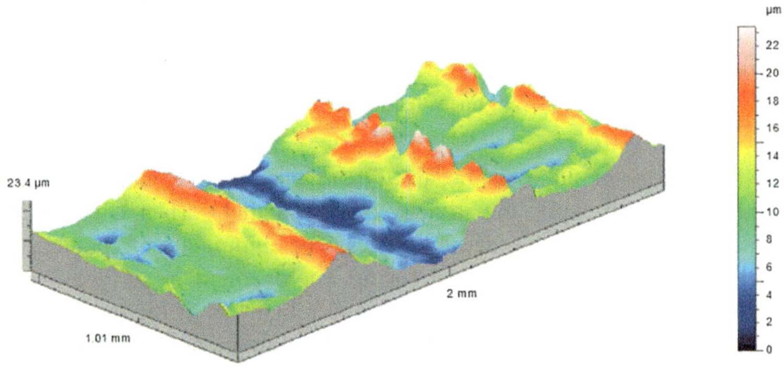

Fig. 34. Surface topography of main bevel made by parallel strategy without coolant.

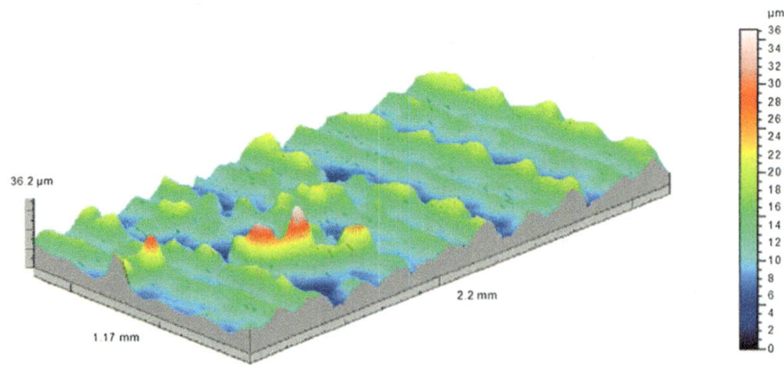

Fig. 35. Surface topography of main bevel made by morph strategy using ball nose end mill.

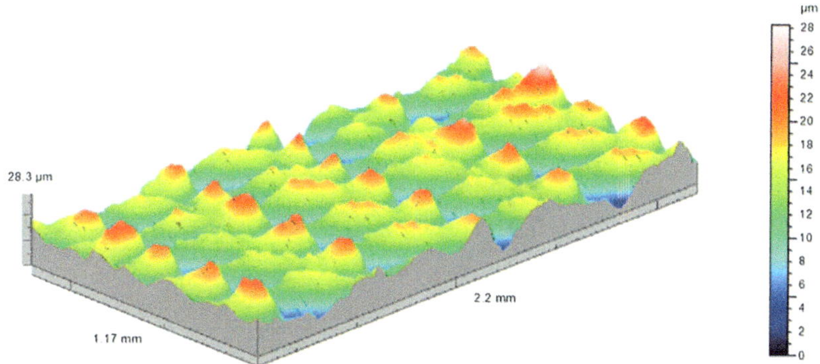

Fig. 36. Surface topography of main bevel made by parallel strategy (B).

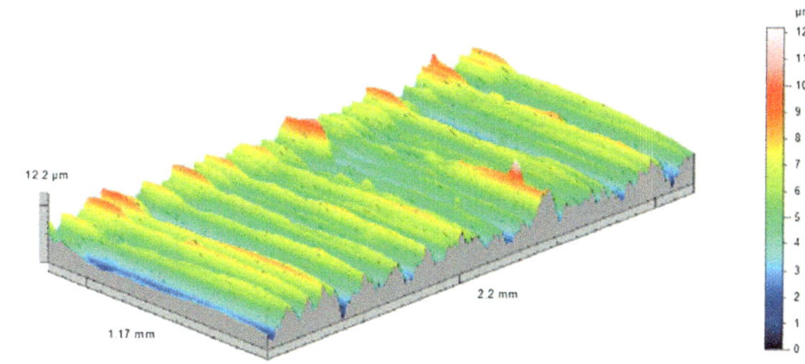

Fig. 37. Surface topography of a main bevel made by hand grinding.

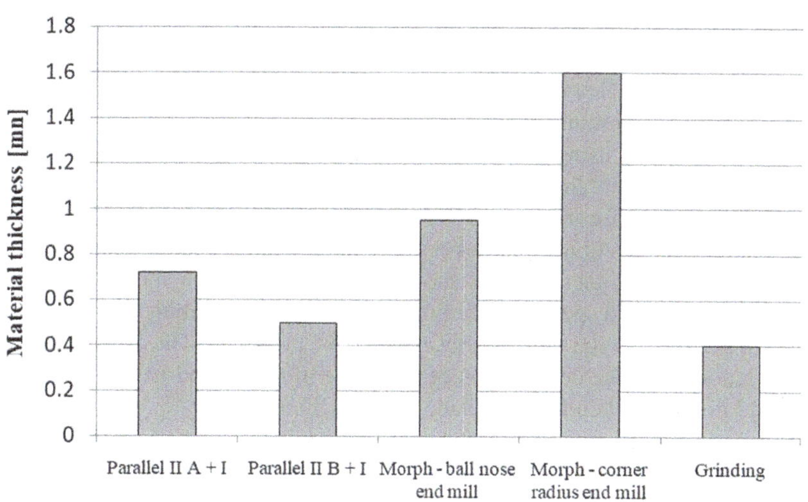

Fig. 38. Comparison of the allowance value for further processing.

Figure 38 shows a summary of the material thickness left on the cutting edge. The ideal value equal to 0.4 mm is shown in Fig. 39 depicting the knife model in the descent of main bevels view.

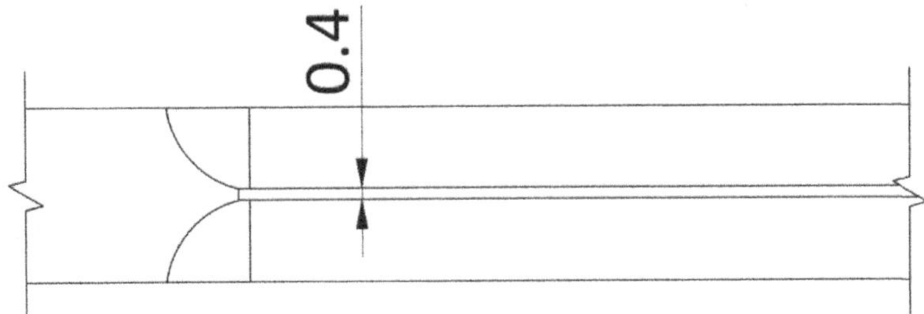

Fig. 39. The view from the forehead to the surface of the knife main bevel descent.

The grounded blade made it possible to obtain the most accurate allowance for further processing, and the bevels on both sides are symmetrical and converge equally to the knife's axis. Manual grinding allows continuous control of the obtained effects and the possibility of correction in the case obtaining too thick allowance.

4   Conclusions

The following conclusions can be made on the basis of the tests and results:

- using the CAM module in Inventor, the machining strategy was programmed with two different methods, the choice of the method itself did not affect the obtained results like the distance between the lines and the choice of the type of mill and its diameter,
- on the basis of the analysis of the tests, it can be stated that manual grinding is the most accurate machining process of tested knives, while among milling methods, the morph strategy using a ball nose end mill with a diameter of $D = 10$ mm,
- the use of concurrent and counter-milling significantly affected the deterioration of the surface quality for the knife made with the parallel strategy, one advantage was the significant reduction of the processing time,
- the most important element that should be improved when milling knives main bevels is the use of a special machining fixture with the possibility of supporting the knife together with the entire surface of one of the bevel. The lack of this in the work has affected the deformation of the shape of the second bevel on each of the milled knives, and bending the blade along its bottom edge,
- when designing the tool path in the area of the bottom main bevel line, additional tool paths should be placed outside the geometry of the model, so as to be sure to obtain the appropriate thickness of the finishing allowance.

References

1. Utkin J (2016) Hunting knives of Poland and Europe. Bellona, Warsaw
2. Bi W, Tan W, Yang Z, Ma A (2018) Study on standards of ceramic knives in contact with foodstuffs part 2: performance requirement. In: IOP conference series: materials science and engineering, vol 381, pp 1–3
3. Janusz OM (2016) An evaluation of modern day kitchen knives: an ergonomic and biomechanical approach, Iowa State University, Digital Repository, pp 16–17
4. Sykut B, Kowalik K, Opielak M (2005) Testing the effects of blade angles and blade putting on food product cutting resistance. Lublin University of Technology, Lublin
5. Hrisoulas J (1987) The complete bladesmith: forging your way to perfection. Paladin Press, Colorado
6. Pimenov DY, Guzeev VI, Krolczyk G, Mia M, Wojciechowski S (2018) Modeling flatness deviation in face milling considering angular movement of the machine tool system components and tool flank wear. Precis Eng 54:327–337
7. Pimenov DY, Bustillo A, Mikolajczyk T (2018) Artificial intelligence for automatic prediction of required surface roughness by monitoring wear on face mill teeth. J Intell Manuf 29(5):1045–1061
8. Antoniadis A, Savakis C, Bilalis N, Balouktsis A (2003) Prediction of surface topomorphy and roughness in ball-end milling. Int J Adv Manuf Technol 21:965–971
9. Wojciechowski S (2014) Cutting forces during ball-end milling of hardened steel. Poznan University of Technology, Poznań
10. Wojciechowski S, Twardowski P, Wieczorowski M (2014) Surface texture analysis after ball end milling with various surface inclination of hardened steel. Metrol Meas Syst 21(1):145–156
11. Ko TJ, Kim HS, Lee SS (2001) Selection of the machining inclination angle in highspeed ball end milling. Int J Adv Manuf Technol 17:163–164
12. Shan C, Lv X, Duan W (2016) Effect of tool inclination angle on the elastic deformation of thin-walled parts in multi-axis ball-end milling. Procedia CIRP 56:311–312
13. Wojciechowski S, Mrozek K (2017) Mechanical and technological aspects of micro ball end milling with various tool inclinations. Int J Mech Sci 134:424–435
14. Wojciechowski S, Wiackiewicz M, Krolczyk GM (2018) Study on metrological relations between instant tool displacements and surface roughness during precise ball end milling. Measurement 129:686–694
15. Bagci E, Yüncüoğlu EU (2017) The effects of milling strategies on forces, material removal rate, tool deflection, and surface errors for the rough machining of complex surfaces. J Mech Eng 63:643–656
16. Miko B, Baranyai G (2016) Comparison of milling strategies in case of free form surface milling. Dev Mach Technol 6:76–86
17. Pahole I, Studenčnik D, Gotlih K, Ficko M, Balič J (2011) Influence of the milling strategy on the durability of forging tools. J Mech Eng 57:898–903
18. Pervaiz S, Deiab I, Rashid A, Nicolescu M (2013) Experimental analysis of energy consumption in milling strategies. In: International conference on computer systems and industrial informatics. IEEE
19. http://www.alfa-tech.com.pl/stale-wysokostopowe-o-specjalnych-wlasnosciach-stalnierdzewna-4h13
20. http://www.oberonrd.pl/?p=main&what=58

Uncertainty of Sine Input Calibration Apparatus for the Air Gauges

Michal Jakubowicz[1](✉), Miroslaw Rucki[2], and Matej Babic[3]

[1] Institute of Mechanical Engineering, Poznan University of Technology,
Poznan, Poland
michal.jakubowicz@put.poznan.pl
[2] Faculty of Mechanical Engineering, Kazimierz Pulaski University
of Technology and Humanities in Radom, Radom, Poland
[3] Jožef Stefan Institute, Ljubljana, Slovenia

Abstract. In the paper, dynamic calibration of air gauges with small chambers is discussed. Proposed sine input test rig is described, and its uncertainty is evaluated. Three main sources of uncertainty are identified as a Mechanical Unit, Pneumatic Unit and Digital Unit. Type B uncertainty estimation is considered, and the most difficult to assess sources are pointed out. Namely, it is the fluctuation of the back-pressure due to the airflow inside the air gauge. Thus, it was proposed to estimate the Type A uncertainty through the repetitions of the time constant measurement. The results provided highly satisfactory level of expanded uncertainty ca. 0.001 s. This result is covering all factors of influence on the dynamic calibration result.

Keywords: Air gauge · Dynamic calibration · Time constant ·
Frequency domain · Uncertainty

1 Introduction

Air gauges are usually mentioned among other precise measuring devices, primarily but not exclusively used for in-process control [1]. Air gauges are able to work as an automatic gauging devices in harsh dynamic conditions like electrochemical honing in-process control [2] and find their applications even in untypical devices like in humanoid robots falling forward experiments [3]. However, usually engineers do not calculate the exact response time or amplitude characteristics of the air gages. For instance, Grandy et al. [4] used the simplified dynamic models with reference to the geometric parameters of an air gage. Liu et al. [5] considered the velocity of the gage head during the form measurement relatively low and ignored its dynamic error. It was proved that the dynamics of the air gages with small volumes (ca. 0.5 to 4.0 cm^3) combined with a piezoresistive pressure transducer could be treated as a first-order dynamic system [6], and be applied for the measurement in dynamic conditions, like in-process measurement or a form errors assessment.

Traditionally, quality control is performed offline, after a part is produced [7]. However, the automated quality control becomes more and more important aspect of modern manufacturing process [8]. Advanced manufacturing technology characterized

© Springer Nature Switzerland AG 2019
M. Diering et al. (Eds.): *Advances in Manufacturing II* - Volume 5, LNME, pp. 82–94, 2019.
https://doi.org/10.1007/978-3-030-18682-1_7

by high accuracy and efficiency is focused on intelligent machining with its online monitoring and optimization [9]. It is crucial to apply the inspection technologies that allow data to be collected online along the process chain in order to increase quality control significantly in current dynamic and modifiable environments [10].

A detailed study on roundness measurement with air gauges proved their ability to work in dynamic conditions and to provide reliable results for further digital processing [11]. However, it was crucial to determine the uncertainty of dynamic calibration in order to gain full understanding of the dynamic accuracy of the air gauges.

2 Sine Input Calibration Apparatus

General rules for measurement uncertainty estimation in calibration process are presented in the document EA-4/02 M: 2013 [12]. The term "dynamic calibration" for a pressure quantity is used to define the process of characterizing a dynamic pressure transducer, determining its properties such as natural frequency, damping, peak time, stabilization time and sensitivity [13]. Unlike well defined by metrology organizations calibration procedures for a sensor under static pressure conditions, dynamic calibration of a piezoelectric pressure transducer has no traceable reference standard. Similarly, while the uncertainty of an air gauge test rig for static characteristics is relatively easy to determine [14], air gauge dynamic calibration equipment require additional analysis. Dynamic variables are time or space dependent in both their magnitude and frequency content. A dynamic calibration determines the relationship between an input of known dynamic behavior and the measurement system output. Usually, such calibrations involve applying either a sinusoidal signal or a step change as the known input signal [15]. In case of piezoelectric pressure transducers, it was demonstrated that the general methodology proposed by GUM (Guide to the Expression of Uncertainty in Measurement) methodology [16] is adequate for uncertainty estimation [17]. In some cases, it is useful to calculate the equipment variation EV according to the widely known procedure described in detail by Dietrich and Schulze [18]. Then it can be assumed that an examined air gauge in the experimental rig constitutes a measurement system itself. This approach was demonstrated to be appropriate by Rucki and Barisic [19].

The investigated dynamic calibration apparatus for the air gauges is presented in Fig. 1. In the analysis, three main units should be viewed separately: Mechanical Unit, Pneumatic Unit and Digital Unit.

The Mechanical Unit consists of the DC electric motor with the belt transmission and the eccentric shaft. Dependent on rotational speed of the shaft, its distance from the air gauge s is periodically changing with an eccentricity value $\pm e$. The rotational speed ω is set from the main computer and provides the sine input $s(\omega t)$ to the air gauge. The unit is able to generate input frequency in the range from 0.2 to 20 Hz, with a respective step of 0.2 Hz.

The Pneumatic Unit is similar to the pneumatic path of the back-pressure experimental rig described in [20]. It consists of a typical back-pressure air gauge with inlet

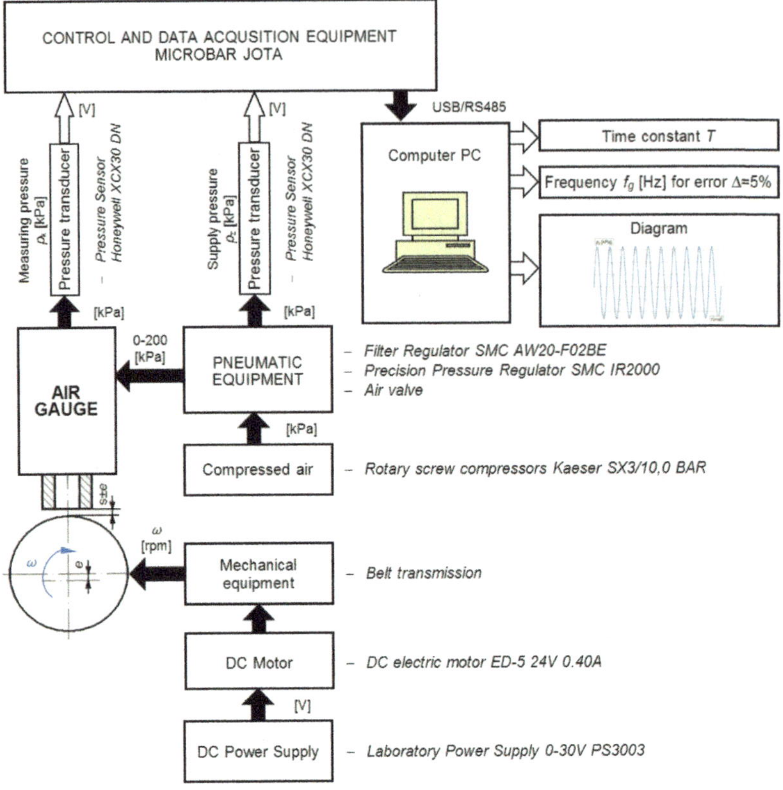

Fig. 1. Air gauge sine input apparatus.

and measuring nozzles d_w and d_p, respectively [21], where the pressure p_k inside the measuring chamber of volume V_k represents the dimension-dependent slot width s:

$$p_k = f(s). \tag{1}$$

Since only small volumes with integrated piezoresistive pressure transducers were proven to have good dynamic properties [6], the experiments were performed with volumes between $V_{min}= 0.5$ and $V_{max}= 4.0$ cm^3.

It is important to keep in mind that the air gauge static characteristics represented by the Eq. (1) are dependent on the geometrical dimensions, especially the diameters of inlet and measuring nozzles, d_w and d_p, respectively. These dimensions determine a range of the thermodynamic phenomena that take place inside the air gauge body and have their impact on the metrological characteristics, just to mention one of fundamental publications by Breitinger [22].

The back-pressure is measured with a pressure transducer Honeywell XCX 30DN (non-linearity <0.5% of span, repeatability 0.1%, compensated temperature range 25 °C). The response time was below $T = 0.1$ ms, so this type of transducers was found appropriate for the sine input calibration of the air gauges that performed at best $T = 2.0$ ms [19].

And, finally, the Digital Unit consists primarily of the dedicated MicroBar device made by JOTA company, which is able to register changes of the pressure with high-frequency sampling (range between 16 Hz and 4 kHz), and the main computer. The data processing software enabled analysis of the recorded changes of back-pressure $p_k = f(t)$ in the measuring chamber of the examined air gage model. The main computer had two functions: control and signal processing. The operator should input information about the examined air gage model (diameters and other characteristics of the nozzles, initial and final displacement, sampling step and so on), and the measurement could be started by the control program.

3 Uncertainty Sources

For indicating instruments, it is sufficient to rate the response time, which is a partial dynamic characteristics [23]. Having the registered sine input response of the system, it is possible to calculate its time constant T. For different frequencies, the amplitude-frequency characteristics are made, so the ratio of amplitudes corresponding with respective frequencies ω_i can be calculated as follows:

$$A(\omega) = \frac{A(\omega_i)}{A(\omega_0)}. \tag{2}$$

Knowing that the ratio of amplitudes is bond with the time constant T (Eq. 3), it is possible to derive T from the series of measurement of $A(\omega_i)$.

$$\frac{A(\omega_i)}{A(\omega_0)} = \frac{1}{\sqrt{1 + T^2 \omega_i^2}}. \tag{3}$$

Thus, it can be written as follows [24]:

$$T = \frac{\sum_{i=1}^{n} \frac{1 - A^2(\omega)}{A^2(\omega)}}{\sum_{i=1}^{n} \left(\omega \sqrt{\frac{1 - A^2(\omega)}{A^2(\omega)}}\right)}. \tag{4}$$

Functional dependence of the final value on the measured amplitude A and rotational speed ω is one of the uncertainty sources. However, the values $A(\omega)$ are themselves affected additionally with the analog to digital conversion [25] with its error of the instant output value:

$$e_i = y_i - x_i \approx \Delta_G x_i + \Delta_{OFF} + \Delta_{INL}(x + n) + n_0 + \Delta_{DNL}(x + n_0) + \Delta_q(z), \tag{5}$$

where:

n – noise,

Δ_G, Δ_{OFF}, Δ_{INL}, Δ_{DNL}, Δ_q – errors of the multiplication, offset, integral non-linearity, differential non-linearity and quantization, respectively.

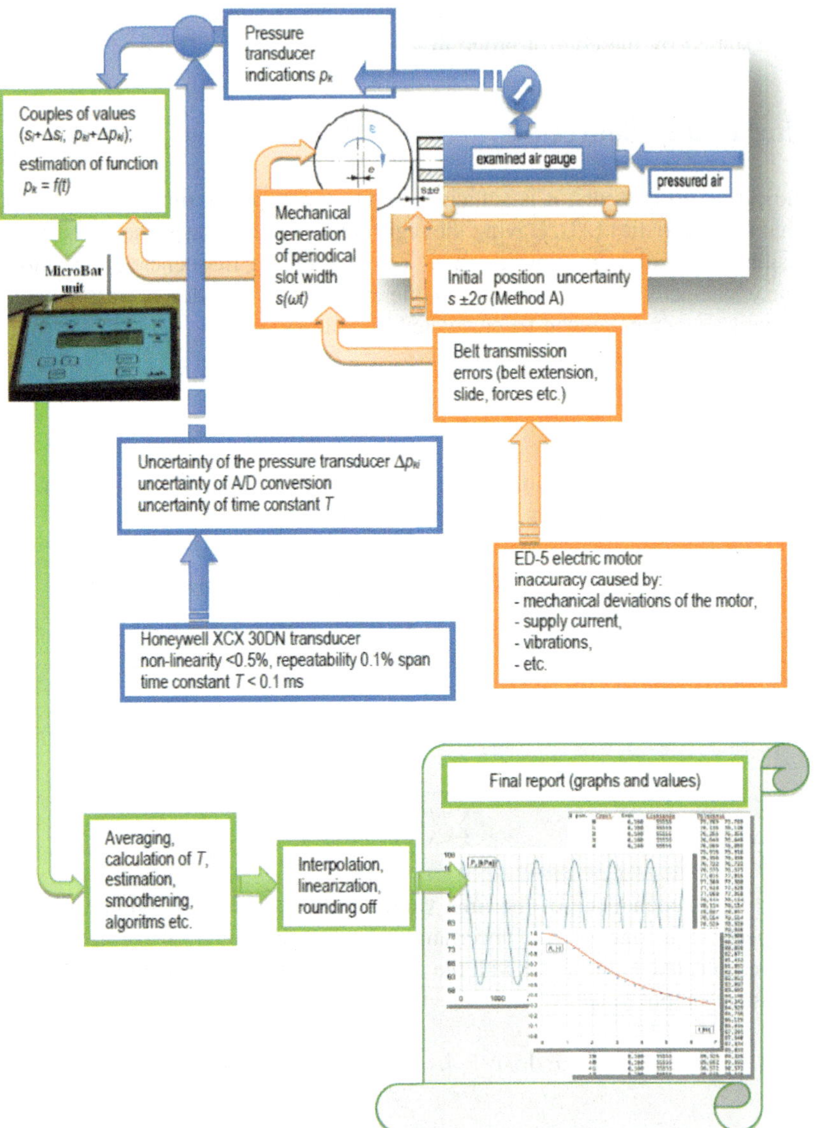

Fig. 2. Sources of uncertainty in the air gauge sine input test rig.

Because the true values of the instant errors are not known, the producers provide the ranges covering those errors:

$$|\Delta_G| \le M_g; \; |\Delta_{OFF}| \le M_{OFF}; \; |\Delta_{INL}| \le M_{INL}; \; |\Delta_{DNL}| \le M_{DNL}. \tag{6}$$

Finally, the combined standard uncertainty of D/A conversion can be calculated as follows:

$$u_C(e_i) = \sqrt{u_G^2 x^2 + u_{OFF}^2 + u_{INL}^2 + u_{n0}^2 + u_{DNL}^2 + u_q^2}. \tag{7}$$

Moreover, both Mechanical and Pneumatic units add numerous uncertainties to the final result, as it is shown in Fig. 2.

Of course, there are numerous other sources of uncertainty apart from those of Mechanical, Pneumatic and Digital Units. Figure 3 presents the diagram of the most important ones that contribute to the uncertainty final results.

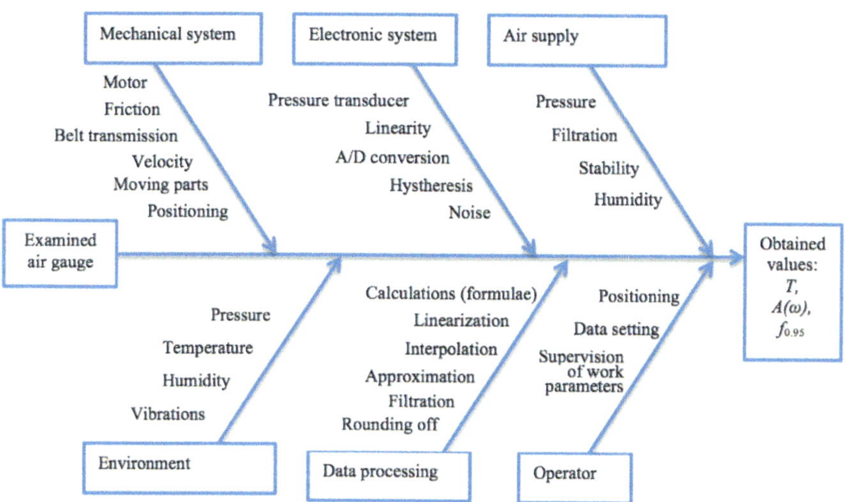

Fig. 3. Uncertainty sources for the air gauge dynamic calibration apparatus.

4 Results of Measurement and Uncertainty Estimation

In the experiments, configuration of a typical air gauge with measuring nozzle $d_p = 1.810$ mm and multiplication $|K| = 1$ kPa/μm was used. Its linear characteristics $p_k = f(s)$ approximated with non-linearity $\delta = 1\%$ was in range of $z_p = 59$ μm (between the slot widths $s_p = 52$ and $s_k = 111$ μm), as it is shown in Fig. 4.

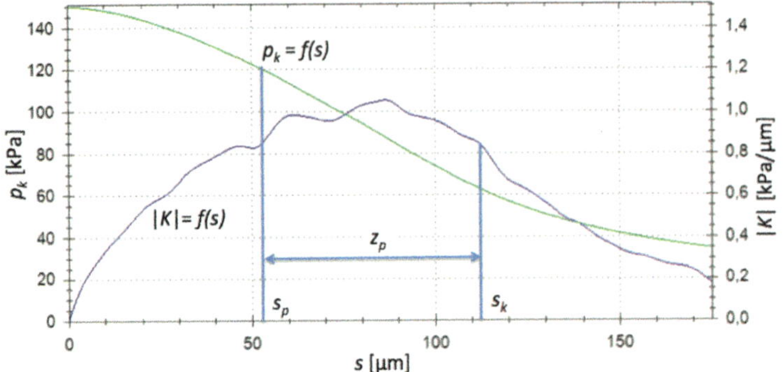

Fig. 4. Static characteristics of the examined air gauge.

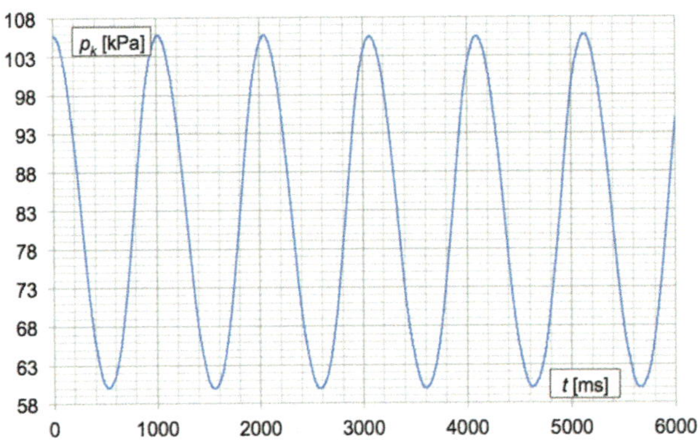

Fig. 5. Sine response graph $p_k = f(t)$ of the examined air gauge.

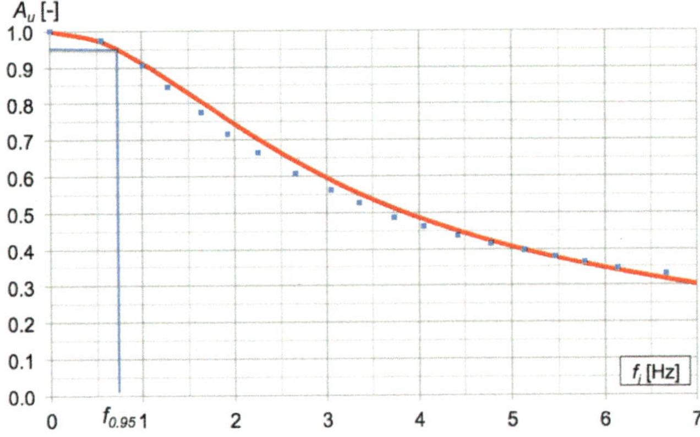

Fig. 6. Example of obtained amplitude-frequency characteristics.

The example results of dynamic calibrations are presented graphically as a sine response of the examined air gauge in Fig. 5 and as an amplitude-frequency characteristic in Fig. 6.

From the amplitude-frequency graph, it is possible to determine the maximal frequency $f_{0.95}$ which generates the dynamic error 5% of the amplitude. As it is shown in Fig. 6, for the examined air gauge it was $f_{0.95} = 0.733$ Hz.

4.1 Back-Pressure Fluctuations

Since many publications have reported the effect of pressure fluctuation in static conditions (when the slot s remains unchanged) [26], it was necessary to evaluate the impact of this phenomenon. Thus, the eccentric shaft was put into position of maximal slot width $s_{\max} = s - e$, and the measurement of back-pressure p_k was performed for 10 s with sampling frequency 1000 Hz. This way 10,000 results were obtained, so the statistical analysis of the p_{ki} results enabled to assess the impact of back-pressure fluctuations inside the measuring chamber. Similarly, the experiment was performed for the minimal slot width $s_{\min} = s + e$. The respective graphs and histograms are shown in the Figs. 7 and 8.

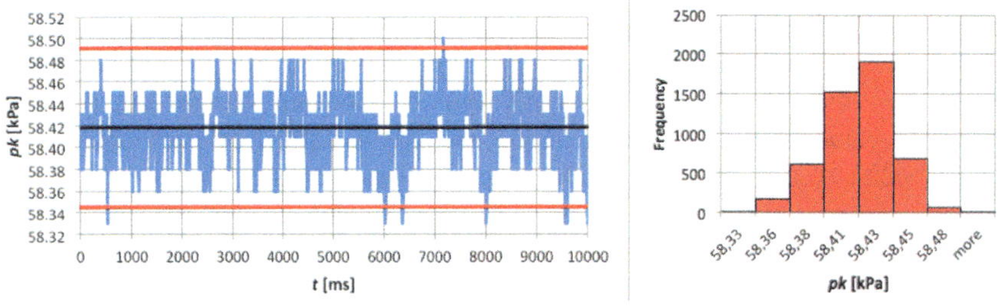

Fig. 7. Fluctuations of back-pressure p_k for unchanged maximal slot $s_{\max}$.

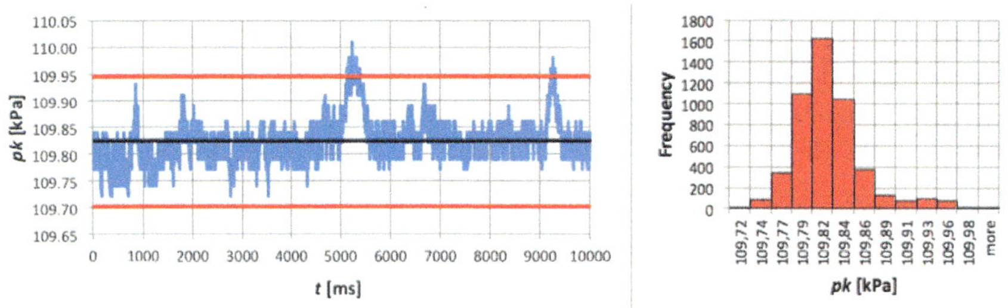

Fig. 8. Fluctuations of back-pressure p_k for unchanged minimal slot $s_{\min}$.

The graphs in Figs. 7 and 8 demonstrate that the values of dispersion range ($R = p_{k\max} - p_{k\min}$) differ substantially. In case of maximal slot, dispersion was $R = 0.17$ kPa, while for minimal slot it was 70% larger, i.e. $R = 0.29$ kPa. Moreover, histograms are skewed in opposite direction, with respective statistical skewness of –0.56 and 1.06. Standard deviations were 0.0243 kPa and 0.0406 kPa, respectively, so it could be assumed that the maximal standard uncertainty component of static p_k measurement caused by airflow fluctuations for any slot between $s_{\max}$ and $s_{\min}$ was $u(p_k) = 0.04$ kPa.

It is worth noting that the fluctuations were registered also in the stabilizer of feeding pressure p_z. Here, for both $s_{\max}$ and $s_{\min}$ the dispersion range was smaller than that measured in the air gauge chamber. Moreover the distribution of values was close to normal, as it is seen in Fig. 9.

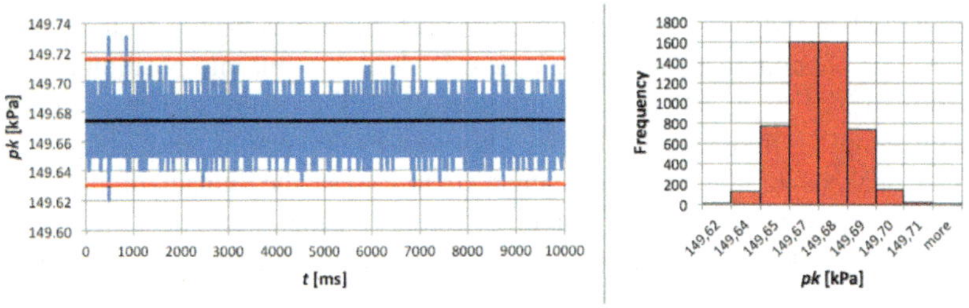

Fig. 9. Fluctuations of feeding pressure p_z for unchanged maximal slot $s_{\max}$.

The feeding pressure p_z in stabilizer was set on the value 150 kPa, but in reality the slot had some impact on its value. This way, minimal slot causes slight drop of p_z down to the mean value 149.67 kPa, and maximal slot causes further pressure drop down to the mean value 149.71 kPa. Statistically, in case of maximal slot, dispersion of feeding pressure p_z was $R = 0.11$ kPa, while for minimal slot it was even smaller, i.e. $R = 0.09$ kPa. Standard deviations were 0.0142 kPa and 0.0170 kPa, respectively, which was much smaller than those for back-pressure p_k.

4.2   Amplitude Approximation

As it is seen in Fig. 6 above, the experimental points do not lay exactly on the approximated curve. To evaluate the approximation error δA, $i = 30$ repetitions were made for amplitude-frequency characteristics obtaining subsequent values of normalized experimental amplitudes A_{ui} and corresponding theoretical values A_{uti}. Then the square measure of the approximation error was calculated from the formula:

$$\delta A = \sqrt{\frac{1}{n}\sum \left(A_{ui} - A_{uti}\right)^2}. \tag{8}$$

The results obtained for 30 repetitions are shown in Fig. 10 as a histogram (left) and as a subsequent measurements (right).

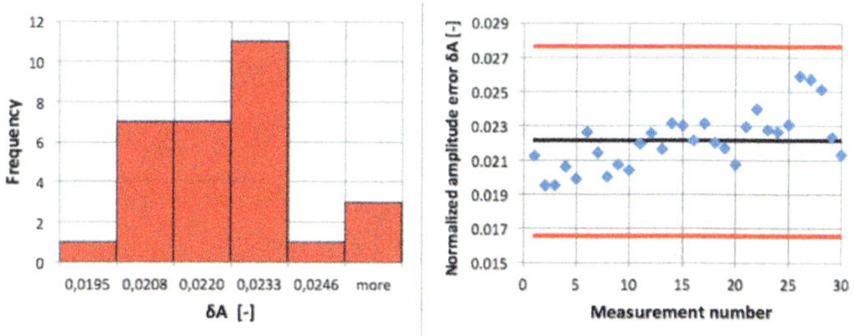

Fig. 10. Errors of amplitude approximation δA for 30 repetitions.

The latter reveals a slight trend towards increasing values of δA. Statistically, for 30 repetitions, its mean value is 0.0221, and maximal $\delta A_{max} = 0.0259$. Hence, it can be assumed that the amplitude is calculated with relative error no larger than 2.6%

4.3   Overall System Uncertainty

Since it is difficult to predict pressure fluctuations due to airflow in the measuring chamber and it is rather impossible to have this phenomenon under control or supervision, it was decided to assess the overall uncertainty statistically as a Type A uncertainty [16]. In order to do that, 30 repetitions under repeatability conditions were performed with supervised values of temperature, humidity and atmospheric pressure. From each full-cycle repetition, final value of time constant T was calculated along with the related frequency $f_{0.95}$ that generated a dynamic error below 5%. Figures 11 and 12 present the respective graphs.

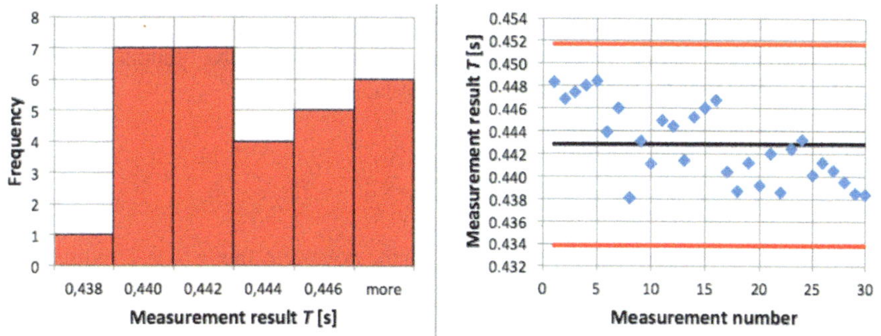

Fig. 11. Time constants T obtained for 30 repetitions.

Some slight declining and rising trends can be noted for subsequent measurements of respective values of time constant T and frequency $f_{0.95}$. In Table 1, there are statistical parameters of the obtained values, including amplitude approximation error δA for the dynamic characteristics.

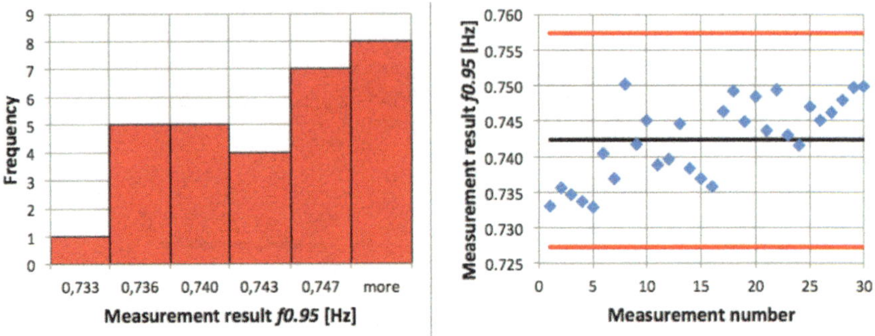

Fig. 12. Dynamic frequency $f_{0.95}$ obtained for 30 repetitions.

Table 1. Statistical analysis for 30 repetitions.

Value	T [s]	$f_{0.95}$ [Hz]	δA [-]
Mean	0.4428	0.7424	0.0221
Maximal	0.4484	0.7502	0.0259
Minimal	0.4381	0.7329	0.0195
R (Max − Min)	0.0103	0.0173	0.0064
St. deviation σ	0.0030	0.0050	0.0018
3σ	0.0089	0.0150	0.0055
Mean + 3σ	0.4517	0.7573	0.0277
Mean − 3σ	0.4338	0.7274	0.0166

For the Gaussian distribution, it can be assumed that the standard uncertainty $u(x)$ is equal to the standard deviation σ. However, the histograms presented in the Figs. 11 and 12 above are closer to the rectangular distribution, so the standard uncertainty should be calculated from the following formula:

$$u(x) = \frac{\frac{1}{2}R}{\sqrt{3}}. \tag{9}$$

Expanded uncertainty U_{95} was calculated for coverage factor $k = 2$, which corresponds with approximately 95% confidence level. The Table 2 presents values of uncertainties:

Table 2. Statistical analysis for 30 repetitions.

Uncertainty	T [s]	$f_{0.95}$ [Hz]
$u(x)$	0.0005	0.0009
U_{95}	0.0010	0.0018

Thus, it can be stated that the examined sine input apparatus enabled to determine the time constant T of air gauge with expanded uncertainty $U_{95} = 0.001$ s, and its working frequency $f_{0.95}$ with expanded uncertainty $U_{95} = 0.0018$ Hz. These uncertainties estimated using Type A method covered all the determinants that had impact on the measurement results.

5 Conclusions

The analysis and experiments described in the paper were performed in order to evaluate uncertainty of the dynamic calibration of an air gauge. The sources of uncertainty were divided to three main groups: Mechanical Unit, Pneumatic Unit and Digital Unit. From 30 repetitions, it was found that amplitude approximation error did not exceed 2.6%.

However, analysis proved that some sources of uncertainty cannot be assessed theoretically, such as back-pressure fluctuations caused by flow instability. It was possible to measure it for this particular air gauge, but in case of other configuration or even shape of flow-through elements, its value might appear significantly different. Thus, it was assumed that the best way was to calculate Type A uncertainty based on statistical analysis of final results.

The final results may be viewed as highly satisfactory. Namely, the time constant T of the air gauge was determined with expanded uncertainty $U_{95} = 0.001$ s, and its working frequency $f_{0.95}$ with expanded uncertainty $U_{95} = 0.0018$ Hz.

Acknowledgments. The research work reported here was performed under the subject of No. 02/22/DSPB/1432, funded with grants for education allocated by the Ministry of Science and Higher Education in Poland.

References

1. Koenig B (2017) Marriage of convenience. Manuf Eng 158(3):59–63
2. Rao PS, Jain PK, Dwivedi DK (2017) Optimization of key process parameters on electro chemical honing (ECH) of external cylindrical surfaces of titanium alloy Ti 6Al 4V. Mater Today Proc 4(2 Part A):2279–2289
3. Zhang Z, Liu H, Yu Zh, et al. (2017) Biomimetic upper limb mechanism of humanoid robot for shock resistance based on viscoelasticity. In: Proceedings of 2017 IEEE-RAS 17th international conference on humanoid robotics (humanoids), pp 637–642
4. Grandy D, Koshy P, Klocke F (2009) Pneumatic non-contact roughness assessment of moving surfaces. CIRP Ann Manuf Technol 58(1):515–518
5. Liu J, Pan X, Wang G, Chen A (2012) Design and accuracy analysis of pneumatic gauging for form error of spool valve inner hole. Flow Meas Instrum 23:26–32
6. Rucki M, Jermak CzJ (2012) Dynamic properties of small chamber air gages. J Dyn Syst Meas Contr 134(1):011001 (6 pages)
7. Gao RD, Tang X, Gordon G, Kazmer DO (2014) Online product quality monitoring through in-process measurement. CIRP Ann Manuf Technol 63(1):493–496

8. Milo MW, Roan M, Harris B (2015) A new statistical approach to automated quality control in manufacturing processes. J Manuf Syst 36:159–167

9. Chen M, Wang Ch, An Q, Ming W (2018) Tool path strategy and cutting process monitoring in intelligent machining. Front Mech Eng 13(2):232–242

10. Brosed FJ, Zaera AV, Padilla E, Cebrian F, Aguilar JJ (2018) In-process measurement for the process control of the real-time manufacturing of tapered roller bearings. Materials 11 (8):1371 (17 pages)

11. Jermak CzJ, Rucki M (2012) Air gauging: static and dynamic characteristics. IFSA, Barcelona

12. European Co-operation for Accreditation (2013) EA-4/02 M: 2013 evaluation of the uncertainty of measurement in calibration

13. Theodoro F, Reis M, d'Souto C, Barros E (2016) Measurement uncertainty of a pressure sensor submitted to a step input. Measurement 88:238–247

14. Jermak C, Jakubowicz M, Derezynski J, Rucki M (2017) Uncertainty of the air gauge test rig. Int J Precis Eng Manuf 18(4):479–485

15. Figliola RS, Beasley DE (2011) Theory and design for mechanical measurements, 5th edn. Wiley, Hoboken

16. BIPM/JCGM 100:2008 (2008) Evaluation of measurement data – guide to the expression of uncertainty of measurement, Joint Committee for Guides in Metrology BIPM

17. Theodoro F, Reis M, d'Souto C, Barros E (2018) An overview of the dynamic calibration of piezoelectric pressure transducers. J Phys Conf Ser 975:012002 (5 pages)

18. Dietrich E, Schulze A (2010) Statistical procedures for machine and process qualification. Hanser Publications, Cincinnati

19. Rucki M, Barisic B (2009) response time of air gauges with different volumes of the measuring chambers. Metrol Meas Syst 16(2):289–298

20. Jermak C, Rucki M, Jakubowicz M (2018) Air gauge back-pressure uncertainty estimation for the advanced test rig. Int J Mater Prod Technol 57(4):274–286

21. Jakubowicz M, Rucki M, Varga G, Majchrowski R (2018) Influence of the inlet nozzle diameter on the air gauge dynamics. In: Advances in manufacturing. Lecture notes in mechanical engineering. Springer, pp 733–743

22. Breitinger R (1969) Fehlerquellen beim pneumatischen Längenmessen. Dissertation, TU, Stuttgart (in German)

23. Rabinovich SG (2018) Evaluating measurement accuracy: a practical approach, 3rd edn. Springer, Cham

24. Soboczyński RJ (1978) Research on metrological properties of pneumatic high-pressure devices. Dissertation, Wroclaw University of Technology, Wrocławska (in Polish)

25. Domanska A (2011) Uncertainty of the analog to digital conversion result. In: Fotowicz P et al (eds) Measurement Uncertainty: Theory and Practice. GUM, Warsaw, pp 22–31 (in Polish)

26. Jermak CzJ (2017) Discussion on flow-through phenomena in the air gauge cascade. Acta Mechanica et Automatica 11(1):38–46

Testing Geometric Precision and Surface Roughness of Titanium Alloy Thin-Walled Elements Processed with Milling

Józef Kuczmaszewski[1], Kazimierz Zaleski[1], Jakub Matuszak[1(✉)], and Janusz Mądry[2]

[1] Department of Production Engineering, Lublin University of Technology, Lublin, Poland
j.matuszak@pollub.pl
[2] Polskie Zakłady Lotnicze Sp. z o.o. - PZL Mielec, Mielec, Poland

Abstract. Many branches of industry aim to minimize the weight of the structure. This can be achieved by selecting a low-density construction material or by making thin-walled structures. The specificity of the aerospace industry requires a combination of these two methods of manufacturing the structure. During machining of thin-walled components, there are a number of problems associated with the reduced rigidity. The main problems include issues with obtaining the required accuracy and assumed quality of the machined surface. One of the causes of these difficulties may include the occurrence of vibrations of thin-walled elements while machining. The paper presents the results of geometric accuracy and surface roughness after the milling process of thin-walled elements using various machining strategies. Titanium alloy Ti6Al4V samples were used for studies. The tests were carried out on a three-axis machining centre using four strategies for making walls with the same dimensions. Fixed cutting technological parameters included: cutting speed v_c, feeding rate v_f, milling width a_e. The variation was based on the application of cutting depths specific for the axial strategies. A significant impact of the machining strategy was shown on the geometric accuracy and quality of the processed surface.

Keywords: Milling · Geometric accuracy · Surface roughness · Thin-walled elements

1 Introduction

Titanium alloys are structural materials widely used in machine construction. The major fields of application of these alloys should include the aerospace, marine, automotive, chemical and metallurgical industries. Titanium alloys are used for manufacturing implants used in medicine, while titanium and fibre composites are also used for making hybrid laminates [1, 2].

High attractiveness of titanium alloys, as construction materials, is associated with their properties, such as high resistance at low density, high resistance to corrosion and high temperature, good tolerance by the human body. Titanium alloys are characterised

© Springer Nature Switzerland AG 2019
M. Diering et al. (Eds.): *Advances in Manufacturing II* - Volume 5, LNME, pp. 95–106, 2019.
https://doi.org/10.1007/978-3-030-18682-1_8

by good fatigue properties. The fatigue resistance of the elements made of these alloys can be greatly enhanced by improving the surface properties of these elements, subjecting them to surface treatment with burnishing or peening [3–5]. The properties of the top layer of titanium alloys also affect the effectiveness of adhesive joints [6].

Production of titanium alloys is often accompanied by problems related to the difficult machinability of this material. When cutting titanium alloys, the tools are worn quickly due to the very low thermal conductivity of these alloys, the tendency for the formation of accretion and friction, and significant cutting forces. The cemented carbides are a tool material which is widely used in the treatment of titanium alloys. Studies on the wear mechanisms of cemented, uncoated carbide plates, with anti-wear layer were conducted by the authors of [7–9]. The increasing wear of the blade along with the increase in the cutting path results in an increase in roughness of the treated surface, while this dependence is highly dependent on the cutting speed [10–12].

In many branches of industry, such as aviation and automotive industries, also thin-walled components are used, which is associated with a desire to reduce the weight of manufactured products. These elements are characterised by low rigidity during machining, they deform due to the influence of cutting forces, resulting in vibration in the cutting process. Testing the stability of machining thin-walled components is carried out using various methods, such as the modal analysis, finite element method, recursive analysis [13–15]. The vibrations occurring during the processing of thin-walled components affect the deterioration of the accuracy and increase the surface roughness of the workpieces [16, 17].

Particularly severe difficulties occur during machining of thin-walled components of titanium alloys, which is associated with high cutting forces, the relatively small Young's module (e.g. for steel, Young's module $E = 210$ GPa, and for titanium alloys $E = 110$ GPa) and the intensive wear of cutting edges. The authors of [18] have found that the vibrations during "dry" milling of titanium alloy thin-walled components depend on the cutter's pitch – the displacement is smaller for variable-pitch cutters than for even ones. The paper [15] presented cutting stability graphs (so-called "bag" curves) for workpieces with different rigidity. Gang [19] showed that the finite element method can be used for the simulation of the titanium alloy thin-walled components cutting process. The paper [20] found that the parameters of the trochoidal milling of thin-walled elements of titanium alloys affect the errors of wall thickness and roughness of their surface.

Previous studies on the treatment of titanium alloy thin-walled components have focused mainly on assessing the impact of cutting parameters, tool geometry and shape of the workpiece on the vibrations of the cutting speed and quality of the processed surface. This paper presents the research results of the influence of different methods for removing the machining surplus in the milling process of thin-walled components of Ti6Al4V titanium alloy on the geometric accuracy of these walls and the roughness of their surfaces.

2 Methodology of Research

Samples made of Ti6Al4V titanium alloy were used for studies. The chemical composition and physical properties of the material, according to the delivery card, are presented in Table 1. The Ti6Al4V titanium alloy, due to its properties, belongs to the group of materials that are difficult to process.

Table 1. Chemical composition and physical properties of the Ti6Al4V titanium alloy.

Chemical composition, %					Mechanical properties		
Al	V	C	Fe	Ti	Rm	HRC	E
6,25–6,31	4,09–4,12	0,026–0,027	0,18–0,21	Rest	1014 MPa	33	120 GPa

The milling process was performed on the three-axis machining centre Avia VMC-800HS, equipped with Heidenhain control. The machine has a high power spindle of 25 kW, and a maximum speed of 24 thousand rpm, so it is dedicated to the HSM processing. The tests were performed using Mitsubishi (VF-JHV 12) cutters with a diameter d = 12 mm. The working length of the tool is 30 mm. Four-blade tools made of cemented carbides are dedicated to processing materials, which are difficult to process.

Coarse processing consisted of making a 1,8 mm thick wall. While tests of geometric accuracy and roughness were performed after producing a 1 mm thick and 29 mm high wall in the finishing process. Wall height (29 mm) was adopted due to the maximum working length of the tool. The cutting speed was set to v_c = 60 m/min, while the feeding rate f_z = 0,07 mm/blade. Both sides of the walls were machined removing the material surplus from each side (a_e = 0,4 mm).

Four finishing strategies were applied in the finishing process. The method of treatment for each strategy is illustrated in Fig. 1. Strategy A consisted of machining one side of the wall in one pass, the cutting depth a_p is equal to the height of the wall. Strategy B assumed alternating machining with a cutting depth a_p = 7 mm (the numbers in Fig. 1 denote subsequent passes). Strategy C is similar to the second one, with a reduction in cutting depth to a_p = 5 mm. Strategy D, in addition to the lower cutting depth a_p = 3 mm, is still different from the other ones with its "spruce" way of embedding on the opposite sides of the walls, as shown in Fig. 1.

The machine times were determined for the presented strategies from the dependence:

$$t_m = \frac{l_d + l_p + l_w}{v_f} * i \tag{1}$$

assuming the runway and overrun equal to half the cutter's diameter $l_d = l_w$ = 6 mm. The feed rate was determined as v_f, while l_p indicated the cutting distance.

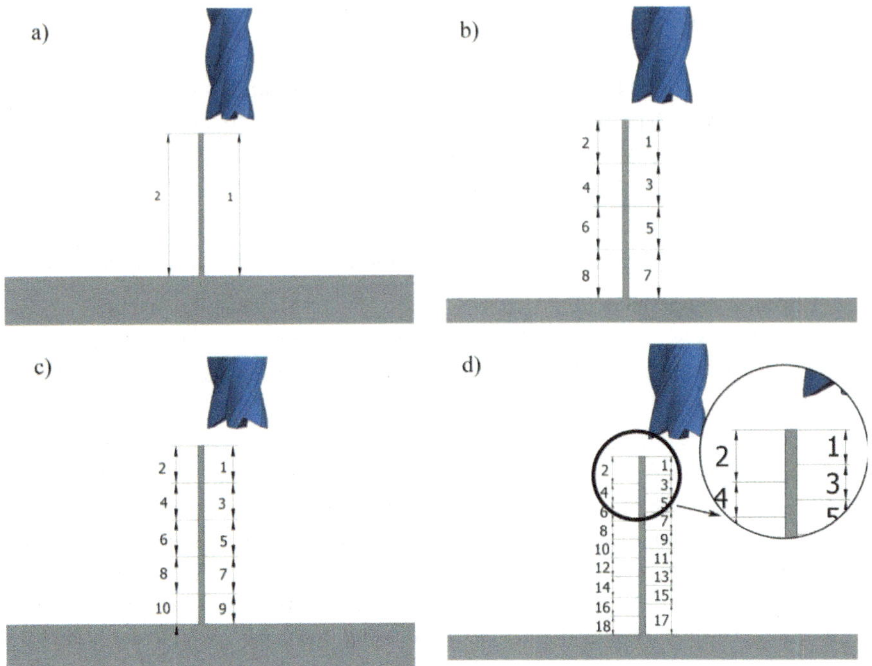

Fig. 1. Wall treatment strategies: (a) strategy A – full depth (a_p = 29 mm); (b) strategy B alternating (a_p = 7 mm), (c) strategy C – alternating (a_p = 5 mm), strategy D – equilateral (a_p = 3 mm).

The number of passes and needed to perform the wall for individual strategies was: strategy A - i = 2, strategy B - i = 8, strategy C - i = 10, strategy D - i = 18.

After machining, the geometric accuracy measurements were performed directly on the machine using the TS640 inspection probe. Figure 2 shows the methodology for measuring the geometric accuracy of the wall. Measurements were taken at five height levels in five points along the wall on the left and right side of the wall. The total number of measurement points for each wall, based on which the wall thickness measurement was taken, was fifty. In addition, the wall shape was illustrated in the Surfer Golden Software program using the data from the measurement probe.

Surface roughness and surface topography tests were carried out on the 3D T8000 RC120-400 roughness measuring device of the Hommel-Etamic company. For each machining strategy the study was repeated five times, from which the means and standard deviations were determined. Bar graphs were prepared using the statistica software.

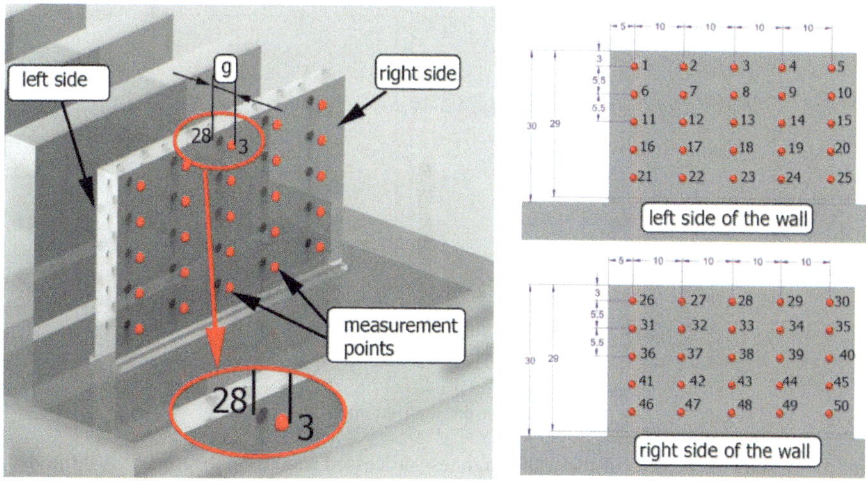

Fig. 2. The measurement method of the geometric accuracy of the wall.

3   Research Results

Figure 3 shows the example thin-walled specimen after milling.

Fig. 3. Thin-walled specimen after machining.

Figure 4 shows the results of the surface roughness measured at different distances from the bottom of the wall. There is a significant increase in roughness noticeable in the upper wall area, indicating the possibility of vibrations both during the treatment of the left and the right side of the wall.

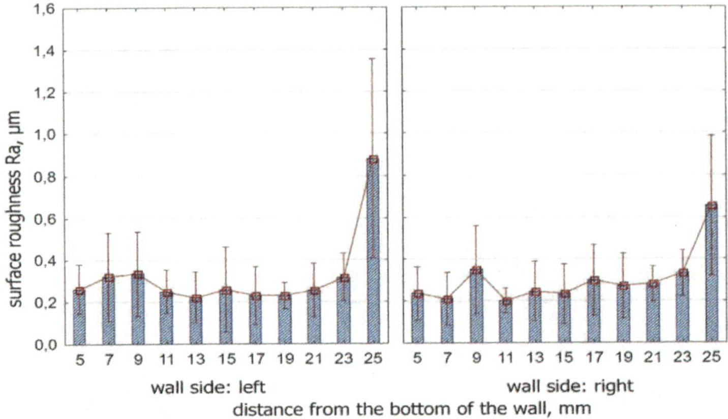

Fig. 4. Results of the tests of the wall roughness processed according to strategy A – full depth: (a) left side of the wall, (b) right side of the wall.

An exemplary topography made in the area of roughness increase is shown in Fig. 5. A distinctive pattern of micro-roughness evoked by the emergence of vibrations can be seen.

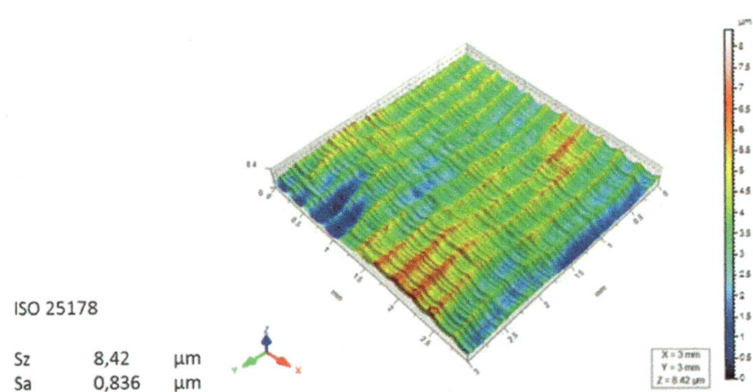

Fig. 5. Topography prepare in the upper area of the milled wall according to strategy A – full depth.

Figure 6 shows the results of roughness measurements according to strategy B. No significant roughness differences in the distance from the wall's depth function were observed in the entire tested area, indicating that the treatment was stable.

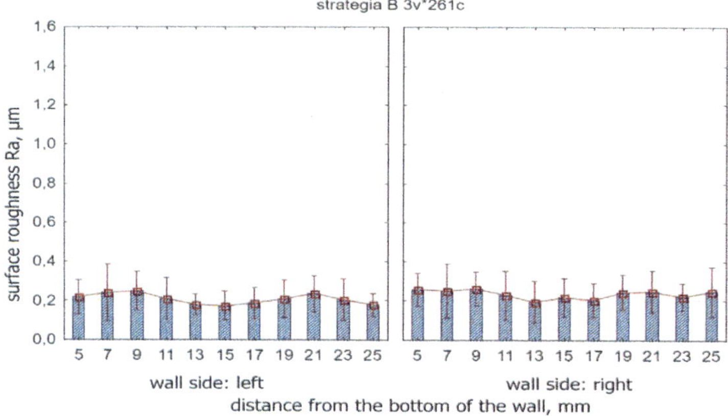

Fig. 6. Results of the roughness tests of the wall treated according to strategy B: (a) left side of the wall, (b) right side of the wall.

Figures 7 and 8 show the results of surface roughness measurements, in the distance from the wall's bottom function, processed according to strategies C and D. No significant surface roughness differences were observed for all strategies, which consisted of the distribution of the entire cutting depth into several alternating machining passes.

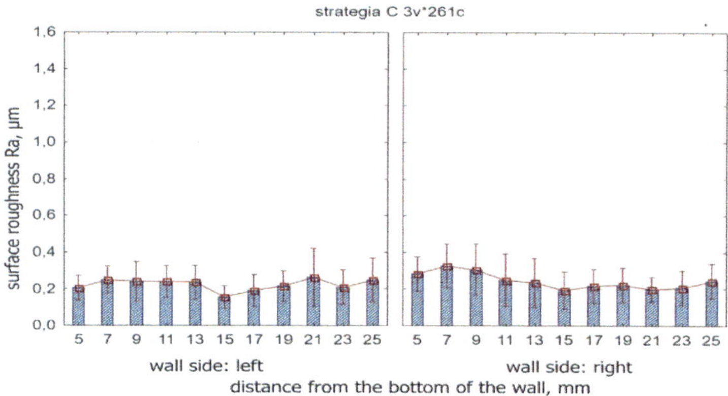

Fig. 7. Results of roughness tests of the wall treated according to strategy C: (a) left side of the wall, (b) right side of the wall.

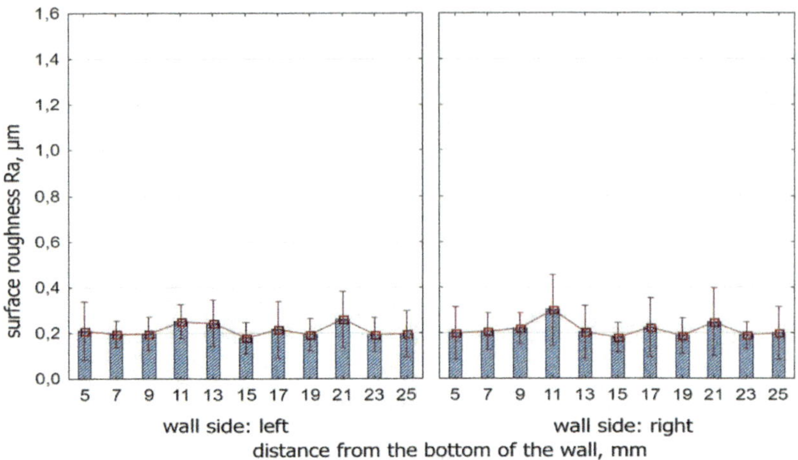

Fig. 8. Results of roughness tests of the wall processed according to strategy D: (a) left side of the wall, (b) right side of the wall.

Figure 9 shows a comparison of surface roughness for the machining strategies studied. Results were averaged over the entire wall area. The highest roughness was obtained after the treatment according to strategy 1, while the surface roughness is comparable for the other strategies.

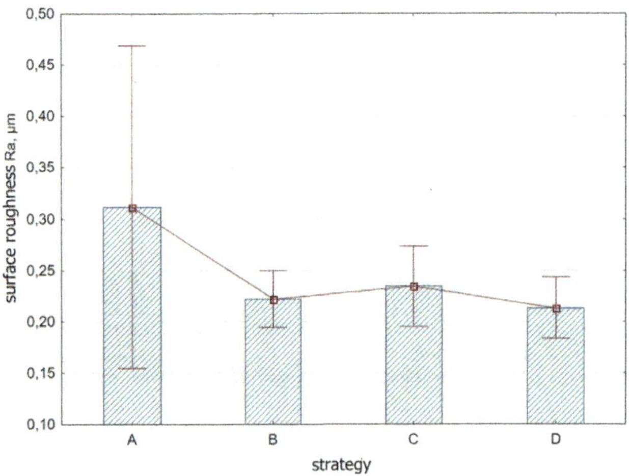

Fig. 9. Comparison of surface roughness after treatment according to four strategies.

Figure 10 shows the comparison of wall thickness after treatment according to four strategies. The thickness was compared in the lower and upper area of the wall. The difference in normal size to the nominal (1 mm) size can be the result of a combination of wall deformations during machining and deformation of the cutter.

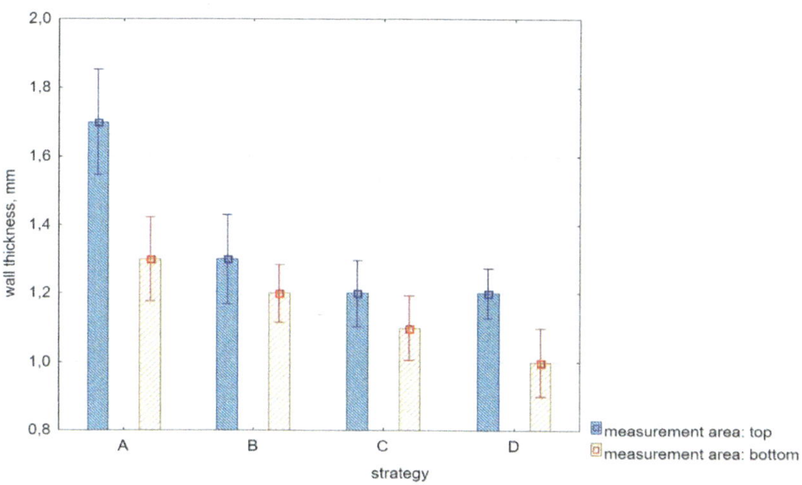

Fig. 10. Comparison of wall accuracy.

The largest deviation from the nominal deviation characterised the walls treated according to strategy A, while in the area of the upper wall surface the wall width was 1,7 mm, while in the bottom one – 1,3 mm. Full depth processing is characterised by greater interaction forces between the tool and the workpiece. Increasing the width at the top of the wall is associated with wall bending. While increasing the width at the base of the wall relative to the nominal dimension may be related to the cutter bending. The combination of these factors makes the wall geometry after treatment complex, as shown in Fig. 11.

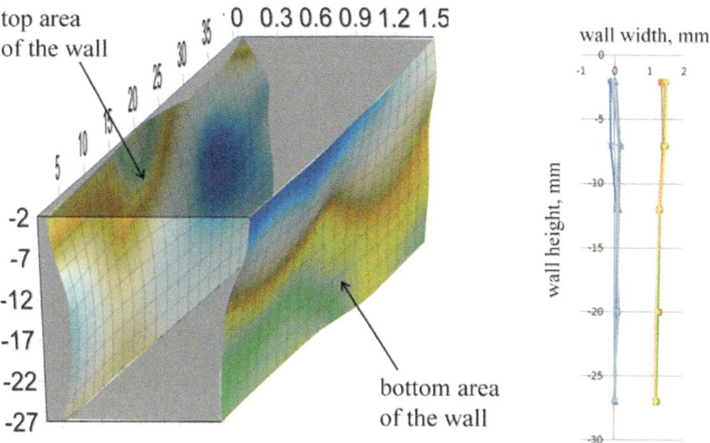

Fig. 11. Geometry of the wall after treatment according to strategy A.

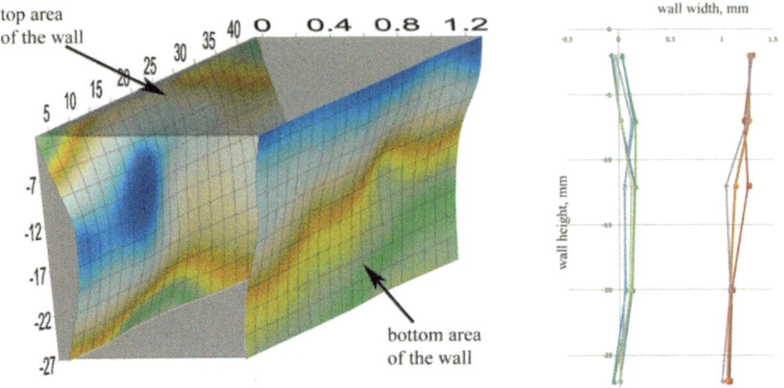

Fig. 12. Geometry of the wall after treatment according to strategy D.

The smallest deviation from the nominal dimension was observed after treatment according to strategy D, which assumed a division of the 30 mm high wall treatment into nine passages on each side of the wall. The geometry of the wall treated according to strategy D is presented in Fig. 12.

A smaller deviation from the nominal dimension can be seen in the area of the upper wall surface caused by smaller cutting forces due to the larger number of cutting workpieces.

Achieving higher accuracy is associated with a significant increase in the number of passes for each strategy. This is connected with an increase in machine time for performing the given path. Figure 13 illustrates the comparison of machine times for the strategies used in the experiment.

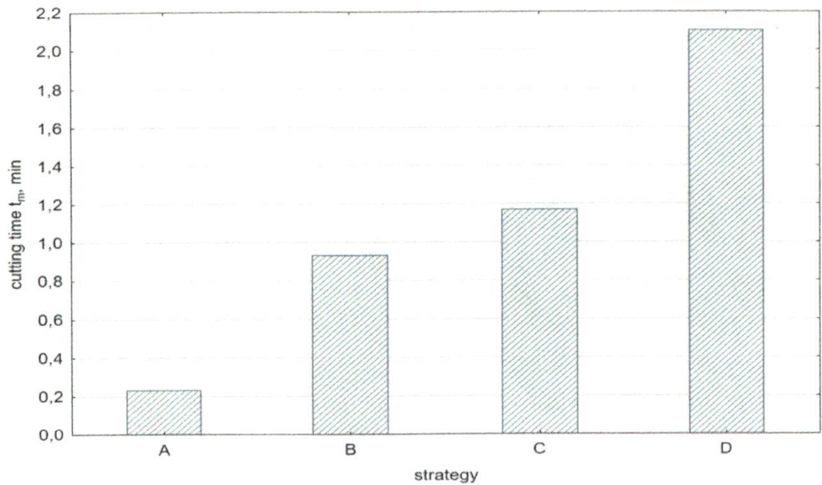

Fig. 13. Comparison of machine times of wall construction

Time for making an entire wall according to strategy D, characterised by the greatest dimensional and shape accuracy, is nine times longer than the time for making the wall according to strategy A (less accurate). The final choice of the strategy should also take into account the aspect related to machine time.

4   Research Results

The performed experimental tests of the impact of the milling machining strategy of thin-walled elements made of Ti6Al4V titanium allow on the geometric accuracy and surface roughness have led to the following conclusions:

- The effect of the machining strategy of walls has been observed on the geometric accuracy and roughness of the treated surface.
- During the treatment, walls in one pass (according to strategy no. A), vibrations have occurred, causing a significant increase in roughness in the area of the upper surface of the wall.
- There were no significant differences in surface roughness in the entire wall area after the thin-walled elements treatment according to the strategies, which divided the total wall height into several passes (strategy b, c and d), which may indicate the lack of vibrations during machining.
- The walls treated according to strategy a were characterised by the smallest geometric accuracy.
- The highest accuracy (the smallest deviation from the nominal dimension) was obtained after the treatment of thin-walled elements according to strategy d.
- Obtaining the highest dimensional and shape accuracy and small roughness (strategy d) requires a longer machine time.

The final choice of the machining strategy should take into account both the analysed geometric accuracy and surface roughness, but also the machine time for conducting the treatment. The execution time of the wall at the specified cutting speed v_c and federate on the blade f_z according to strategy no. D is the longest, however, the machining results in terms of obtaining the best accuracy and the smallest roughness are the best.

Thin-walled elements are typical for the aviation industry. These elements are often components in the assembly process of larger structures. Machining accuracy in such cases is important in order to avoid additional assembly stresses. The choice of D strategy may allow to facilitate the further assembly process.

Acknowledgments. Research carried out within the INNOLOT (Acronym AMpHOra) sectoral project entitled "Research on incremental hybridization treatment technologies for the developmental purposes of the innovative aviation production", coordinated by Polskie Zakłady Lotnicze Sp. z o.o. - PZL Mielec, co-financed by the National Research and Development Centre and the European Union within the framework of the European Regional Development Fund for the Operational Program of Innovative Economy Priority I., Measure 1.5 PO IG., contract no. INNOLOT/I/6/NCBR/2013.

References

1. Jakubczak P, Surowska B, Bienias J (2016) Evaluation of force-time changes during impact of hybrid laminates made of titanium and fibrous composite. Arch Metall Mater 61(2):689–693
2. Sieniawski J (2003) Nickel and titanium alloys in aircraft turbine engines. Adv Manuf Sci Technol 27(3):23–33
3. Ferri OM, Ebel T, Borman R (2009) High cycle fatigue behavior of Ti-6Al-4V fabricated by metal injection moulding technology. Mater Sci Eng A 504:107–113
4. Wagner L (1999) Mechanical surface treatments on titanium, aluminum and magnesium alloys. Mater Sci Eng A 263:210–216
5. Zaleski K (2009) The effect of shot peening on the fatigue life of parts made of titanium alloy Ti-6Al-4V. Eksploatacja i Niezawodnosc - Maintenance and Reliability 4(44):65–71
6. Kwiatkowski MP, Klonica M, Kuczmaszewski J, Satoh S (2013) Comparative analysis of energetic properties of Ti6Al4V titanium and EN-AW-2017 (PA6) aluminum alloy surface layers for an adhesive bonding application. Ozone-Sci Eng 35(3):220–228
7. Abdel-Aaal HA, Nouari M, El Mansori M (2009) Influence of thermal conductivity on wear when machining titanium alloys. Tribol Int 42:359–372
8. Grzesik W, Małecka J, Zalisz Z, Żak K, Niesłony P (2016) Investigation of friction and wear mechanisms of TiAlN coated carbide against Ti6Al4V titanium alloy using pin-on-discs tribometer. Arch Mech Eng LXIII(1):113–127
9. Jawaid A, Sharif S, Koksal S (2000) Evaluation of wear mechanisms of coated carbide tools when face milling titanium alloy. J Mater Process Technol 99:266–274
10. Ribeiro MV, Moreira MRV, Ferreira JR (2003) Optimization of titanium alloy (6Al-4V) machining. J Mater Process Technol 143–144:458–463
11. Sharif S, Rahim EA (2007) Performance of coated- and uncoated-carbide tools when drilling titanium Alloy-Ti-6Al-4V. J Mater Process Technol 185:72–76
12. Zębala W, Gawlik J, Matras A, Struzikiewicz G, Ślusarczyk Ł (2014) Research of surface finish during titanium alloy turning. Key Eng Mater 581:409–414. https://doi.org/10.4028/www.scientific.net/KEM.581.409 ISSN 1662-9795
13. Bolsunovskij S, Vermel V, Gubanov G, Kacharava I, Kudryashov A (2013) Thin-walled part machining process parameters optimization based on finite-element modeling of workpiece vibrations. Procedia CIRP 8:276–280
14. Rusinek R, Zaleski K (2016) Dynamics of thin-walled element milling expressed by recurrence analysis. Meccanica 51:1275–1286
15. Sun C, Shen X, Wang W (2016) Study on the milling stability of titanium alloy thin-walled parts considering the stiffness characteristics of tool and workpiece. Procedia CIRP 56:580–584
16. Kuczmaszewski J, Pieśko P, Zawada-Michałowska M (2016) Surface roughness of thin-walled components made of aluminium alloy EN AW-2024 following different milling strategies. Adv Sci Technol Res J 10(30):150–158
17. Michalik P, Zajac J, Hatala M, Mital D, Fecova V (2014) Monitoring surface roughness of thin-walled components from steel C45 machining down and up milling. Measurement 58:416–428
18. Huang PL, Li JF, Sun J, Zhou J (2014) Study on performance in dry milling aeronautical titanium alloy thin-wall components with two types of tools. J Clear Prod 67:258–264
19. Gang L (2009) Study on deformation of thin-walled part in milling process. J Mater Process Technol 209:2788–2793
20. Polishetty A, Goldberg M, Littlefair G, Puttaraju M, Patil P, Kalra A (2014) A preliminary assessment of machinability of titanium alloy Ti 6Al 4V during thin wall machining using trochoidal milling. Procedia Eng 97:357–364

Influence of Cutting Conditions in the Topography of Texturized Surfaces on Aluminium 7075 Plates Produced by Robot Machining

Alejandro Pereira[1]($\boxtimes$), M. T. Prado[1], M. Fenollera[1],
Michał Wieczorowski[2], Bartosz Gapiński[2], and Thomas Mathia[3]

[1] Faculty of Industrial Engineering of Vigo, Lagoas Marcosende,
36210 Vigo, Spain
apereira@uvigo.es

[2] Faculty of Mechanical Engineering and Management, Poznan University
of Technology, Pl. M. Sklodowskiej-Curie 5, 60965 Poznan, Poland

[3] Ecole Centrale de Lyon, Laboratoire de Tribologie et Dynamique
des Systèmes, 69134 Ecully, France

Abstract. Current manufacturing processes attempt to improve productivity by reducing idle time in the system, without being detrimental to quality requirements. Robotic machining is increasingly being introduced because of the advantages compared to CNC machines, for instance better adaptability to use the same equipment to perform different products or the feasibility to machining higher dimensional parts. But nonetheless robots have less rigidity, which means a loss of accuracy in geometrical dimensions. The research carried out in this study is focused on the introduction of robots in the machining of surfaces to be joined by adhesives. The main objective is to study the surface quality by the analysis of geometrical dimensions of textures made in aluminium 7075 by robotic machining. The experiments consist on reproducing different texturing patterns by robotic machining, where machining parameters such as axial and radial depth of path and angle between paths have been considered as variable parameters. Results have been analysed by different dimensional metrology devices, such as a Nikon SMZ800 microscope, a Trimek Coordinate Measuring Machine and an Alicona 3D interferometer with variable focal length.

Keywords: Aluminium 7075 · Robot machining · Mechanical texturized · 3D topography

1 Introduction

The automotive industry is including adhesive bonding into the manufacturing processes due to some advantages comparing to welding processes. On one hand, in the adhesive joints the ratio weight/power decreases and consequently more environmental and sustainable vehicles are obtained. Furthermore, welding joints present several problems in the welding of dissimilar materials. Then, the significance of adhesives joining technology is increasing in the field of manufacturing due to the advantages

© Springer Nature Switzerland AG 2019
M. Diering et al. (Eds.): *Advances in Manufacturing II* - Volume 5, LNME, pp. 107–121, 2019.
https://doi.org/10.1007/978-3-030-18682-1_9

compared with other joining methods, such as welding, riveted or threaded. The present work is part of a wider project focused on the inclusion of adhesive bonding into the automotive industry. It is focused on the topographical analysis of the textured Aluminium 7075 used in the automotive industry in order to joint with steel and other polymeric composites.

Uehara and Sakurai studied the influence of the surface roughness on the bonding strength of the adhesion. They concluded that, for specific materials, an optimum surface roughness exists in the tensile strength of adhesion [1]. Adhesive bonding is a preferred joining technique for manufacturing fibre reinforced composite metal hybrids. Khan et al. analysed the behaviour of 7xxx series extruded aluminium alloy surface, treated using different conventional surface treatment methods, and bonded with glass fibre reinforced tape/mat (FRP). They observed that surface treatment improve adhesion between aluminium and epoxy composite [2]. Seo et al. evaluated the effect of surface treatments on the adhesive strength of aluminium/policarbonate single lap joints. The adhesive strength shows linear relationship with surface roughness and loading speed [3].

In order to improve productivity in the automotive industry, a robotic machining was selected to reduce cycle time during the machining of the different textured surfaces. A great investment is required to develop robotic machining technology to a state where it can be implemented in high tolerance applications using a variety of materials [4]. Robotic machining was used to manufacture great prototypes but one of the inconveniences is the tolerance of parts. Chen et al. studied different strategies using the robot for rough machining. This method can generate tool paths for rough machining with different tolerances. As a verification of the proposed algorithm, a number of prototypes have been produced, which have demonstrated the feasibility and advantages of the algorithm [5].

The furthest objective of the study carried out in this paper is to minimize the time during the generation of a mechanical texture. In previous research, in order to characterize the texture and to study the tolerance of machining, the topographic surface was analysed [6, 7] considering different measurement methods and parameters to be used [8–10]. The objective of this research is to analyse the geometrical dimensions of textures obtained in aluminium 7075 by robotic machining.

2 Methodology

2.1 Materials and Equipment

Machining was performed on a robotic cell composed by a robot ABB IRB 6640/235, equipped with a rotary axis table and a Peron PS-TCV1 machining spindle mounted on the sixth axis of the robot, with an HSK DIN fixture system. This spindle allow speeds up 60.000 rpm. A force control system was installed as showed in Fig. 1. The machining was performed in aluminium 7075 plates, a high strength aluminium alloy used for aerospace applications, and nowadays also used in the automotive industry. Its relevance is reflected in the number of research on machining and optimal cutting conditions of this material [11–14].

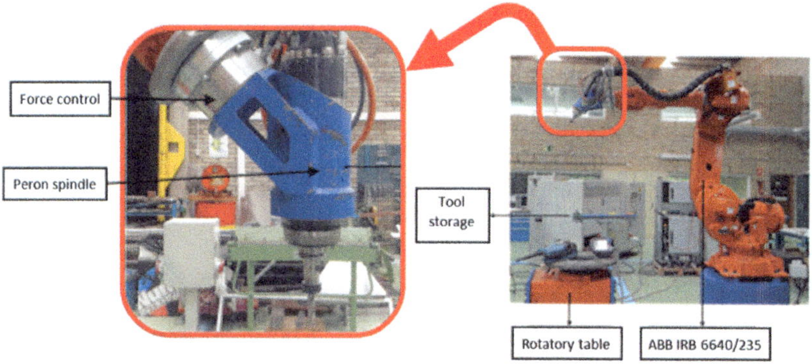

Fig. 1. Machining robot ABB 6640 with the Peron Spindle.

The 120 mm × 100 mm × 4 mm plates of aluminium 7075 have been prepared in order to carry out the machining tests, Fig. 2.

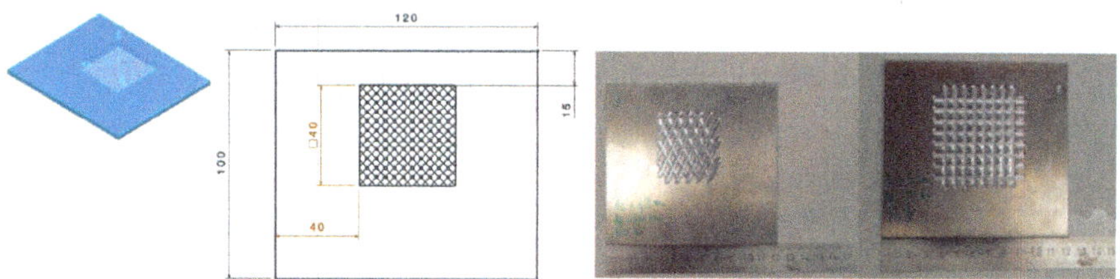

Fig. 2. Dimensions of aluminium 7075 plates for testing.

Regarding to avoid both vibrations and displacements due to flexion in the plates, a special fixture device has been designed and developed, Fig. 3. Thanks to this device, flatness accuracy was 0.06 mm, lower than repeatability of robot.

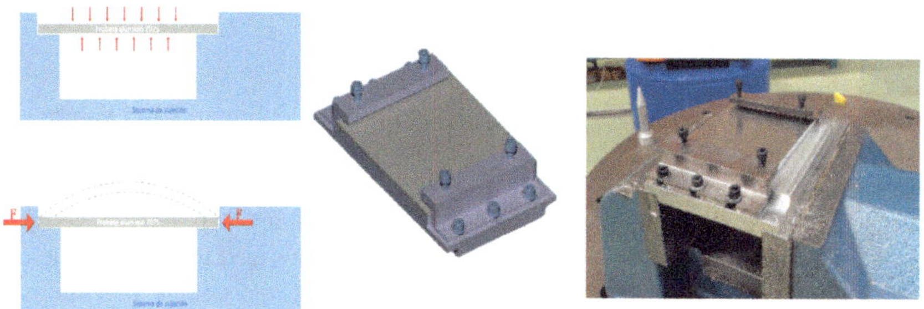

Fig. 3. Designed fixture device.

The different machining paths have been programmed with the software Power-MILL from Delcam. First of all, the calibration of the work cell has been done in order to parameterize and to reference all the positions in the cell. This operation involves the definition of the Tool Center Point (TCP) and the work-object, related to the fixture system as it is showed in the Fig. 4. Both operations have been done hand-operated with the flexpendant of the ABB Robot, positioning the end of the machining tool in different positions.

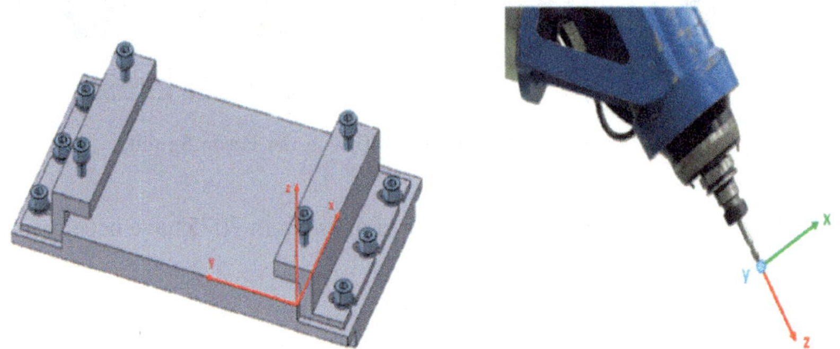

Fig. 4. Work-object in the fixture device and Tool Center Point (TCP).

2.2 Experimental Design

The tool applied was a 12 mm HACHENBACH HSS ECO-8 with two flutes, Fig. 5.

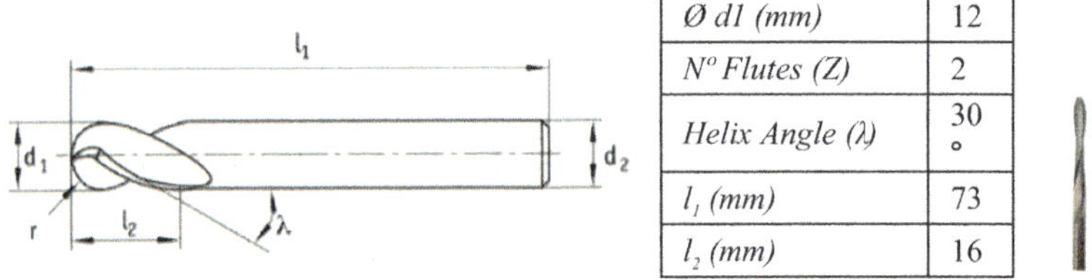

Ø d1 (mm)	12
N° Flutes (Z)	2
Helix Angle (λ)	30°
l₁ (mm)	73
l₂ (mm)	16

Fig. 5. Hachenbach tool.

Previous experiments were performed to stablish the optimum fixed values of the machining parameters (which are going to stay fixed all along the different experiments). These experiments were carried out in 9 samples with the cutting conditions presented in Table 1 and keeping the seventh external rotary axis fixed. The optimal selection was based on maximum deviation and machining feed to optimize the cycle. After running and assessing these previous experiments, the cutting conditions presented in Table 2 have been selected. The variable parameters introduced for the design of experiments are listed in Table 2 and showed in Fig. 6.

During the experiments, the effective diameter of the cutting tool has been considered to evaluate the results. This effective diameter depends on the axial depth (a_p).

Table 1. Previous experiments to choose cutting conditions.

Samples	Spindle speed (rpm)	Feed (mm/min)
1	10000	1800,00
2		2400,00
3		3000,00
4	20000	1800,00
5		2400,00
6		3000,00
7	30000	1800,00
8		2400,00
9		3000,00

Table 2. Fixed and variable parameters selected.

Fixed parameters		Variable parameters
Spindle speed	20.000 rpm	Radial strategy or step-over (a_c)
Feed	2.400 mm/min	Axial strategy or axial depth (a_p)
		Angle of path ($\alpha = \beta$)

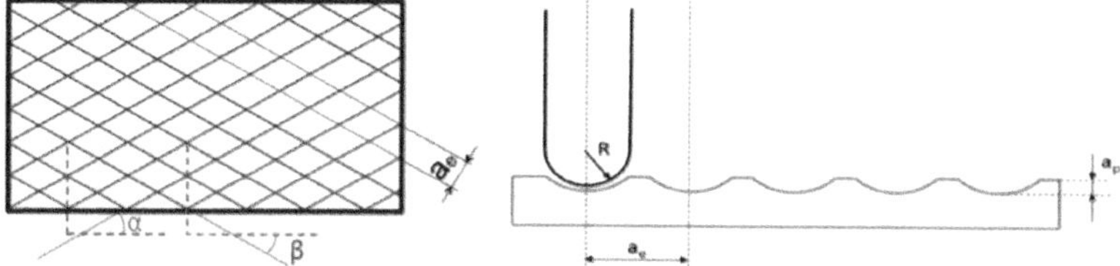

Fig. 6. Machining parameters.

If a_p varies and the spindle speed and the feed remain constants, the cutting speed varies due to the effective cutting diameter.

Two replicas of each experiment conditions have been machined. This has provided enough data to analyse the repeatability of the ABB Robot and more measurements to process afterwards.

Two kinds of measurements (samples 1 and 2) have been planned; the geometrical orientation deviation and the topographic measurement of surface. With regard to the orientation deviation, the inclination and the parallelism of path have been chosen. The value of angle deviation is calculated as showed in (1). This measurements have been made in an optical microscope Nikon SMZ 800, Fig. 7. The parallelism Dα was the maximum difference between step-over measured directly with the software Nikon of the optical microscope (2).

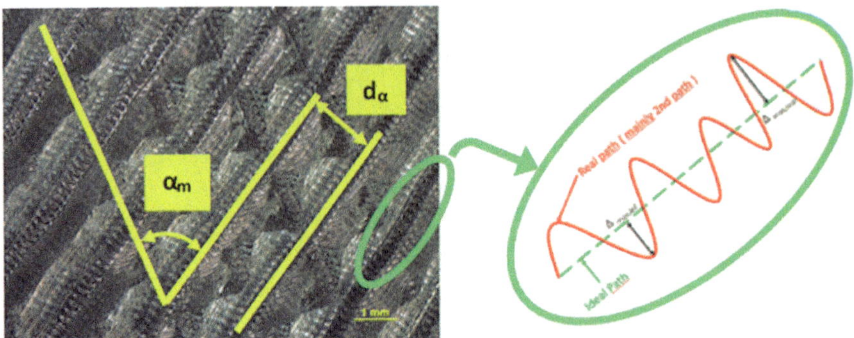

Fig. 7. Inclination α_m and parallelism of paths by d_α.

$$\Delta\alpha(^{\circ}) = \alpha_m - \alpha \tag{1}$$

Where α_m = Measured angle of path and $\Delta\alpha$ (°) = Angle Deviation

$$D\alpha(mm) = Max\{d_\alpha\} \tag{2}$$

Where d_α = Measured step-over.

The 3D topographic measurements haven been realised by an Alicona interferometer. The Infinite Focus Alicona interferometer is a system that combine a 3D micro coordinate measurement machine and a surface roughness measurement device. With InfiniteFocus, Alicona provides a vertical resolution measurement system equal to 100 nm. The obtained files have been analysed with Mountain Maps® software according to the methodology shown in Fig. 8. The parameters have been obtained according ISO 25178. The robust Gaussian regression filter with cut-off value equal to 2.5 mm was applied [15].

The analysis of the surface was carried out taking into account the following parameters, according to ISO 25178:

- Ssk (Skewness): represents the degree of symmetry of the surface.
- Sku (Kurtosis): Highlights the presence of peaks or valleys with same values (when Sku > 3), or shows when the surface is almost normal (Sku < 3).
- Sz: It is total height and is equal to the addition of maximum peak Sp and minimum valley Sv.
- Sa: It is the mean surface roughness.
- Vv: (Void Volume) It is the volume of space bounded by the surface texture from a plane at a height corresponding to a chosen "p%" value to the lowest valley.
- Std: (Texture direction) This parameter calculates the main angle for the texture of the surface. The angle is given between 0° and 360° counterclockwise, from a reference angle.

In order to set out a relationship between input (α, a_p y a_e) and output (orientation deviation and 3D topography), 14 experimental tests were conducted as outlined in Table 3.

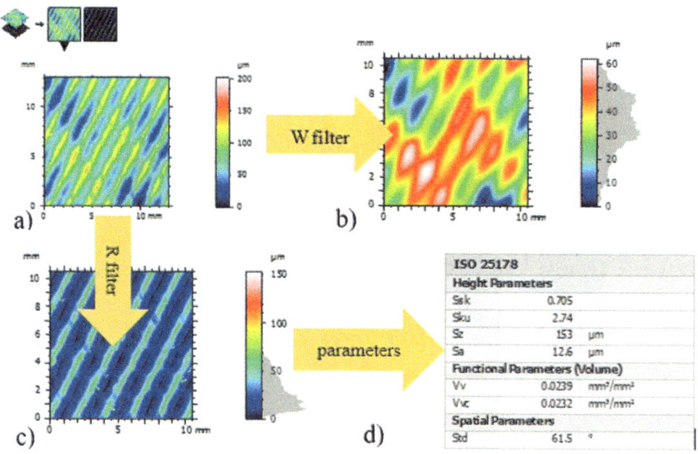

Fig. 8. (a) Primary surface (b) Waveness surface (c) Roughness surface (d) Parameters.

Table 3. Design of experiments.

Sample 1	Sample 2	α	a_p (mm)	a_e (mm)	Measurements		
					Orientation	Deviation	3D Topography
1	15	60°	0,20	2,00	$\Delta\alpha(°)$	$D\alpha$(mm)	Ssk Sku
2	16			3,00			Sz, Sa
3	17			5,00			Vv, Vc
4	18		0,40	1,80			Std
5	19			3,30			
6	20			4,30			
7	21			6,30			
8	22	90°	0,20	2,00			
9	23			3,00			
10	24			5,00			
11	25		0,40	1,80			
12	26			3,30			
13	27			4,30			
14	28			6,30			

3   Test Results and Analysis

The measurements of orientation deviation such as inclination, $\alpha_m(°)$, and measurements of step–over, d_α(mm), are shown in Fig. 9. The parallelism, $D\alpha$ (mm) should be the maximum difference between the measurements of step-over.

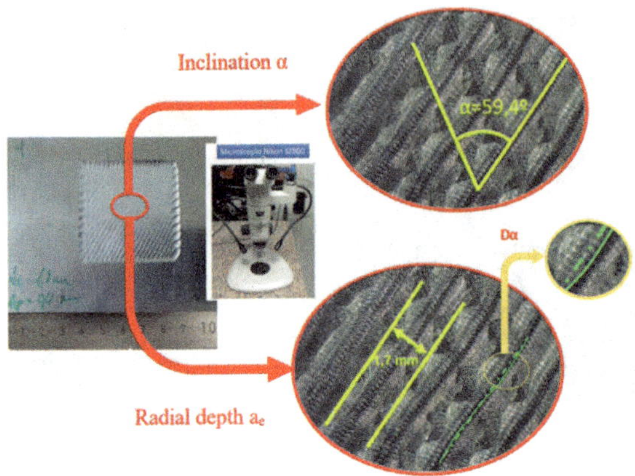

Fig. 9. Sample 1 - test 4; Measurement of inclination (αm) and measurement of step-over (dm).

The representation of the inclination versus the step-over is showed in Fig. 10. In this graph it can be noted the significant average inclination $\Delta\alpha = -0.775°$ for the case of axial depth $a_p = 0.2$ mm. In the case of axial depth $a_p = 0.4$ mm the average inclination is $-0.4°$. Also in Fig. 9 the lineal error $D\alpha$ of the paths can be observed on the microscope image.

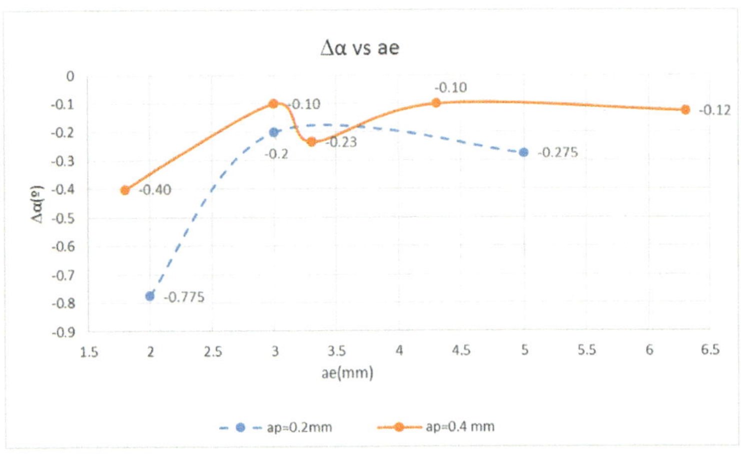

Fig. 10. Inclination deviation $\Delta\alpha$ versus step-over.

In Fig. 11, the parallelism Dα (mm) in front of the step-over (radial depth) are presented. It can be observed that the maximum variability is 0.41 mm (0.3 mm + 0.11 mm) in case of axial depth a_p = 0.2 mm, smaller than in the case of a_p = 0.4 mm with a maximal variability of 0.5 mm (0.20 mm + 0.3 mm).

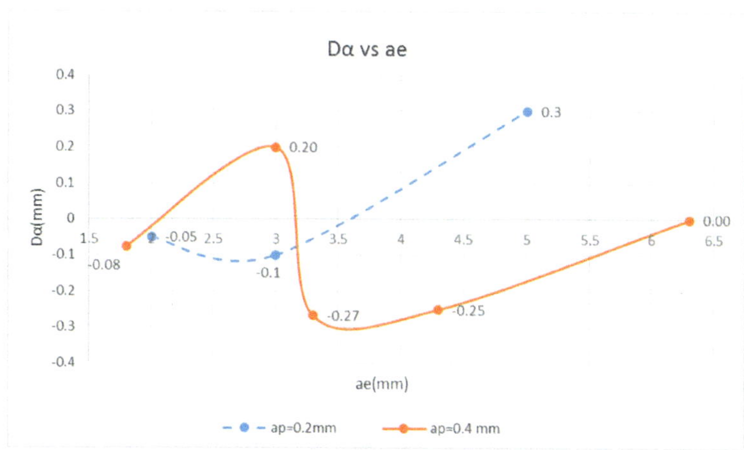

Fig. 11. Parallelism Dα versus step-over.

In relation to the topography measurements, it can be noted the proportional relation between the average surface of primary profile, Pa, and the radial depth a_e, as it can be shown in Fig. 12. There are no variance between different angles and the proportionality is similar in each angle.

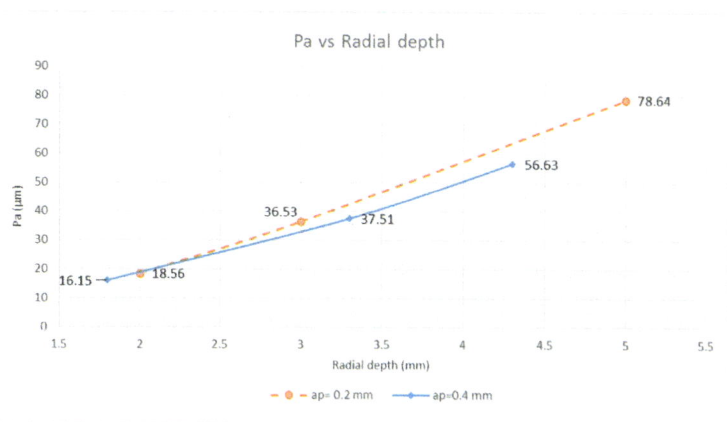

Fig. 12. Average surface of primary profile Pa versus radial depth.

116   A. Pereira et al.

After applying the Gaussian roughness filter, and cutt-off of 2.5 mm, the behaviour of average surface Sa value follows lower proportionality to the radial depth, as it can be observed in Fig. 13. Also, there is a significant variability of Sa between 8.43 μm (Fig. 18c) and 17.1 μm (Fig. 18b), caused by the cutting conditions and the limited rigidity of robot.

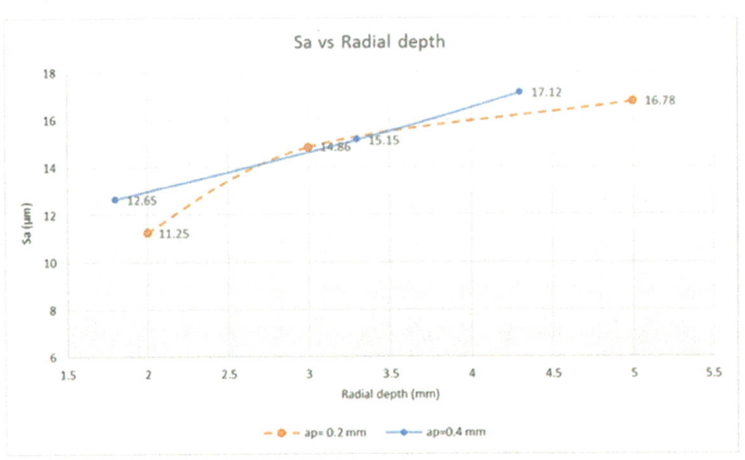

Fig. 13. Average surface Sa versus radial depth.

In relation to skewness parameter, the values of Ssk versus the radial depth are presented in the Fig. 14. All the values are positive and for that reason there are more significant peaks. In the case of axial depth equal to 0.4 mm the values are higher than the values of case with axial depth equal to 0.2 mm.

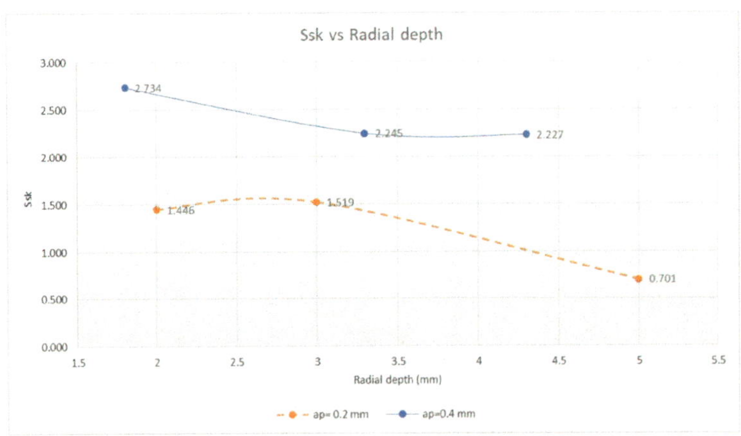

Fig. 14. Skewness Ssk versus radial depth.

The Fig. 15 presented the Void Volume in front of the radial depth. In this case the parameter has been calculated with a plane at 10% from valley. It can be observed that there are no significant differences respect the axial depth. It can be noticed the higher is the radial depth the higher is the parameter Vv.

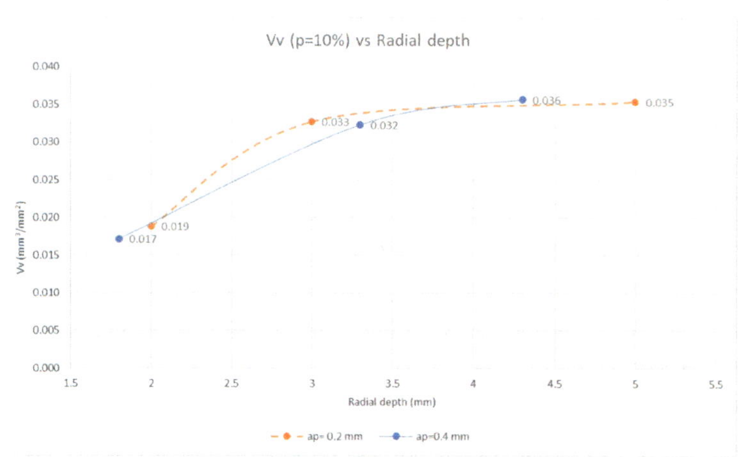

Fig. 15. Void Volume Vv versus radial depth

The parameter texture direction Std, case of angle α equal to 60° are presented in the Fig. 16. This parameter has been chosen with a reference of angle equal to 0°. It can be observed that there are a significant deviation over the designed texture, in both cases of axial depth.

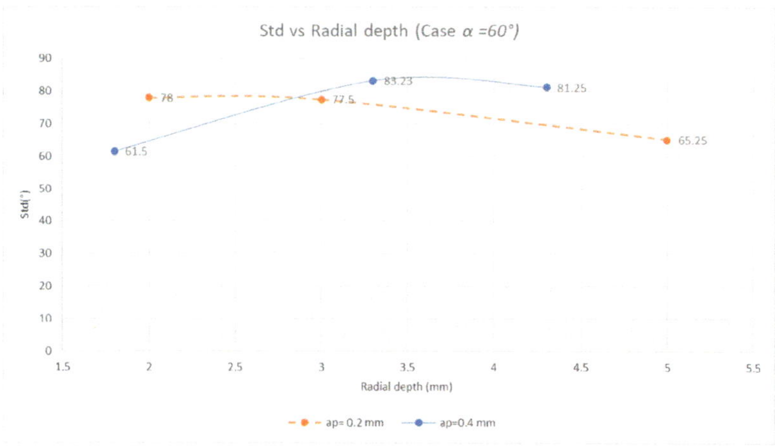

Fig. 16. Texture direction Std versus radial depth, case of angle α equal to 60°.

The Fig. 17 shows the texture direction in the case of angle α equal to 90°.

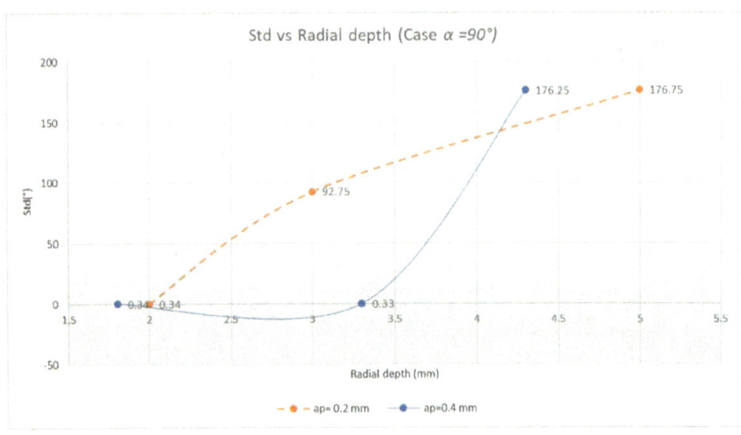

Fig. 17. Texture direction Std versus radial depth, case of angle α equal to 90°.

Taking into account that 0° and 180° could be the same direction, depending the methodology of measurement, there are lower differences in the parameter than in the case of α equal to 60°.

In Fig. 18, it can be observed the comparative surface between minimum and maximum average surface Sa in cases of $\alpha = 60°$ and $\alpha = 90°$. It can be noted the non-uniform rhombus, which could be caused by the second path of machining inclined 60°. This result could be due to the different cutting forces, obtained by machining areas where there are no material to remove. This is one of the future work, to be carried out in order to introduce the correction after measuring with force control of spindle.

In case of $\alpha = 90°$, studying high radial depth, particularly in the 13rd test, the texturize surfaces are more uniform, than tests which were machined with $a_e = 4.3$ mm. The behaviour of the values of parameter Sz is proportionality to the axial depth.

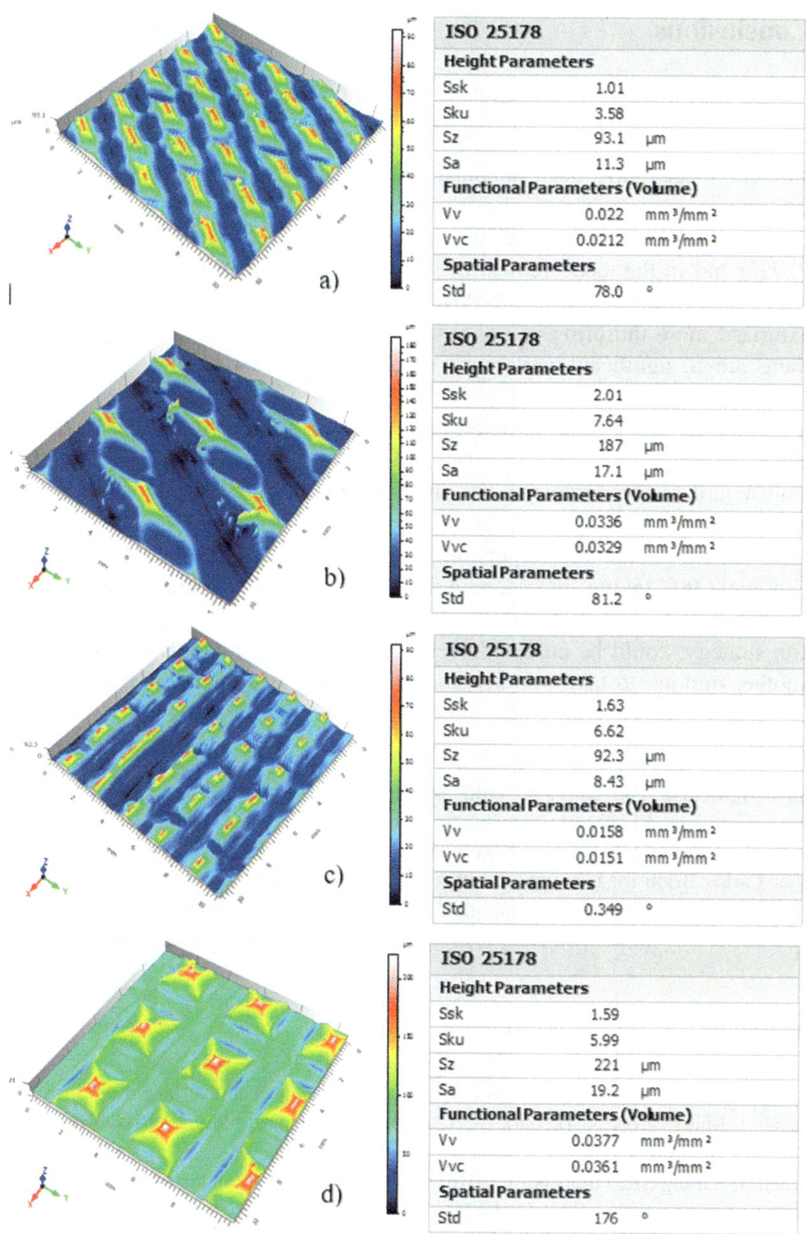

a)

ISO 25178		
Height Parameters		
Ssk	1.01	
Sku	3.58	
Sz	93.1	µm
Sa	11.3	µm
Functional Parameters (Volume)		
Vv	0.022	mm³/mm²
Vvc	0.0212	mm³/mm²
Spatial Parameters		
Std	78.0	°

b)

ISO 25178		
Height Parameters		
Ssk	2.01	
Sku	7.64	
Sz	187	µm
Sa	17.1	µm
Functional Parameters (Volume)		
Vv	0.0336	mm³/mm²
Vvc	0.0329	mm³/mm²
Spatial Parameters		
Std	81.2	°

c)

ISO 25178		
Height Parameters		
Ssk	1.63	
Sku	6.62	
Sz	92.3	µm
Sa	8.43	µm
Functional Parameters (Volume)		
Vv	0.0158	mm³/mm²
Vvc	0.0151	mm³/mm²
Spatial Parameters		
Std	0.349	°

d)

ISO 25178		
Height Parameters		
Ssk	1.59	
Sku	5.99	
Sz	221	µm
Sa	19.2	µm
Functional Parameters (Volume)		
Vv	0.0377	mm³/mm²
Vvc	0.0361	mm³/mm²
Spatial Parameters		
Std	176	°

Fig. 18. (a:) 1st Test, 60°, (b) 6th test, 60° (c) 8th test, 90° (d) 13th test, 90°.

4 Conclusions

After presentation the obtained results, the next conclusions can be explained:

- The texturized of aluminium 7075 with robotic machining is easy and very fast.
- Different shapes have been obtained by easy path strategy and common spherical end tool.
- There are a significant variability in the geometrical inclination (from $-0.10°$, $-0.77°$) and in the measurement of parameter parallelism (± 0.3 mm).
- During the machining with the ABB 6640 robot, the higher is the step-over, the texturized more uniform is.
- There are a significant variability of average surface heights Sa in the tested conditions.
- In relation to skewness parameter, the higher is the axial depth the higher are the values of Ssk.
- With regard to the parameter Vv, there are no significant differences respect the axial depth.
- There are no remarkable dispersion in the values of Std in the case of machining with angle 90°, taking into account that the value 180° is equal to 0°
- As a future line of work, the study of the influence of the cutting force in the second path strategy, could be analysed, in order to avoid geometrical deviations.
- Another outlook to take into account is the study of the adhesive strength results with this texturized samples.

Acknowledgments. The authors would appreciate the support of Juan González from the company Vigotec and Francisco Sánchez from the Automotive Technical Center (CTAG). Also the authors would like to thanks to company Vigotec, and particularly with his operational manager Carlos Brion for his support and advices. This work is supported by project "Ecoboss Development of new solutions of ecological structures of body work for sustainable automotive industry", in collaboration with company Vigotec S.L. CDTI (10/9/2015 – 30/09/2017).

References

1. Uehara K, Sakurai M (2002) Bonding strength of adhesives and surface roughness of joined parts. J Mater Process Technol 127(2):178–181. https://doi.org/10.1016/S0924-0136(02)00122-X Art. no. Pii
2. Khan S, Sarang SK, Hiratsuka I (2016) Study of bending strength for aluminum reinforced with epoxy composite. SAE Int J Mater Manuf 9(3):781–787
3. Seo DW, Yoon HC, Lee JY, Lim JK (2004) Effects of surface treatment and loading speed on adhesive strength of aluminum to polycarbonate lap joints. Key Eng Mater 261–263:405–410
4. Barnfather JD, Goodfellow MJ, Abram T (2016) A performance evaluation methodology for robotic machine tools used in large volume manufacturing. Robot Comput. Integr Manuf 37:49–56
5. Chen YH, Hu YN (1999) Implementation of a robot system for sculptured surface cutting. Part 1. Rough machining. Int J Adv Manuf Technol 15(9):624–629

6. Pereira A, Hernandez P, Martinez J, Perez JA, Mathia T (2013) Study of morphology wear model of moulds from alloys of aluminium EN AW-6082 in injection process. In: Current state-of-the-art on material forming: numerical and experimental approaches at different length-scales, Pts 1–3, vol 554–557, pp 844–849
7. Pereira A, Martínez J, Prado MT, Perez JA, Mathia T (2014) Topographic wear monitoring of the interface tool/workpiece in milling. Adv Mater Res 966–967(2014):152–167
8. Mathia TG, Pawlus P, Wieczorowski M (2011) Recent trends in surface metrology. Wear 271(3–4):494–508
9. Gapinski B, Wieczorowski M, Marciniak-Podsadna L, Dybala B, Ziolkowski G (2014) comparison of different method of measurement geometry using CMM, optical scanner and computed tomography 3D. In: 24th Daaam international symposium on intelligent manufacturing and automation, vol 69, pp 255–262
10. Gapinski B, Wieczorowski M, Marciniak-Podsadna L, Pereira Domínguez A, Cepova L, Martinez Rey A (2018) Measurement of surface topography using computed tomography. In: Lecture notes in mechanical engineering, no 201519, pp 815–824
11. Davim JP, Maranhao C, Jackson MJ, Cabral G, Gracio J (2008) FEM analysis in high speed machining of aluminium alloy (Al7075-0) using polycrystalline diamond (PCD) and cemented carbide (K10) cutting tools. Int J Adv Manuf Technol 39(11–12):1093–1100
12. Rivero A, de Lacalle LNL, Penalva ML (2008) Tool wear detection in dry high-speed milling based upon the analysis of machine internal signals. Mechatronics 18(10):627–633
13. Maranhao C, Davim JP, Jackson MJ (2011) Physical thermomechanical behavior in machining an aluminium alloy (7075-O) using polycrystalline diamond tool. Mater Manuf Process 26(8):1034–1040
14. Raykar SJ, D'Addona DM, Mane, AM (2015) Multi-objective optimization of high speed turning of Al 7075 using grey relational analysis. In: 9th CIRP conference on intelligent computation in manufacturing engineering - CIRP ICME 2014, vol 33, pp 293–298
15. Seewig J (2005) Linear and robust Gaussian regression filters. In: 7th symposium on measurement technology and intelligent instruments, Univ Huddersfield, Huddersfield, England, 2005, vol 13, pp 254–257

The Effect of Dimple Distortions on Surface Topography Analysis

Przemysław Podulka[(✉)]

The Faculty of Mechanical Engineering and Aeronautics, Rzeszow University
of Technology, Powstancow Warszawy 12 Str, 35959 Rzeszow, Poland
p.podulka@prz.edu.pl

Abstract. In this paper the influence of improper selection of reference plane
on dimples distortions was taken into consideration. The effect of application of
various procedures (cylinder fitting algorithm; polynomial approximation; digital filtering: regular Gaussian regression filter, Robust Gaussian regression filter) were compared and discussed. Plateau-honed cylinder liner surfaces were
studied (more than 20 measured and/or 20 with digitally added dimples surfaces
were studied). They were measured by stylus instrument Talyscan 150 or white
light interferometer Talysurf CCI Lite. The influence of usage of commonly
applied algorithms on surface topography parameters (from ISO 25178 standard) was also taken into account. It was assumed that dimples distortion is of a
great importance for calculation of surface topography parameters (especially of
Sk parameters) of plateau-honed cylinder liners with additionally added oil
pockets created by burnishing techniques. False estimation of areal form
removal procedure can cause of classification of properly made parts as a lack
and its rejection. The dimples size (width D_W and depth D_D) was also taken into
consideration for proposal of selection of areal form removal procedure.

Keywords: Surface topography · Surface topography parameters · Dimples ·
Oil pockets

1   Introduction

Surface topography (ST) is established in the final stage of a machining process. Its
measurement and/or analysis is directly relevant in assessments of functional properties, such as material contact, lubricant retention or wear resistance. The evaluation of
ST of internal combustion engines is of a great importance due to their wide applications; since they consume fuel and contribute to air pollution, reducing their fuel
consumption and emissions has an overriding importance. Therefor the process of
surface metrology should be subjected to strict control.

The errors in ST assessment can be classified in measurement errors [1, 2], the
measured object errors [3], software and measuring method errors [4]. The measuring
uncertainty can be also divided in errors: typical for measuring approaches, resulted by
digitization process, obtained while data processing and other errors [5–8].

Usually the ST parameters of car engine cylindrical elements are calculated after
form removal (error of form, waviness due to imperfections in manufacturing process).

© Springer Nature Switzerland AG 2019
M. Diering et al. (Eds.): *Advances in Manufacturing II* - Volume 5, LNME, pp. 122–133, 2019.
https://doi.org/10.1007/978-3-030-18682-1_10

Selection of reference plane is often proposed by: cylinder fittings [9], polynomials [10] and plenty of filtering methods: Gaussian themes [11] (regression filter [12], robust regression filter [13]), morphological schemes and its modifications [14]; spline filters [15], wavelets [16] and others algorithms and/or procedures [17]. Modal filtering, wavelet filtering and curvature-scale analyses were also presented in [18].

The cylinder plateau honed surface is an example of surface texture which consist of smooth plateaus with valleys and is characterized by highly acceptable sliding properties and lubricant maintains. Two-process cylindrical surfaces, containing wide and/or deep valleys, have clear advantage over one-process surfaces [19]. ST of cylindrical elements with additionally added oil pockets created by burnishing techniques were taken into consideration in [20]; the influence of areal form removal on ST parameters was also studied [21]. However there are many problems in measurement and analysis of stratified-called surface textures [22, 23]. One of the problem is to select the proper reference line (plane) when profile (surface) exploration is taken into account. Areal form removal of plateau-honed cylinder liners with deep and wide edge-located oil pockets was taken in consideration in [24, 25]. Erroneous selection of reference plane can cause allotment of properly mad parts as a lacks and its rejection [26].

There were many paper considering the areal form removal of cylindrical elements by various digital processing methods [1, 27–30]. However, the influence of distortion of oil dimples on ST parameters was not taken into consideration; false estimation of procedures for reference plane selection was not comprehensive studied with valley distortion analysis.

2 Materials and Methods

The plateau-honed cylinder surfaces with additionally added oil pockets created by burnishing techniques (in some cases dimples were added in computing process) were taken into consideration. The average width (D_W) and depth (D_D) of oil pockets were between 0.3 mm and 1 mm and between 10 μm and 100 μm respectively. More than 20 measured and 20 with digitally added valleys surfaces were analysed but only few of them were discussed and showed in details. The isometric view, material ratio curves and selected parameters of examples of details from analysed surfaces were presented in Fig. 1.

They were measured by stylus instrument Talyscan 150 (nominal tip radius about 2 μm, height resolution about 10 nm) and/or white light interferometer Talysurf CCI Lite (height resolution 0.01 nm). Measured area was 5 mm by 5 mm, however, many results were presented only in extracted details. The spacing was between 5 μm and 15 μm. The measurement was repeated three times and average values was taken into account (A method for surface measurement uncertainty determination was applied).

The effect of dimples distortion on the following parameters (from ISO 25178 standard) was taken into consideration: root mean square height Sq, skewness Ssk, kurtosis Sku, maximum surface peak height Sp, maximum valley depth Sv, maximum

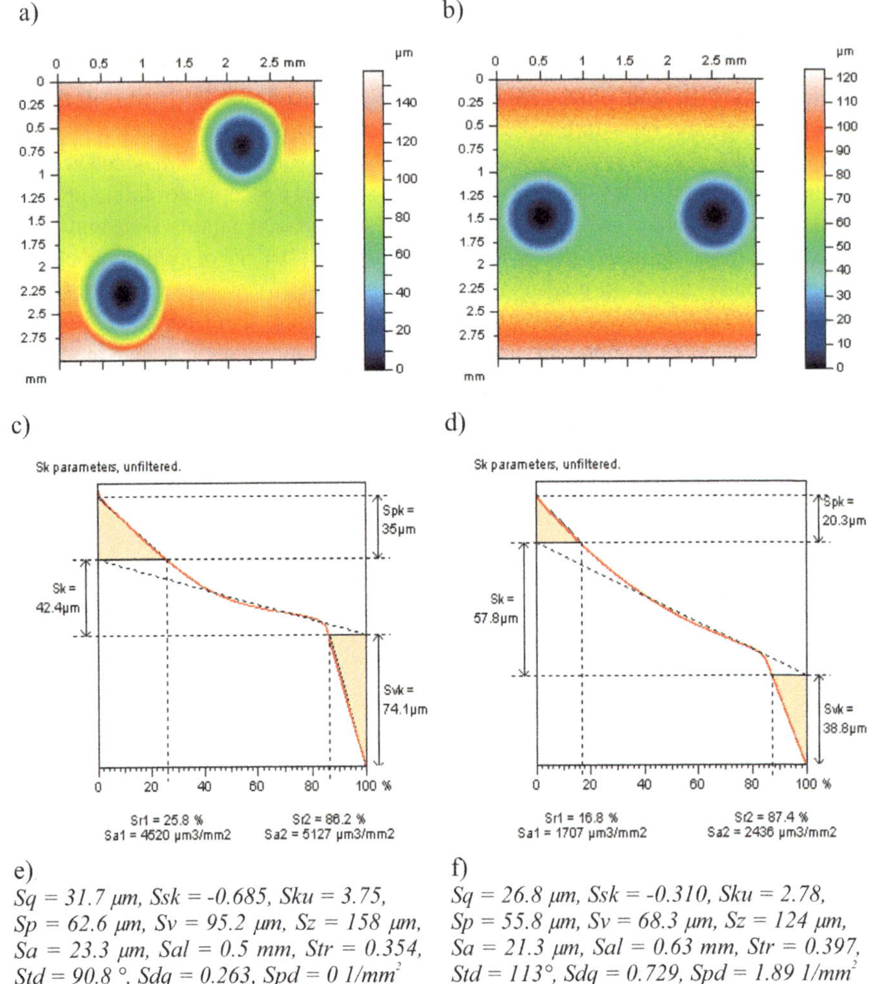

e)
Sq = 31.7 µm, Ssk = -0.685, Sku = 3.75,
Sp = 62.6 µm, Sv = 95.2 µm, Sz = 158 µm,
Sa = 23.3 µm, Sal = 0.5 mm, Str = 0.354,
Std = 90.8 °, Sdq = 0.263, Spd = 0 1/mm²

f)
Sq = 26.8 µm, Ssk = -0.310, Sku = 2.78,
Sp = 55.8 µm, Sv = 68.3 µm, Sz = 124 µm,
Sa = 21.3 µm, Sal = 0.63 mm, Str = 0.397,
Std = 113°, Sdq = 0.729, Spd = 1.89 1/mm²

Fig. 1. Isometric views of examples of extracted details from measured (a) and modelled (b) surfaces, their material ratio curves (c, d) and selected parameters (e, f) correspondingly.

height Sz, arithmetic mean height Sa, auto-correlation length Sal, texture-aspect ratio Str, texture direction Std, root mean square slope Sdq, interfacial area ratio Sdr, summit density Spd, arithmetic mean peak curvature Spc; Sk group parameters: reduced summit height Spk, reduced valley depth Svk, core roughness depth Sk, upper bearing area $Sr1$ and lower bearing area $Sr2$.

For areal form removal the following (commonly used) algorithms and/or procedures were applied: cylinder fitted by the least square method (C_{LSM}); polynomial approximation (P_{2ND}, P_{4TH}); digital filters: regular Gaussian regression filter (GRF), Robust Gaussian regression filter ($RGRF$). In some cases the influence of dimples distribution (dimple-to-edge D_{DTE} and/or dimple-to-dimple D_{DTD} distances) on distortion of valleys (dimples) was taken into consideration. The effect of filter cut-off (bandwidth) value (F_{CO}) on areal form removal and/or values of parameters was also studied.

3 Results and Discussions

From the results of previous research by author of this paper [10, 21, 24] it was assumed that application of C_{LSM} for selection of reference plane of cylindrical surface containing wide and/or deep valleys did not allow to remove form correctly, especially when the $D_D > 10$ μm and/or $D_W > 0.5$ mm. It was also suggested to select the reference plane by minimization of Sk parameter value and the distortion of dimples (for two-process surfaces). It was also recommended to select the reference plane by use of polynomials from 2nd to 4th degree.

From the analysis of the isometric view of surfaces it was assumed that application of polynomial (P_{2ND} or P_{4TH}) did not allow to remove form entirely; the biggest degree of polynomial was applied the biggest not eliminated errors of form were observed (it was indicated by the arrows in Fig. 2).

Application of GRF caused the increase of dimples distortions (compared with polynomial and/or $RGRF$); valleys were flattened and values of Sk parameters were falsely estimated, the biggest underestimation was observed for Svk parameter – more than 80% (parameters are presented in Fig. 3b). The smallest (biggest) value of Sk (Sv) parameter was noticed when $RGRF$ was used. However the value of Svk parameter decreased (increased) after $RGRF$ usage in accordance with polynomial (GRF) appliance; it can be caused by the distortion of dimples, indicated in Fig. 3d and f.

From the analysis of extracted profiles, application of polynomials might seem the best solution. However the value of maximum height of profile Pt was smaller than after usage of $RGRF$; the dimples flattening occurred.

For two-process surfaces digital filtering (GRF and $RGRF$) was proposed. Application of GRF caused a distortion of dimples regardless of D_D (even the $D_D < 10$ μm) when $D_W > 0.4$ mm; oil pockets were flattened and the near-dimple areas were displaced (Fig. 4a, c, e). When the D_{DTE} decreased the distortion of dimples increased; especially when $D_{DTE} < F_{CO}$. Moreover, when scratches occurred in near-dimple areas then reference plane displacement has a tendency to grow (Fig. 4c). The distortion of dimples and/or reference plane has also a tendency to increased when $D_{DTD} < D_W$ and/or $D_{DTD} < F_{CO}$; the highest distortion (displacement) of dimples (reference plane) was observed when $D_{DTD} < D_W$ occurred.

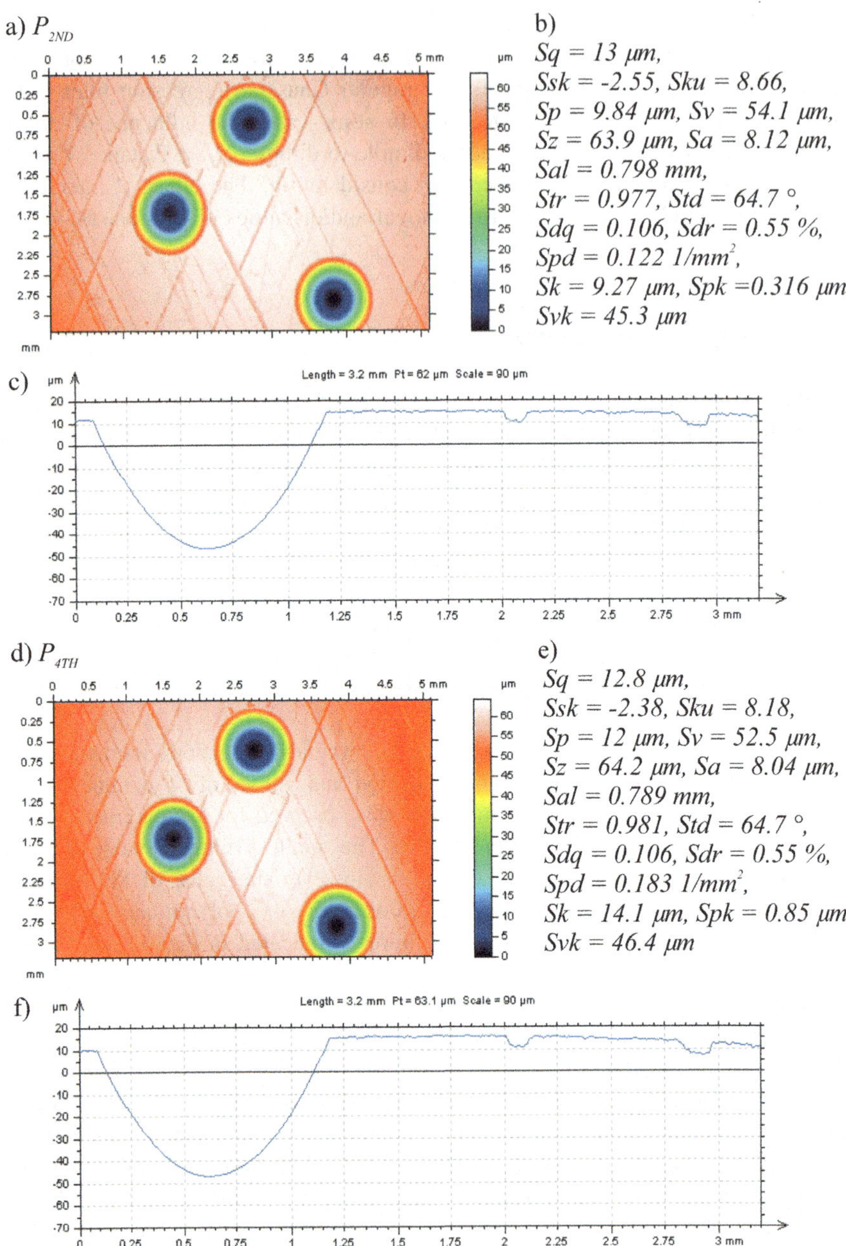

Fig. 2. Isometric views (a, d), selected parameters (b, e) and extracted profiles (c, f) from surface after preprocessing by polynomial appliance.

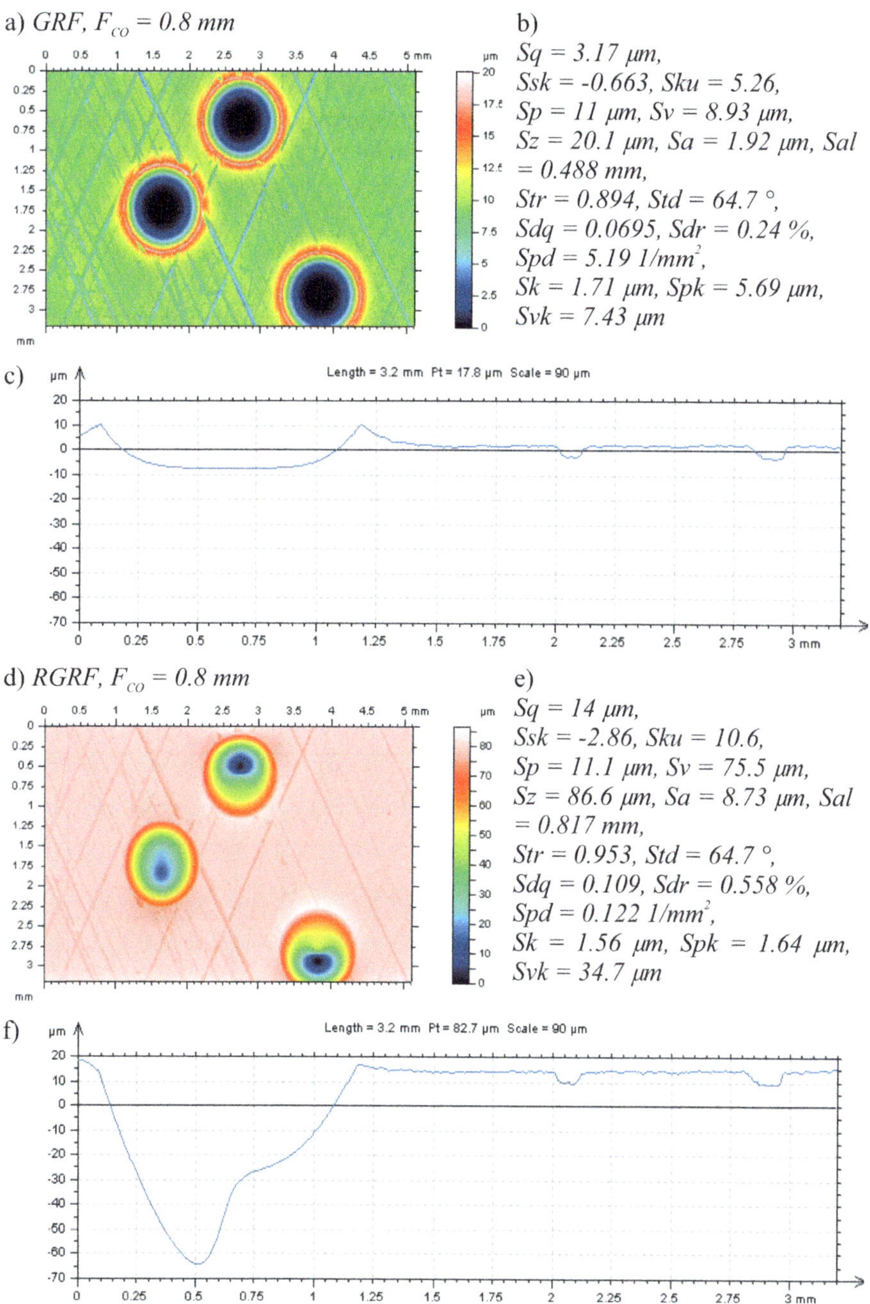

Fig. 3. Isometric views (a, d), selected parameters (b, e) and extracted profiles (c, f) from surface after form removal by Gaussian filtering methods.

When *RGRF* was applied, the distortion of valleys decreased (Fig. 4b, c, d) in accordance to *GRF*; F_{CO} = 0.8 mm. However some areas of surface situated between dimples and edges of analysed detail were displaced (Fig. 4b, d) when $D_{DTE} < F_{CO}$. Moreover, when distance between wide scratches and dimples (and D_{DTD}) was smaller than D_W and/or F_{CO} than displacement of reference plane has also a tendency to increase.

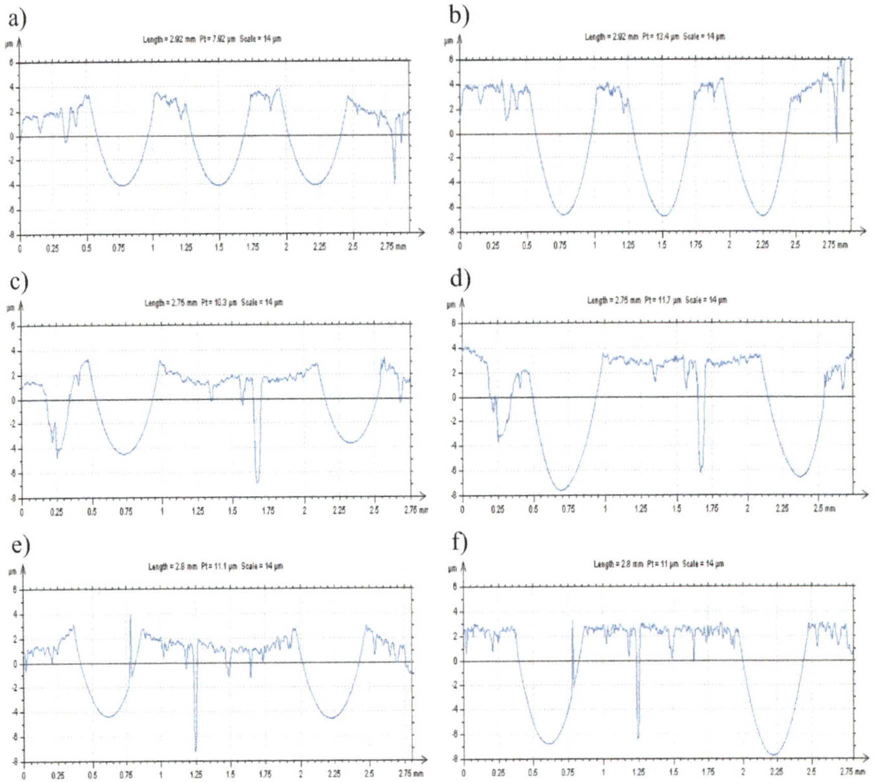

Fig. 4. Extracted profiles from surfaces after form removal by *GRF* (a, c and e) and *RGRF* (b, d and f); F_{CO} = 0.8 mm.

In Figs. 5 and 6 extracted details, profiles (and their parameters) from surface were presented. Two types of dimples distribution were taken into consideration: center-located (Fig. 5) and edge-located (Fig. 6).

From the analysis of isometric view of surface it was assumed that when valleys were located in the center of studied detail then distortion of oil pocket was negligible or did not occur when the P_{2ND} or P_{4TH} were applied. However the form was not entirely removed (profiles from e and j example). Increase of the degree of the polynomial cause the increase (decrease) of *Sk* (*Svk*) parameter; the variation of *Spk* parameter was very small or it was indiscernible.

a)

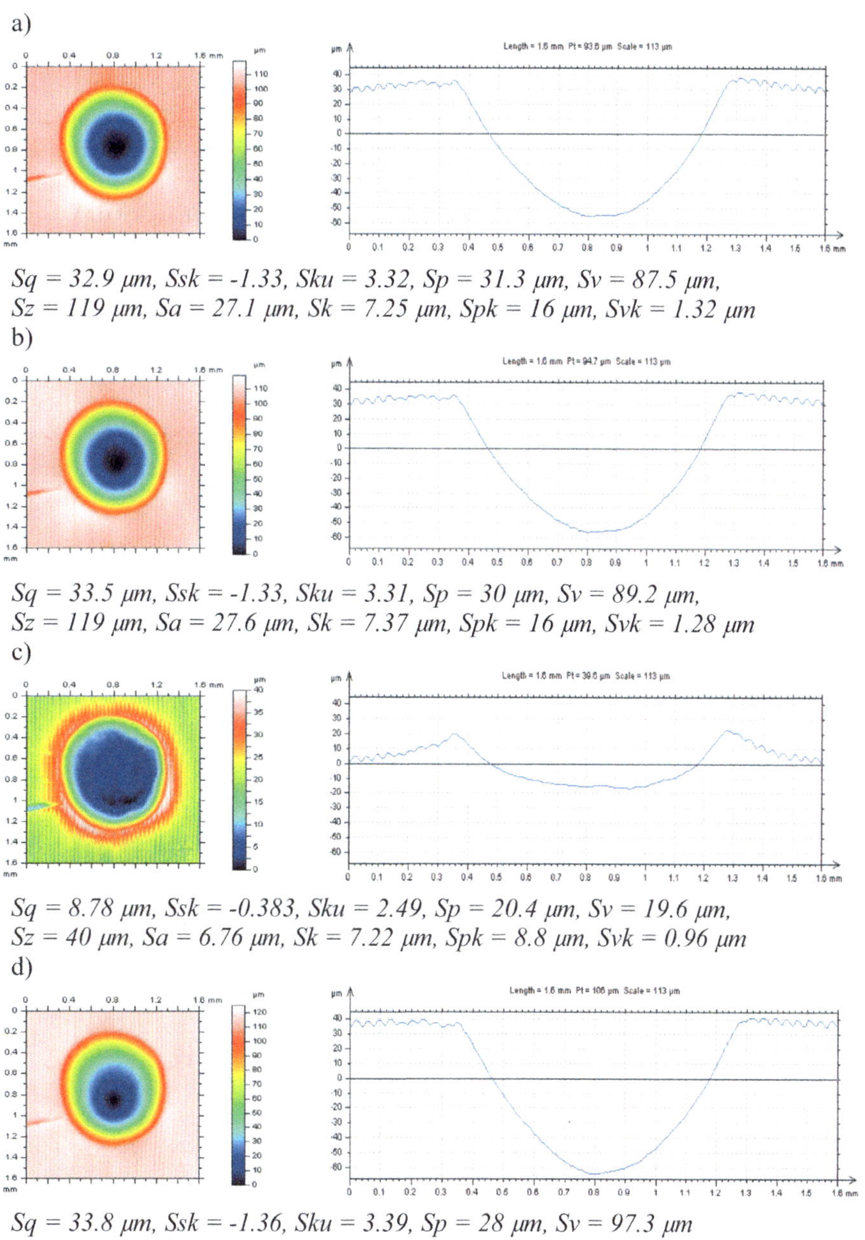

Sq = 32.9 μm, Ssk = -1.33, Sku = 3.32, Sp = 31.3 μm, Sv = 87.5 μm,
Sz = 119 μm, Sa = 27.1 μm, Sk = 7.25 μm, Spk = 16 μm, Svk = 1.32 μm

b)

Sq = 33.5 μm, Ssk = -1.33, Sku = 3.31, Sp = 30 μm, Sv = 89.2 μm,
Sz = 119 μm, Sa = 27.6 μm, Sk = 7.37 μm, Spk = 16 μm, Svk = 1.28 μm

c)

Sq = 8.78 μm, Ssk = -0.383, Sku = 2.49, Sp = 20.4 μm, Sv = 19.6 μm,
Sz = 40 μm, Sa = 6.76 μm, Sk = 7.22 μm, Spk = 8.8 μm, Svk = 0.96 μm

d)

Sq = 33.8 μm, Ssk = -1.36, Sku = 3.39, Sp = 28 μm, Sv = 97.3 μm
Sz = 125 μm, Sa = 27.9 μm, Sk = 9.19 μm, Spk = 15.7μm, Svk = 6.28 μm

Fig. 5. Details with center-located dimples, their profiles and parameters respectively, extracted from surface after form removal by: P_{2ND} (a) or P_{4TH} (b), *GRF* (c) and *RGRF* (d); F_{CO} = 0.8 mm

Application of *GRF* caused the flatteners of dimples, therefor *RGRF* was proposed for selection of reference plane in ST analysis. Usage of this type of form removal method allowed to minimization of values of parameters describing plateau-part (free-of-dimple part) of surface: *Sp* and *Spk*; the value of *Svk* parameter increased with accordance to P_{2ND}, P_{4TH} and/or *GRF*; *Sk* parameter increased. Moreover, the *Pt* parameter (when profiles were taken into account) increased when *RGRF* was applied; the depth of the dimples was flattened with polynomials application.

When oil pockets were edge-situated areal form removal was hampered. When the degree of polynomial increased, *Sq*, *Sv* and *Sa* decreased but *Sv* and *Svk* parameters increased. The smallest value of *Sk* parameter was obtained when *GRF* was used. However, the dimples were flattened (*Svk* parameter decreased much more than *100%*). The highest value of *Svk* parameter was obtained when *RGRF* was applied. Moreover, the *Sq* and *Sa* parameters were minimalized with *RGRF* application (in some cases values were smaller after use of *GRF*). However, the maximum height of surface (*Sz*) has also obtained the biggest value. Moreover, when the displacement of reference plane (dimples) increased, the value of *Ssk* and *Sku* parameters also increased (in some cases more than 300%); evaluation of skewness and kurtosis parameters is of a great importance in ST of car engine parts assessments.

It was assumed that: when D_{DTE} decreased, the distortion of dimples increased; the values of *Sk* parameter increased; when valley were edge-situated the value of *Sk* parameter was twice as large than obtained with center located dimples (when *RGRF* was used). When polynomials (P_{2ND} or P_{4TH}) were used, the value of *Sk* parameter increased around three times (value was overestimated).

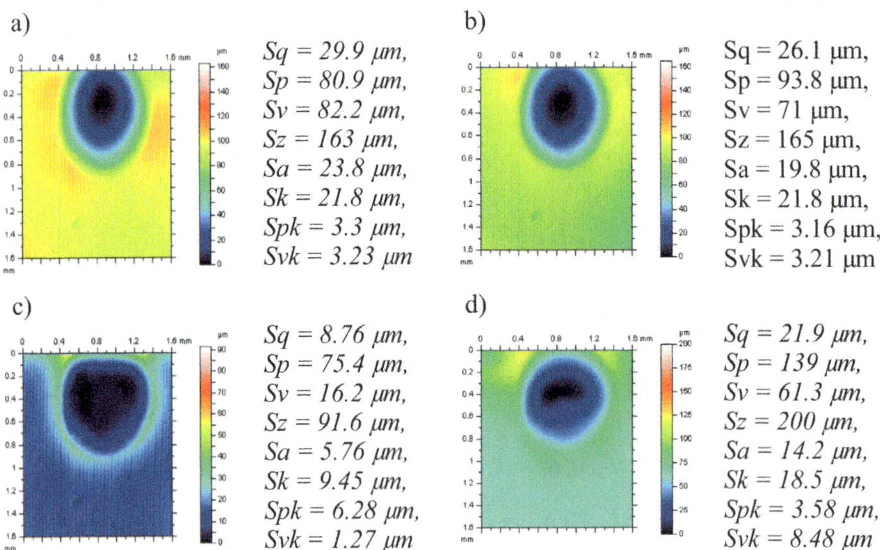

Fig. 6. Details (and their parameters correspondingly) with edge-located dimples, extracted from surface after form removal by: P_{2ND} (a) or P_{4TH} (b), *GRF* (c) and *RGRF* (d); F_{CO} = 0.8 mm

Usually, when distortion of oil pockets increased, the values of Sv and/or Svk parameters decreased; it can be caused by the flattens of dimples. Moreover, when $D_{DTD} < F_{CO}$ and/or $D_{DTE} < F_{CO}$ then displacement of reference plane also increased.

4 Conclusions

The following conclusions and/or remarks were noticed:

1. Application of C_{LSM} did not allow to remove form correctly of cylindrical surfaces containing dimples; displacement of position of reference plane occurred when the $D_D > 10$ μm and/or $D_W > 0.5$ mm.
2. When polynomials (P_{2ND} or P_{4TH}) were applied form was not entirely eliminated. The biggest degree of polynomial was applied the biggest not eliminated errors of form were obtained.
3. When $D_W > 0.4$ mm application of GRF caused a distortion of dimples regardless of D_D (even the $D_D < 10$ μm); dimples were flattened and the neardimple areas were displaced. Moreover, when the D_{DTE} decreased (especially when $D_{DTE} < F_{CO}$) then the distortion of dimples increased.
4. When GRF was applied, displacement of reference plane has a tendency to grow when scratches were located with near-dimple distribution. Moreover, the distortions of dimples increased when $D_{DTD} < D_W$ and/or $D_{DTD} < F_{CO}$; when $D_{DTD} < D_W$ the highest distortion (displacement) of dimples (reference plane) was observed.
5. After usage of $RGRF$ with $F_{CO} = 0.8$ mm the distortion of valleys decreased (in accordance to GRF). However, when $D_{DTE} < F_{CO}$ some dimple-to-edge areas of analysed surface were displaced. When $D_{DTD} < D_W$ and/or $D_{DTD} < F_{CO}$ then displacement of reference plane increased.
6. When D_{DTE} decreased, the distortion of dimples increased (the values of Sk parameter increased when valley were edge-situated – the value of Sk parameter was twice as large than obtained with center-located dimples) when $RGRF$ was applied.
7. After application of P_{2ND} or P_{4TH} the value of Sk parameter was overestimated (increased around three times) for details with edge-situated valleys (compared with its center-located distribution).
8. Usually, when distortion of oil pockets increased, the values of Sv and/or Svk parameters decreased; it can be caused by the flattens of dimples. Moreover, when $D_{DTD} < F_{CO}$ and/or $D_{DTE} < F_{CO}$ then displacement of reference plane also increased.
9. For areal form removal of plateau-honed cylinder surfaces containing deep and/or wide oil pockets it is suggested to application of digital filtering such as $RGRF$.
10. Selection of reference plane should be defined with valley distribution analysis, especially when dimples are edge-located. False estimation of procedure for areal form removal can cause a grave errors in ST parameter calculations. Moreover, the isometric view analysis of surface should be taken into consideration in dimples distortion exclusion.

References

1. Blunt L, Jiang X (eds) (2003) Advanced techniques for assessment of surface topography. Kogan Pages, London
2. Podulka P, Pawlus P, Dobrzański P, Lenart A (2014) Spikes removal in surface measurement. J Phys: Conf Ser 483(1):012025
3. Stout KJ (ed) (1994) Three-dimensional surface topography – measurement, interpretation and application. Penton Press, London
4. Thomas TR (1999) Rough surfaces, 2nd edn. Imperial College Press, London
5. Miller T, Adamczak S, Świderski J, Wieczorowski M, Łętocha A, Gapiński B (2017) Influence of temperature gradient on surface texture measurements with the use of profilometry. Bull Pol Acad Sci Tech Sci 65(1):53–62
6. Magdziak M (2016) An algorithm of form deviation calculation in coordinate measurements of free-form surfaces of products. Strojniški Vestnik - J Mech Eng 62(1):51–59
7. Wdowik R, Magdziak M, Porzycki J (2014) Measurements of surface roughness in ultrasonic assisted grinding of ceramic materials. Appl Mech Mater 627:191–196
8. Magdziak M (2017) The influence of a number of points on results of measurements of a turbine blade. Aircr Eng Aerosp Technol 89(6):953–959
9. Forbes AB (1989) Least squares best fit geometric elements. NLP report DITC 40 (89), Teddington, UK
10. Podulka P, Dobrzański P, Pawlus P, Lenart A (2014) The effect of reference plane on values of areal surface topography parameters from cylindrical elements. Metrol Meas Syst 21 (2):247–256
11. Janecki D (2011) Gaussian filters with profile extrapolation. Precis Eng 35(4):602–606
12. Brinkmann S, Bodschwinna H, Lemke H-W (2001) Accessing roughness in three-dimensions using Gaussian regression filtering. Int J Mach Tools Manuf 41(13–14):2153–2161
13. Brinkman S, Bodschwinna H (2000) Advanced gaussian filters. In: Advanced techniques for assessment surface topography. Elsevier
14. Lou S, Jiang X, Scott PJ (2014) Applications of morphological operations in surface metrology and dimensional metrology. J Phys Conf Ser 483:012020
15. Muralikrishnan B., Raja J (2009) Computational surface and roundness metrology. Springer
16. De Chiffre L, Lonardo P, Trumpold H, Lucca DA, Goch G, Brown CA, Raja J, Hansen HN (2000) Quantitative characterization of surface texture. CIRP Ann Manuf Technol, 49 (2):635–642, 644–652
17. Mainsah E, Greenwood JA, Chetwynd DG (2001) Metrology and properties of engineering surfaces. Springer, London
18. Brown CA, Hansen HN, Jiang XJ, Blateyron F, Berglund J, Senin N, Bartkowiak T, Dixon B, Le Goic G, Quinsat Y, Stemp WJ, Thompson MK, Ungar PS, Zahouani EH (2018) Multiscale analyses and characterizations of surface topographies. CIRP Ann 67(2):839–862
19. Jeng Y (1996) Impact of plateaued surfaces on tribological performance. Tribol Trans 39 (2):354–356
20. Godi A, Grønbæk J, De Chriffre L (2017) Characterisation and full-scale production testing of multifunctional surfaces for deep drawing applications. CIRP J Manuf Sci Technol 16:64–71
21. Podulka P (2018) Problem of selection of reference plane with deep and wide valleys analysis. J Phys Conf Ser 1065:072017
22. Whitehouse DJ (2010) Handbook of surface and nanometrology. CRC Press, Boca Raton

23. Forbes AB (2013) Areal form removal. In: Leach RK (ed) Characterisation of areal surface texture. Springer, pp 107–128
24. Podulka P (2018) Proposal of edge-area form removal of cylindrical surfaces containing wide dimples by application of various robust processing techniques. J Phys Conf Ser 1065:072018
25. Podulka P (2019) Edge-area form removal of two-process surfaces with valley excluding method approach. In: MATEC web of conferences, vol 252, p 05020
26. Stout KJ, Sullivan PJ, Dong WP, Mainsah E, Luo N, Mathia T, Zahouani EH (1993) Publication EUR 15178 EN Commission of the European Communities
27. Podulka P (2016) Selection of reference plane by the least squares fitting methods. Adv Sci Technol Res J 10(30):164–175
28. Podulka P (2018) The effect of valley location in two-process surface topography analysis. Adv Sci Technol Res J 12(4):97–102
29. Yuan Y-B, Qiang X-F, Song J-F, Vorburger TV (2000) A fast algorithm for determining the Gaussian filter mean line in surface metrology. Precis Eng 24(1):62–69
30. Dobrzański P, Pawlus P (2013) Modification of robust filtering of stratified surface topography. Metrol Meas Syst 20(1):107–118

Errors of Surface Topography Parameter Calculation in Grinded or Turned Details Analysis

Przemysław Podulka[(✉)]

The Faculty of Mechanical Engineering and Aeronautics, Rzeszow University of Technology, Powstancow Warszawy 12 Str, 35-959 Rzeszow, Poland
p.podulka@prz.edu.pl

Abstract. In this paper the influence of errors of pre-processing methods on surface topography parameter calculation was taken into consideration. Two types of surfaces were analysed: grinded and/or turned details. Various procedures for separation of form and/or waviness were proposed: polynomials, regular Gaussian filters and splines. More than 20 surfaces, measured by stylus instrument Talyscan 150, were taken into account. The effect of errors of surface topography parameter calculation (from ISO 25178 standard) was taken into consideration. It was assumed that application of commonly used algorithms of polynomial fittings did not always provided a reasonable results; in some cases digital filtering (e.g., Gaussian regression filter) was required. It was also suggested to extract some irregularities with spline pre-processing appliances.

Keywords: Surface topography analysis · Surface topography parameter · Turning · Grinding

1 Introduction

Surface topography measurement and/or analysis is particularly significant in functional properties assessments, lubricant retention or wear resistance, as a material contact issues. Studies of surface topography in engine's piston-piston rings-cylinder liner system is absolutely essential in functional analysis. Since internal combustion engines are widely used in a variety of their applications, the fuel consumption contributed to the air pollution is conducted to the lower emissions. According to the decisive influence of surface texture on functional properties and/or parameters of machined elements, consideration of measurement uncertainty in surface topography analysis seems to be of a non-declining importance.

Many papers are referred to the cylinder liner [1–6] and piston skirt [7–11] assemblies and its surface topography analysis. The influence of surface texture on the functionality of implants, ceramic femoral heads were also studied in details [12, 13]. However not many research are concerned to the errors of surface topography parameter calculation by incorrect pre-processing techniques appliances. Errors of surface texture assessments can be classified to the measurement errors (e.g., the measured object errors, software and/or measuring methods errors [14, 15]), the

© Springer Nature Switzerland AG 2019
M. Diering et al. (Eds.): *Advances in Manufacturing II* - Volume 5, LNME, pp. 134–146, 2019.
https://doi.org/10.1007/978-3-030-18682-1_11

measuring uncertainty errors (e.g., errors caused by digitation process, errors obtained with data processing [16–19]) or other errors [20]. Moreover, various factors affecting machined surface texture was presented in [21]. In some cases the results of surface morphology were obtained with Power Spectrum Density analysis [22]. Parametric and nonparametric description of surface topography after turning in the dry and minimum quantity cooling lubrication conditions were presented in [23]; longitudinal turning tests of duplex stainless steel was taken into account. The problem of precise turning of the mould parts with variable compliance a topographic inspection of the machined surface quality was comprehensive studied in [24].

Usually surface topography parameters are calculated after digital preprocessing by application of various areal form removal algorithms or procedures: fitting algorithms [25–27], filters [28–30] or splines [31–33]. The degree of polynomial approximation for turned piston skirt details was proposed in [34]. However, the influence of falsely estimated pre-processing techniques was not precisely defined according to the errors in surface topography parameter calculations when grinded and/or turned details were taken into account.

2 Materials and Methods

The turned and/or grinded cylindrical surfaces were taken into account. More than 20 surfaces were measured and analysed but only few of them were processed in details. They were measured by stylus instrument Talyscan 150 (nominal tip radius about 2 μm, height resolution about 10 nm). Measured area was around 5 mm by 5 mm, however, usually results were presented only in expanded details. The spacing was up to 5 μm. The measurement was repeated three times and average values was taken into account (A method for surface measurement uncertainty determination was applied). The following pre-processing techniques were applied: least-square polynomial fitting of 4th ($P_{L\text{-}SF4}$), 8th ($P_{L\text{-}SF8}$) or 12th ($P_{L\text{-}SF12}$) degrees, regular Gaussian regression (GF) and robust Gaussian regression (RF) filters or Spline appliance (SF). The value of filter bandwidth (F_{BDW}) was also defined. The effect of pre-processing errors on the surface texture parameter calculation from ISO 25178 standard was taken into consideration.

3 Results and Discussions

From the analysis of isometric views of studied surfaces it was assumed that application of least-square polynomial fitting of nth degree caused the markedly different results of surface topography assessments. In Fig. 1a the various trace of machining process (defined in details: $A1$, $A2$, $A3$ and $A4$) were taken into consideration. It was noticed that application of $P_{L\text{-}SF4}$ and/or $P_{L\text{-}SF12}$ caused the significant growth of heights evaluated between $A3$ and $A4$ details (in accordance to the $P_{L\text{-}SF8}$ appliance); analysed detail seemed to be less flattened. Moreover, when P_{LSF12} was used, the height of the $A2$ detail was overestimated (according to the P_{LSF4} application).

The minimum value of core roughness depth Sk was obtained when $P_{L\text{-}SF12}$ was applied (according to the $P_{L\text{-}SF4}$ and $P_{L\text{-}SF8}$). However, the values of maximum surface

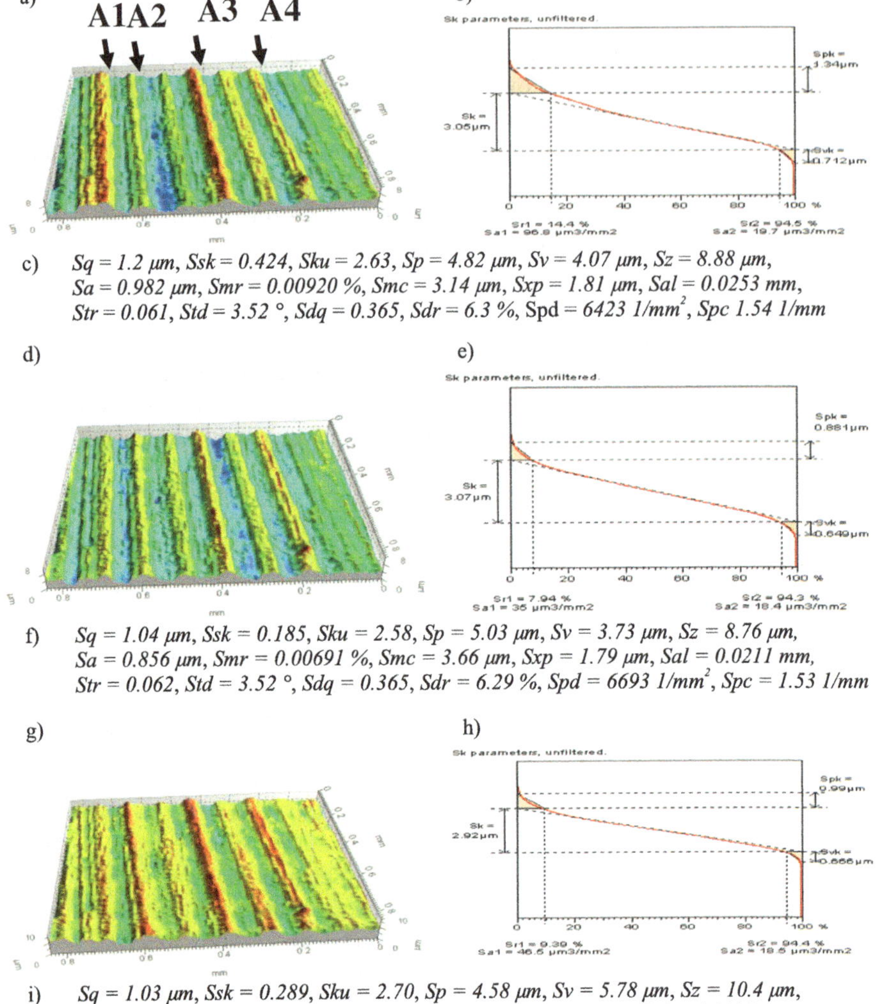

c) *Sq = 1.2 μm, Ssk = 0.424, Sku = 2.63, Sp = 4.82 μm, Sv = 4.07 μm, Sz = 8.88 μm, Sa = 0.982 μm, Smr = 0.00920 %, Smc = 3.14 μm, Sxp = 1.81 μm, Sal = 0.0253 mm, Str = 0.061, Std = 3.52 °, Sdq = 0.365, Sdr = 6.3 %,* Spd = 6423 1/mm², *Spc 1.54 1/mm*

f) *Sq = 1.04 μm, Ssk = 0.185, Sku = 2.58, Sp = 5.03 μm, Sv = 3.73 μm, Sz = 8.76 μm, Sa = 0.856 μm, Smr = 0.00691 %, Smc = 3.66 μm, Sxp = 1.79 μm, Sal = 0.0211 mm, Str = 0.062, Std = 3.52 °, Sdq = 0.365, Sdr = 6.29 %, Spd = 6693 1/mm², Spc = 1.53 1/mm*

i) *Sq = 1.03 μm, Ssk = 0.289, Sku = 2.70, Sp = 4.58 μm, Sv = 5.78 μm, Sz = 10.4 μm, Sa = 0.841 μm, Smr = 0.0124 %, Smc = 3.19 μm, Sxp = 1.7 μm, Sal = 0.0207 mm, Str 0.063, Std = 3.52 °, Sdq = 0.365, Sdr = 6.3 %, Spd = 4548 1/mm², Spc = 1.63 1/mm*

Fig. 1. Details from turned piston skirt surface A after $P_{L\text{-}SF4}$ (a), $P_{L\text{-}SF8}$ (d) and $P_{L\text{-}SF12}$ (g) preprocessing, their material ratio curves (b, e, h) and parameters (c, f, i) respectively.

height Sz and maximum valley depth Sv increased. Moreover, the minimum value of reduced summit height Spk was obtained with $P_{L\text{-}SF8}$ application; kurtosis (Sku) and skewness (Ssk) also decreased. Usually, the autocorrelation length (Sal) decreased when the degree of polynomial approximation increased. The influence of polynomial degree on calculation of texture-aspect ratio Str, texture direction Std, root mean square slope Sdq or interfacial area ratio Sdr parameters did not occur or was negligible. The biggest (smallest) value of summit density Spd (arithmetic mean peak curvature Spc) was obtained when $P_{L\text{-}SF8}$ was applied. All values of parameters were included in Fig. 1.

The minimum values of peak surface area $Sa1$ and valley surface area $Sa2$, upper bearing area $Sr1$ and lower bearing area $Sr2$ were received when $P_{L\text{-}SF8}$ was proposed. In some cases errors of surface topography parameter calculation (from material ratio curves) were greater than *150%* (when $P_{L\text{-}SF4}$ or $P_{L\text{-}SF8}$ were recommended for detail properties extraction). Usually the influence of *A1*, *A2* and *A3* detail location in surface *A* properties extraction was negligible or it not occurred irrespective of pre-processing methods (P_{LSF4}, $P_{L\text{-}SF8}$, $P_{L\text{-}SF12}$); it was assumed that near-edge-distributed upper-part details of analysed surfaces were not significantly vulnerable to flatness when least-square fitting procedures were submitted.

In Fig. 2 lower-part of detail *B* (extracted from turned surface) was taken into account. It was noticed that near-edge (first-edge) located lower-part (*B1*) of studied detail was falsely estimated when $P_{L\text{-}SF4}$ was confirmed (compared with P_{LSF8}). When detail *B1* was improperly evaluated the values of Sq and Ssk (Sku) parameters increased (decreased); differences were small. For Sp, Sv, Sz and Sa parameters they were calculated erroneously (their values were overestimated) after pre-processing by P_{LSF4} according to $P_{L\text{-}SF8}$.

Usually, the value of Sk parameter decreased more than 10% when $P_{L\text{-}SF8}$ was applied instead of P_{LSF4}; the variation (overestimating) of Svk parameter was smaller than 5%. The value of Spk parameter increased more than 30%, however, the sum of Sk and Spk parameters decreased. When the degree of polynomial fitting was enlarged (from $P_{L\text{-}SF4}$ to P_{LSF8} for instance), the values of $Sa1$ ($Sa2$) and $Sr1$ ($Sr2$) parameters raised (decreased) between *10%* and *20%* (less than *1%*) or *50%* and *60%* (*20%* and *30%*) correspondingly. When upper-part (*B2*) of studied detail *B* was taken into consideration, the substantial displacement of pre-processed form was detected in near-edge locations.

Moreover, frequently the first-edge lower-parts of studied detail were vastly exaggerated in association with near-edge upper-parts. Near-edge detail analysis is particularly significant for small-scale surface part analysis; the biggest displacement of near-edge parts of studied detail occurs the biggest errors of calculation of areal surface topography parameters are noticeable. It was also found that displacement of some upper-part of analysed detail was considerably expanded when wear was noticed (it was indicated by the component *B3* in Fig. 2a).

For extraction/calculation of surface topography quantities/parameters the following procedures were proposed: Gaussian filter, Robust Gaussian filter and Spline appliance. The results of its application were compared with commonly used leas-square fitting polynomial algorithms ($P_{L\text{-}SF4}$ and $P_{L\text{-}SF8}$). Upper-parts (*C3*) and lower-parts (*C1* and *C2*) of studied detail *C* were taken into consideration; some description were presented in Fig. 3. Application of *GF*, *RF* and/or *SF* caused a minimization of values of Sq, Ssk and Sku parameters according to the $P_{L\text{-}SF4}$ and/or $P_{L\text{-}SF8}$ appliances.

The errors (values) of Sz parameter calculation increased (decreased) when *RF* was applied in accordance with *GF* and/or *SF* ($P_{L\text{-}SF4}$ or $P_{L\text{-}SF8}$) usage. The values of Sp, Sv, Sz and Sa (Sk, Spk, Svk) parameters were minimized after pre-processing by *GF* (*SF*). From the analysis of isometric views of surfaces it was assumed that application of filtering methods caused that lower-part of studied detail (it was plainly evident with *C2* component evaluation) was flattened; according to the $P_{L\text{-}SF4}$ or P_{LSF8} usage, biggest (smallest) flatten was observed after *RF* (*GF* or *SF*) appliances.

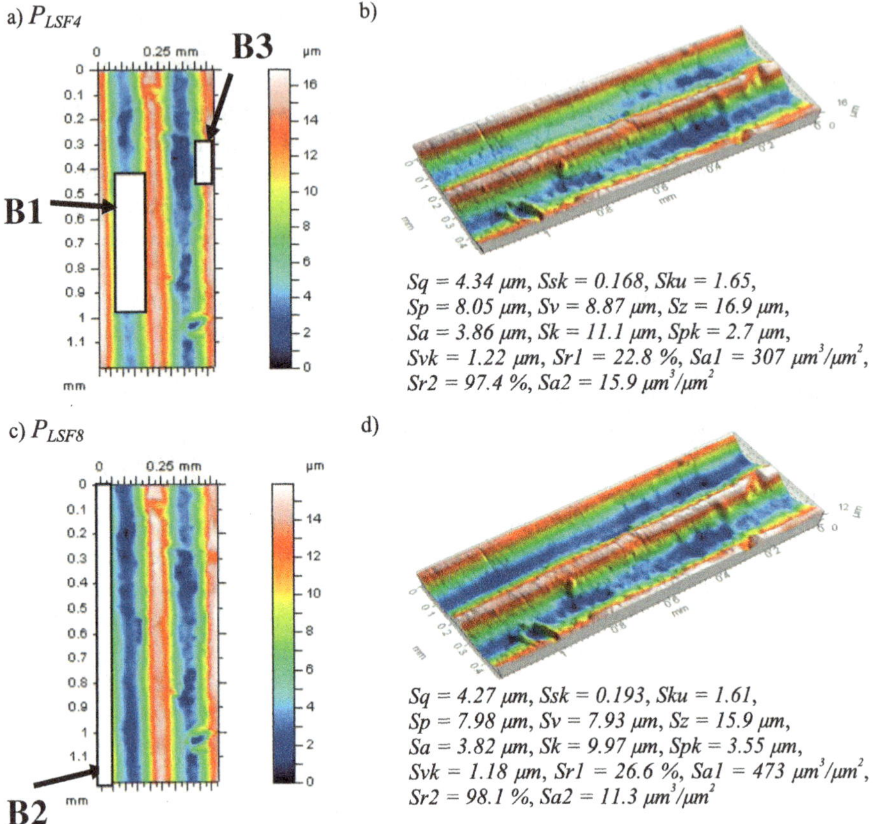

Fig. 2. Details from turned cylindrical element *B*, their contour plots (a, c), isometric views (b, d) and parameters correspondingly, after polynomial fitting method appliance.

Most of the surface topography (especially height) parameters were minimized after *GF* pre-processing in comparison to the $P_{L\text{-}SF4}$ or $P_{L\text{-}SF8}$ schemes. However, when grinded details were taken into account, application of *GF* caused a displacement of near-edge upper-part of studied details; in Fig. 4 differences of dislocation of this type of details (*D1*, *D2*) were presented for *GF* and *SF* applications. Dislocation of *D1* and *D2* details can cause the errors in surface topography parameters; most of the values of surface topography parameters were overestimated when *GF* was applied instead of *SF*. *Sq* parameter was resistant for upper-part detail displacement. However values of *Ssk* and *Sku* parameters were changed more than *200%* and *10%* respectively; values of height parameters (*Sp*, *Sv*, *Sz*) were enlarged ordinarily more than *10%* (*12%*, *10%* and *9%* correspondingly). The displacement of near-edge areas was noticed irrespective of regression quality of applied algorithm.

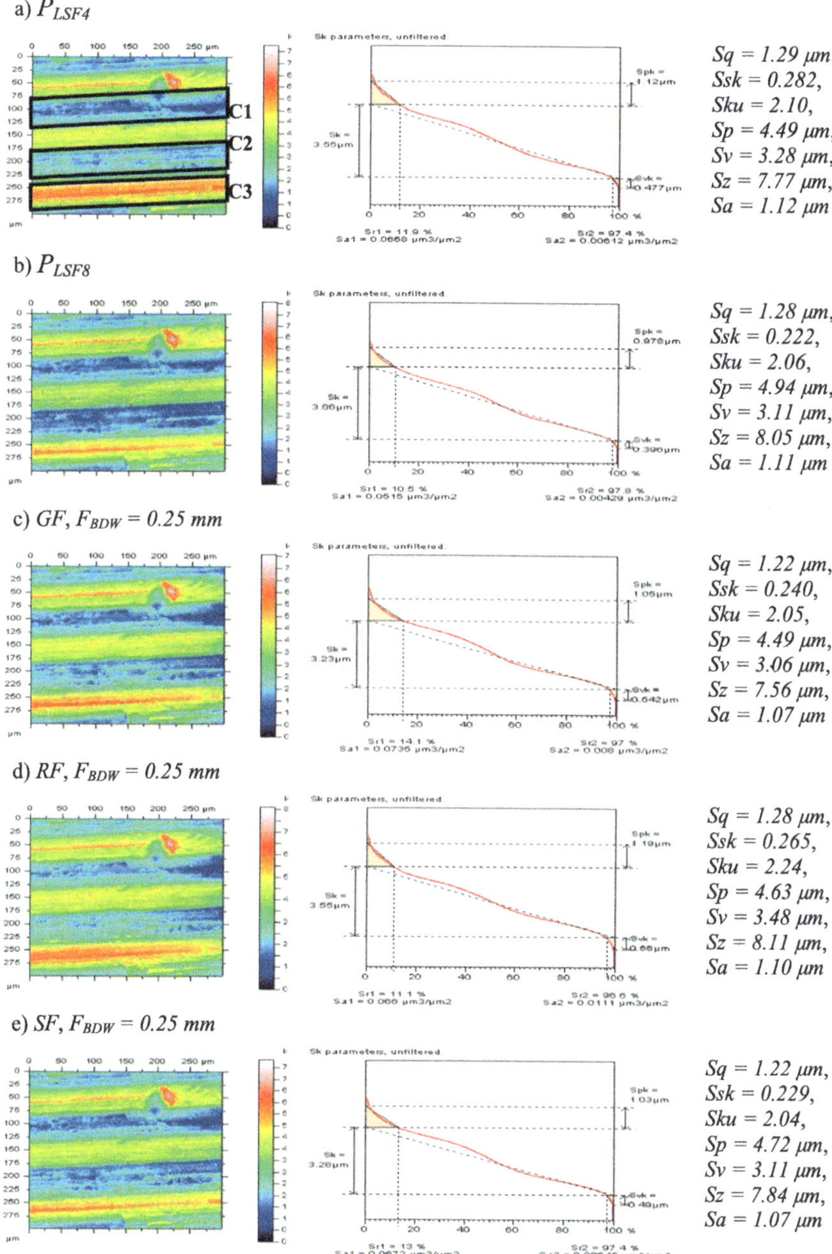

Fig. 3. Contour plots, material ratio curves and selected parameters (correspondingly) of grinded detail C after application of various pre-processing techniques.

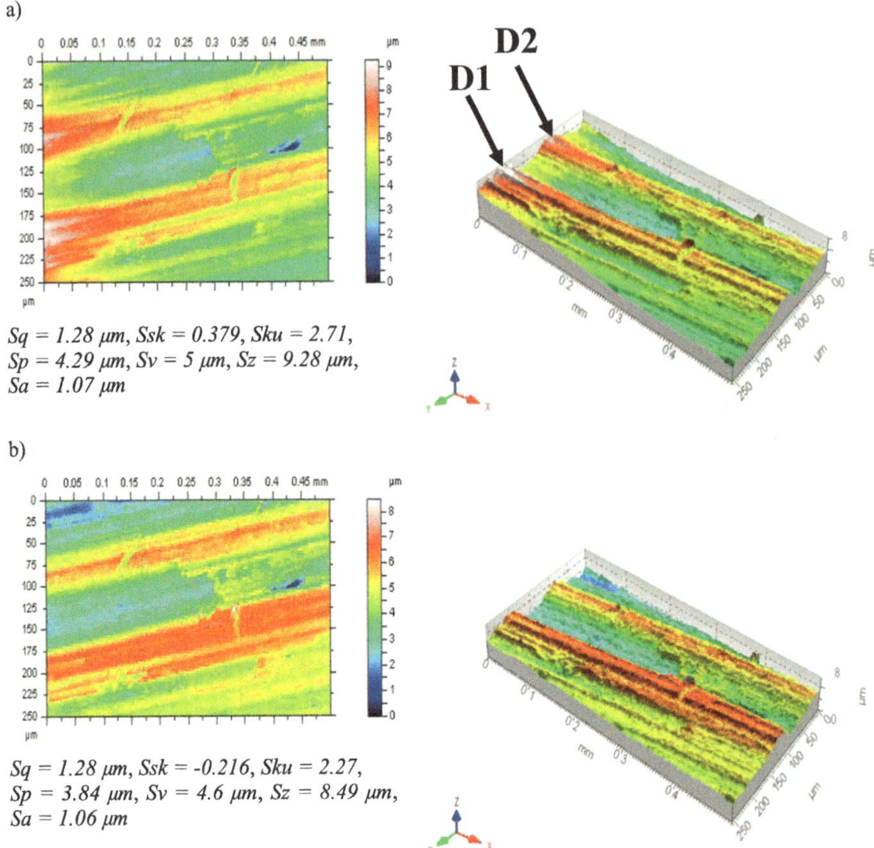

Fig. 4. Contour plots, isometric views and selected parameters of grinded surface D after preprocessing by: GF (a) and SF (b); FBDW = 0.25 mm.

Application of *RF* (compared with the use of *GF*) as a pre-processing technique did not allow to minimize the values of (height) parameters. SF was proposed for surface topography values/parameters extraction/calculation. After usage of *SF* the values of *Sq, Ssk, Sku, Sp, Sv, Sz*, Sa parameters decreased *9%, 13%, 7%, 3%, 5%, 5%, 11%* and *15%, 48%, 1%, 12%, 11%, 11%, 16%* according to *GF* or *RF* application respectively; parameters are included in Fig. 5.

The isometric view of studied surfaces were taken into consideration; in Fig. 5c detail *E4* was given into meticulous attention. It was noticed that *GF* and *RF* caused a displacement of near-edge/near-corner area of analysed detail; in some cases upper-part detail of studied surface was flatten. Moreover, some near-edge located areas were exaggerated (e.g., details *E1, E2* and *E3* presented in Fig. 5); error of shape/waviness were not entirely eliminated.

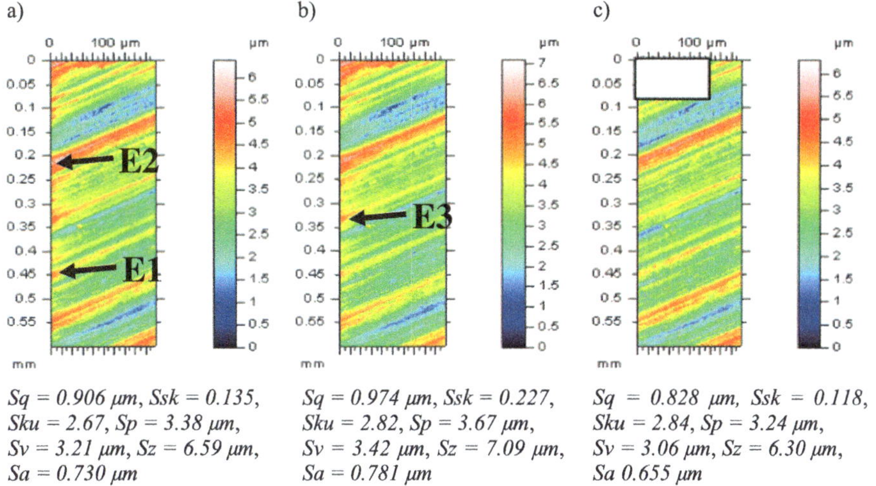

$Sq = 0.906 \, \mu m$, $Ssk = 0.135$, $Sku = 2.67$, $Sp = 3.38 \, \mu m$, $Sv = 3.21 \, \mu m$, $Sz = 6.59 \, \mu m$, $Sa = 0.730 \, \mu m$

$Sq = 0.974 \, \mu m$, $Ssk = 0.227$, $Sku = 2.82$, $Sp = 3.67 \, \mu m$, $Sv = 3.42 \, \mu m$, $Sz = 7.09 \, \mu m$, $Sa = 0.781 \, \mu m$

$Sq = 0.828 \, \mu m$, $Ssk = 0.118$, $Sku = 2.84$, $Sp = 3.24 \, \mu m$, $Sv = 3.06 \, \mu m$, $Sz = 6.30 \, \mu m$, $Sa \; 0.655 \, \mu m$

Fig. 5. Contour plots and parameters of grinded detail E after pre-processing by GF (a), RF (b) and SF (c); FBDW = 0.25 mm.

Application of SF for parameter calculation of turned piston skirt surfaces caused a noticeable improvement according to the *GF* and/or *RF* usage. The values of *Sq*, *Ssk* (*Sku*), *Sp*, *Sv*, *Sz* and *Sa* parameters were minimized (maximized) when *SF* was applied. *Sk* parameter usually decreased more than *5%*, parameter *Spk* increased, parameter *Svk* decreased more than *65%*; however sum of the *Spk* and *Svk* parameters decreased more than *20%*. Parameters *Sr1*, *Sa1* and *Sr2* (*Sa2*) increased (decreased) after *SF* appliance; parameters were presented in Fig. 6.

In many cases upper-part edge-located details were displaced; the position of detail *F1* was lowered when *GF* or *RF* was applied, usage of *SF* allowed to eliminate near-edge and/or near-corner errors of shape/waviness precisely; another remarkable example was presented with *F2* detail assessment. Some displacement were especially (only) noticeable with near-edge analysis.

In Fig. 7 some irregularities of turned piston skirt surface was presented (detail *G1*). Displacement of this type of detail did not occurred or it was negligible regardless of *GF*, *RF* or *SF* appliance. The values of *Sp*, *Sv* and *Sz* parameters were minimized when *GF* was used. However, in some cases displacement of both details (upper-part and lower-part) was noticed after its appliance; false estimation of details was also found when *RF* was purposed; it was indicated with the detail *G2*. After application of *SF* displacement of nearedge located details of turned piston skirt surface were virtually imperceptible or did not exist. Moreover, the values of *Sk*, *Spk* and *Svk* parameters were minimized when *SF* was applied.

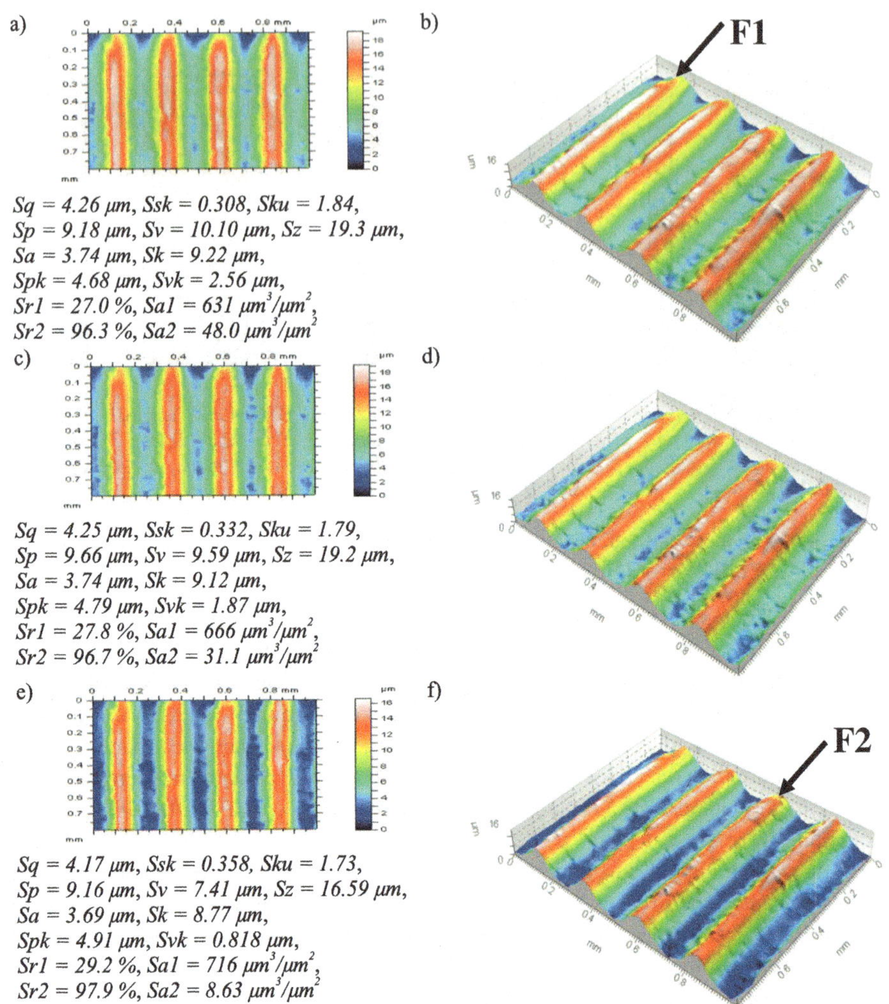

$Sq = 4.26\ \mu m$, $Ssk = 0.308$, $Sku = 1.84$,
$Sp = 9.18\ \mu m$, $Sv = 10.10\ \mu m$, $Sz = 19.3\ \mu m$,
$Sa = 3.74\ \mu m$, $Sk = 9.22\ \mu m$,
$Spk = 4.68\ \mu m$, $Svk = 2.56\ \mu m$,
$Sr1 = 27.0\ \%$, $Sa1 = 631\ \mu m^3/\mu m^2$,
$Sr2 = 96.3\ \%$, $Sa2 = 48.0\ \mu m^3/\mu m^2$

$Sq = 4.25\ \mu m$, $Ssk = 0.332$, $Sku = 1.79$,
$Sp = 9.66\ \mu m$, $Sv = 9.59\ \mu m$, $Sz = 19.2\ \mu m$,
$Sa = 3.74\ \mu m$, $Sk = 9.12\ \mu m$,
$Spk = 4.79\ \mu m$, $Svk = 1.87\ \mu m$,
$Sr1 = 27.8\ \%$, $Sa1 = 666\ \mu m^3/\mu m^2$,
$Sr2 = 96.7\ \%$, $Sa2 = 31.1\ \mu m^3/\mu m^2$

$Sq = 4.17\ \mu m$, $Ssk = 0.358$, $Sku = 1.73$,
$Sp = 9.16\ \mu m$, $Sv = 7.41\ \mu m$, $Sz = 16.59\ \mu m$,
$Sa = 3.69\ \mu m$, $Sk = 8.77\ \mu m$,
$Spk = 4.91\ \mu m$, $Svk = 0.818\ \mu m$,
$Sr1 = 29.2\ \%$, $Sa1 = 716\ \mu m^3/\mu m^2$,
$Sr2 = 97.9\ \%$, $Sa2 = 8.63\ \mu m^3/\mu m^2$

Fig. 6. Contour plots (a, c, e) with selected parameters and isometric views (b, d, f) of turned piston skirt surface F after GF (a, b), RF (c, d) and SF (e, f) pre-processing analysis; FBDW = 0.8 mm.

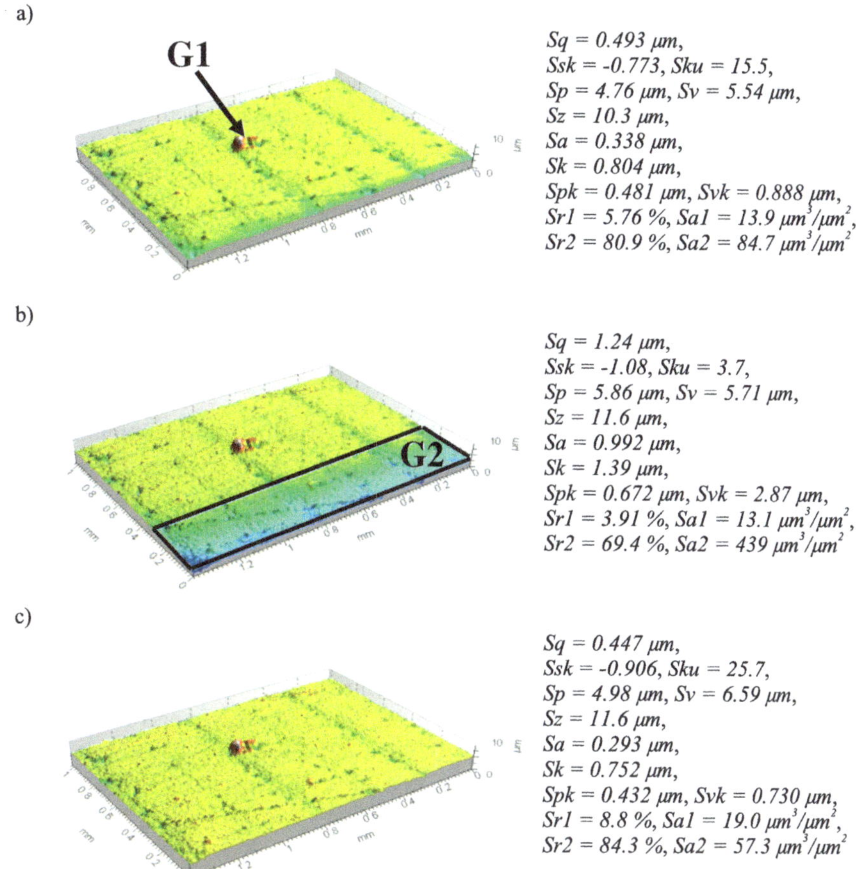

Fig. 7. Isometric views of turned piston skirt surface G (with its height parameters respectively) after assessment by: GF (a), RF (b) and SF (c) applications; FBDW = 0.8 mm.

4   Conclusions

From the analysis of isometric views of studied details it was found that application of least-square fitted polynomials of 4th, 8th or 12th degrees caused the strikingly different results. It was noticed that application of $P_{L\text{-}SF4}$ and/or $P_{L\text{-}SF12}$ caused the considerable growth of some upper-part of analysed surface; analysed detail seemed to be less flattened. Moreover, application of too high degree (e.g., P_{LSF12}) caused overestimating the height of upper-part details considerably. It was suggested to use the $P_{L\text{-}SF8}$ as a pre-processing method instead of $P_{L\text{-}SF4}$ or P_{LSF12} appliance. Generally, near-edge-located upper-part details were not deeply sensitive for flatness process when least-square fitting algorithms (e.g., P_{LSF4}, $P_{L\text{-}SF8}$) were applied. However, in some cases near-edge upper-parts of studied detail were assessed improperly according to the first-edge lower-parts exaggeration. Near-edge detail study is particularly important when small-scale parts of surface are taken into consideration; the biggest errors of near-edge parts position of studied detail occurs the biggest inaccuracy of areal surface topography parameters calculation are found. Moreover, displacement of some upper-part of analysed detail was significantly increased when wear was recognized.

Application of least-square fitted polynomials was compared with *GF*, *RF* and *SF* methods. Usage of *GF*, *RF* and/or *SF* caused a minimization of values of *Sq*, *Ssk* and *Sku* parameters according to the $P_{L\text{-}SF4}$ and/or $P_{L\text{-}SF8}$ applications. The errors of *Sz* parameter calculation increased when RF was applied in accordance with *GF* and/or *SF* appliance. The values of *Sp*, *Sv*, *Sz* and Sa parameters were minimized after pre-processing by *GF*; however the smallest values of *Sk*, *Spk* and *Svk* parameters were obtained by *SF* application. It was also found that application of filtering methods caused the flattening of lower-part of studied detail according to the least-square fitting methods of pre-processing (e.g., $P_{L\text{-}SF4}$ or P_{LSF8}); biggest (smallest) flatten was observed after *RF* (*GF* or *SF*) appliances.

Generally, surface topography (especially height) parameters were minimized after *GF* application in accordance to the least-square fitting methods (e.g., $P_{L\text{-}SF8}$). However, when grinded details were taken into consideration, usage of *GF* caused a displacement of near-edge upper-part of studied details. Most of the surface topography parameters were overestimated when *GF* was applied; the near-edge areas displacement was found regardless of regression aspect of applied procedure.

Application of *RF* as a pre-processing technique did not allow to minimize the values of (height) parameters (according to the *GF* appliance); *SF* was proposed for surface topography values (parameters) extraction (calculation). After usage of *SF* the values of height parameters (*Sq*, *Ssk*, *Sku*, *Sp*, *Sv*, *Sz*, *Sa*) decreased according to *GF* or *RF* application. It was also noticed that *GF* and *RF* caused a displacement of near-edge/near-corner area of analysed surface; in some cases upper-part detail of studied detail was flatten. Moreover, some near-edge located areas were exaggerated; error of shape/waviness were not completely excluded.

When *SF* was applied for parameter calculation of turned piston skirt surfaces a substantial advance (according to the *GF* and/or *RF* appliance) was improved; the values of *Sq*, *Ssk* (*Sku*), *Sp*, *Sv*, *Sz* and *Sa* parameters were minimized (maximized) when SF was proposed as a pre-processing technique. Application of *SF* scheme allowed to eliminate near-edge and/or near-corner errors of shape/waviness; some irregularities and/or wears were not displaced or changes were negligible. After application of *SF* displacement of near-edge located details of turned piston skirt surface were relatively unnoticeable or did not occurred. Moreover, the values of *Sk*, *Spk* and *Svk* parameters were minimized when *SF* was applied. Therefore, *SF* scheme is suggested for surface topography assessment of grinded and/or turned details as an alternative to the application of commonly used least-square fitted polynomials of nth degrees (e.g., $P_{L\text{-}SF8}$).

References

1. Johansson S, Nilsson PH, Ohlsson R, Anderberg C, Rosen B-G (2008) New cylinder liner surfaces for low oil consumption. Tribol Int 41:854–859
2. Pawlus P (1993) Effects of honed cylinder surface topography on the wear of piston-piston ring-cylinder assemblies under artificially increased dustiness conditions. Tribol Int 26:49–56

3. Podulka P (2018) Problem of selection of reference plane with deep and wide valleys analysis. J Phyics Conf Ser 1065:072017
4. Krzyzak Z, Pawlus P (2006) 'Zero-wear' of piston skirt surface topography. Wear 260:554–561
5. Koszela W, Pawlus P, Galda L (2007) The effect of oil pockets size and distribution on wear in lubricated sliding. Wear 263:1585–1592
6. Podulka P (2016) Selection of reference plane by the least squares fitting methods. Adv Sci Technol Res J 10(30):164–175
7. Rao VDN, Kobat DM, Yeager D, Lizotte B (1997) Engine studies of solid film lubricant coated pistons. SAE Paper, Paper 18, p 20–31
8. Knoll GG, Peeken HJ (1982) Hydrodynamic lubrication of piston skirts. ASME J Tribol 104:504–509
9. Oh KP, Li CH, Goenka PK (1987) Elastohydrodynamic lubrication of piston skirts. ASME J Tribol 109:356–361
10. Zhu D, Cheng HS, Arai T, Hamai K (1992) A numerical analysis for piston skirts in mixed lubrication Part I: Basic modeling. J Tribol 114(3):553–562
11. Ye Z, Zhang C, Wang Y, Cheng HS, Tung S, Wang QJ, He X (2004) An experimental investigation of piston skirt scuffing: a piston scuffing apparatus, experiments and scuffing mechanism analyses. Wear 257:8–31
12. Niemczewska-Wojcik M (2014) The measurement and analysis of surface geometric structure of ceramic femoral heads. Scanning 36:105–114
13. Niemczewska-Wójcik M (2011) The influence of the surface geometric structure on the functionality of implants. Wear 271:596–603
14. Pawlus P, Wieczorowski M, Mathia T (2014) The errors of stylus methods in surface topography measurements. ZAPOL, Szczecin
15. Podulka P, Pawlus P, Dobrzański P, Lenart A (2014) Spikes removal in surface measurement. J Phys: Conf Ser 483(1):012025
16. Podulka P (2019) Edge-area form removal of two-process surfaces with valley excluding method approach. In: MATEC web of conferences, vol 252, p 05020
17. Podulka P (2018) The effect of valley location in two-process surface topography analysis. Adv Sci Technol Res J 12(4):97–102
18. Pawlus P, Smieszek M (2005) The influence of stylus flight on change of surface topography parameters. Precis Eng 29:272–280
19. Thomas TR (1999) Rough surfaces, 2nd edn. Imperial College Press, London
20. Magdziak M (2017) The Influence of a number of points on results of measurements of a turbine blade. Aircr Eng Aerosp Technol 89(6):953–959
21. Wojciechowski S, Twardowski P, Wieczorowski M (2014) Surface texture analysis after ball end milling with various surface inclination. Metrol Meas Syst 21(1):145–156
22. Krolczyk GM, Maruda RW, Nieslony P, Wieczorowski M (2016) Surface morphology analysis of Duplex Stainless Steel (DSS) in clean production using the power spectral density. Measurement 96:464–470
23. Krolczyk GM, Maruda RW, Krolczyk JB, Nieslony P, Wojciechowski S, Legutko S (2018) Parametric and nonparametric description of the surface topography in the dry and MQCL cutting conditions. Measurement 121:225–239
24. Nieslony P, Krolczyk GM, Wojciechowski S, Chudy R, Zak K, Maruda RW (2018) Surface quality and topographic inspection of variable compliance part after precise turning. Appl Surf Sci 434:91–101
25. Muralikrishnan B, Raja J (2009) Least-squares best-fit line and plane. In: Computational surface and roundness metrology. Springer

26. Forbes AB (1989) Least-squares best-fit geometric elements, NPL report DITC 140/89, National Physical Laboratory. Teddington, UK
27. Dhanish PB, Shunmugam MS (1991) An algorithm for form error evaluation – using the theory of discrete and linear Chebyshev approximation. Comput Methods Appl Mech Eng 92:309–324
28. Brinkman S, Bodschwinna H, Lemke H-W (2001) Accessing roughness in three-dimensions using Gaussian regression filtering. Int J Mach Tools Manuf 41:2153–2161
29. Friis KS, Godi A, De Chiffre L (2011) Characterisation of multifunctional surfaces with robust filters. In: 4th international swedish production symposium, pp 525–532
30. Janecki D (2011) Gaussian filters with profile extrapolation. Precis Eng 4:602–606
31. Janecki D (2009) A generalized L2-spline filter. Measurement 42(6):937–943
32. Zeng W, Jiang X, Scott P (2011) A generalised linear and nonlinear spline filter. Wear 271:544–547
33. Dobrzanski P, Pawlus P (2010) Digital filtering of surface topography: Part I. Separation of one-process surface roughness and waviness by Gaussian convolution, Gaussian regression and spline filters. Precis Eng 34(3):647–650
34. Podulka P, Pawlus P, Dobrzanski P, Lenart A (2014) The effect of reference plane on values of areal surface topography parameters from cylindrical elements. Metrol Meas Syst 21 (2):247–256

The Effect of a Stylus Tip on Roundness Deviation with Different Roughness

Jan Zelinka[1], Lenka Čepová[1], Bartosz Gapiński[2], Robert Čep[1(✉)],
Ondřej Mizera[1], and Radek Hrubý[1]

[1] Faculty of Mechanical Engineering, VŠB - Technical University of Ostrava,
17. listopadu 15/2172, 708 33 Ostrava-Poruba, Czech Republic
`robert.cep@vsb.cz`
[2] Institute of Mechanical Technology, Poznan University of Technology,
Piotrowo 3, 60-965 Poznań, Poland

Abstract. This article deals with the measurement of roundness deviations with examined samples of different roughnesses. The intent of this research is to determine a proper method for measuring different roughnesses. The experiment was measured on coordinate measuring machine WENZEL LH65 X3M Premium and using the software Metrosoft QUARTIZ R6 located in a laboratory of the Technical University of Ostrava. The final measured values of roundness deviation using a filter were lower than the values measured without a filter. The MZC method showed slightly lower circularity deviations than the LSC method. When using the LSC and MZC method of scanning with a filter, it was found that when using a ball with a diameter ø 2 mm, the deviation is smaller than when using a ball diameter ø 5 mm. Research has shown the dependence of roughness along with the size of the ball on the deviation of circularity.

Keywords: Roundness · Roughness · Coordinate measuring machine · Measuring methods

1 Introduction

Nowadays progress in research and development is getting better every day. To global trend is to make production faster and more qualitative. An increase in the quality of manufactured products has caused an increase in the amount of research on techniques [1]. Therefore, the technology used in production is still being developed. Those changes affect the customer, who is increasingly demanding the best quality at a lower cost. With this in view, companies are trying to provide faster, more accurate and reliable devices, which they can measure and determine if the component meets their requirements.

One of the most frequently measured production parameters is roundness. It is defined by the standard EN ISO 12181-1 [2]. Generally the measurement of roundness is given special attention, because the life of the devices depends on these measurements, which impacts safety and handling.

To measure roundness deviation it is necessary to have not only a technically appropriate and reliable machine, but chosen suitable measurement strategy and the applicable conditions of the environment as well [3].

© Springer Nature Switzerland AG 2019
M. Diering et al. (Eds.): *Advances in Manufacturing II* - Volume 5, LNME, pp. 147–157, 2019.
https://doi.org/10.1007/978-3-030-18682-1_12

Currently, the trend in verifying roundness is through a coordinate measuring machine (CMM). These measuring machines are necessary equipment in most manufacturing companies.

However, it is important to gain knowledge on the measurement process strategy and how to process measurement results, to use a CMM effectively. Scanning and touch methods are used for CMM measurements. There are some differences between these methods depending on the number of scanned points. Also, a very important element for the evaluation of roundness deviation is the filter used, which refers to the standard EN ISO 4291 [4, 5].

Coordination Measuring Machines (CMM) perform measuring on a plane or space. The basis of the calculated geometrical data are position data. From these positions there are determined substitute part geometry. For all elements there are determined a minimal number of points, which is necessary for alternative geometry generation [6]. The CMM probe is one of the most important systems of dimensional measuring instruments and accounts for coordinate measurement accuracy. The fast response and accurate detection of a probe that can be computer controlled represents the current trend for the next generation of coordinate metrology. The stylus tip contact with a detected surface is the source of a signal that will develop the pattern on the working objects. The performance of the whole CMM system is very much dictated by the motion precision of the probe tip and its actuator. So, the probe stylus tip is laterally at the centre of CMM operation and a key element of coordinate measurements. Detection probes branch into two main categories: contact probes and non-contact probes [7]. This issue is not very common in scientific work, so the Literature Review chapter has not been included.

2 Defining the Deviation of Roundness

A deviation of circularity belongs to the group of geometric shape tolerances. The roundness of a single tolerated element is considered to be correct, if the element is between a pair of concentric circles in a plane, whose radial distance is most equal to the prescribed tolerance value [8].

The tolerated element is considered to be appropriate under the condition that the radial distance is equal to the prescribed tolerance value or is less. It is necessary that the radial distance between the two concentric circles be as small as possible (see Fig. 1) [8, 9].

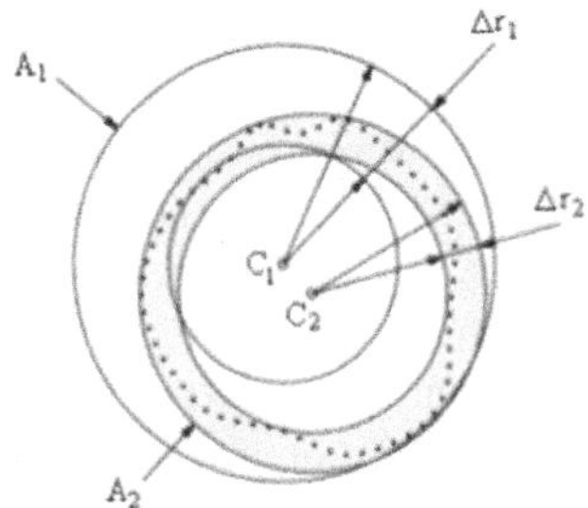

Fig. 1. Placing the centres of two concentric circles [8, 9].

Based on the figure, it is possible to determine that the correct tolerance condition is $\Delta r_2 < \Delta r_1$. The marked out circle A_2 meets the concentricity of circles. The radial distance Δr_2 is equal to or less than the prescribed circular tolerance. For evaluating the deviation of roundness the correct method of attaching reference circle must be chosen.

The standard ISO 6318 gives four suitable elements for a circle: Least Square Circle (LSC), Minimal Circumscribed Circle (MCC), Maximal Inscribed Circle (MIC) and the Minimal Zone Circle (MZC) (see Fig. 2).

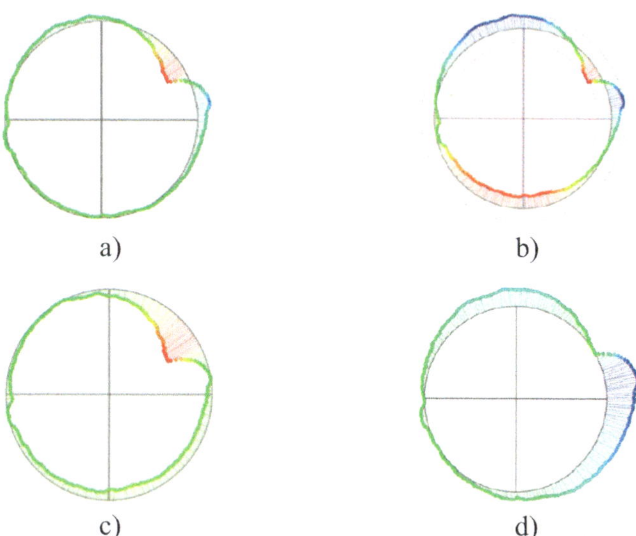

Fig. 2. Suitable elements according ISO 6318: (a) Least Square Circle (LSC), (b) Minimal Circumscribed Circle (MCC), (c) Maximal Inscribed Circle (MIC), (d) Minimal Zone Circle (MZC) [10].

The selected method, the chosen machine tool, and the clamping and cutting tool are the most often reasons for a deviation in roundness. They could be the result of an incorrectly set machine causing vibrations or by an improperly clamped workpiece or tool, causing a misalignment of the workpiece relative to the tool.

The resulting deviation value of the roundness can be reduced with the filter used. Filtration represents a very important element of the surface texture analysis process that affects the parameter values. Filtration is used for many reasons, e.g. a morphological filter is commonly applied to correct the effect of the probe on unprocessed data, or to remove distant protrusions by filtering when the measured data is processed by an optical sensor, etc. The main reason for the filters being used is the separation of the shortwave and longwave components of the surface profile. This mean a separation of waviness from roughness and the subsequent identification of the corresponding characteristic parameters [11].

The standard values of the filters commonly used are 1-15 UPR, 1-50 UPR, 1-150 UPR, 1-500 UPR. These filter values are internationally recognized as a standard UPR range of values for circularity deviation analysis. Today's modern devices already have built-in software that can manually set filter values such as 17-38 UPRs. The second possible division of filters is according to the method of the transfer and processing of the measured data. They can be divided into the 2CR filter and Gauss filter [12].

2CR is an electric filter that consists of two capacitors and two resistors. After selecting the filter, the harmonic wave amplitude at the selected UPR values is reduced by 25%. This type of filter is becoming less and less used, and the Gauss filter, which is known for its better response and sharper curvature, is more widely used [13].

The Gauss filter belongs to a group of mathematical filters. For preselected UPR parameters, the amplitude is reduced by 50% compared to the 2CR, which decreases by only 25%. To use a filter, we must have at least 7 points per wave. This requires a minimum amount of scattered points depending on the shape of the surface. The larger the number of waves per perimeter with the same diameter, the less filtered it is. The size of the W/C filter for a circle or cylinder determines the intensity of the filtration [14].

We can further divide filters into low bandwidth, bandwidth and bandwidth throughput. With low-band filtering, high frequencies are lowered and there are low passes through the filter. In the evaluation, we have obtained waviness parameters and other long wool components. The effects of roughness are suppressed. The high bandwidth filter eliminates waves that occur at high frequencies and low frequencies do not pass. The wrinkle effects are suppressed, and the surface roughness of the component is displayed in the evaluation. Bandwidth is a combination of a low-band and a high-band filter, eliminating the waviness and surface roughness profile that extend beyond the interfaces of the specified boundaries of the individual sections [15].

If multiple points are scanned, it is appropriate to eliminate the remote values. These values distort the resulting deviation of circularity. If the less measured points are to be eliminated, the measured values that define the variance can also be deleted. If we use the elimination of remote values together with filtration, the software first makes an elimination of the remote values and then performs filtration. Prior to filtration, it is recommended to first override the values of the incorrect values before filtering them. Metrosoft QUARTIS software, which was used in this research, also performs the removal of remote values [14].

Table 1. Cutting conditions [15].

Diameter [mm]	Only for evaluation		For evaluating form and waviness	
	Periphery waves [W/U] [-]	No. of points for amplitude [-]	Periphery waves [W/U] [-]	No.of points for amplitude [-]
to 25	15	>105	50	>350
>25 to 80	50	>350	150	>1050
>80 to 250	150	>1050	500	>3500
>250	500	>3500	1500	>10500

3   Research Methodology Work

The samples used for experimental purposes were machined on a DMG Mori Seiki NLX 2500/700. For the preparation of samples there were used the cutting conditions shown in Table 1. All samples with different roughnesses were measured for a deviation of roundness by using the scanning and touch method. The gripping method used the Least Square Circle (LSC) and Minimum Zone Circles (MZC).

Table 2. Cutting conditions [9].

Require roughness [µm]	v_c [m/min]	F [mm]	a_p [mm]
Ra 0,8	300	0,1	1,5
Ra 1,6	120	0,2	1,5
Ra 3,2	120	0,33	1,5
Ra 6,3	120	0,54	1,5

The scanning method was evaluated by using points at a distance 0,1 mm apart. The touch method was measured using 38 points. Measurement was performed with the filter WC150 and without a filter also. The samples that were examined for roughness are in Table 2. A Mitutoyo 410 verified the roughness of samples using a surface roughness measurement. The measured values were then rounded off.

The deviation of roundness was measured by using two different types of stylus tips, see Figs. 3 and 4. The touches of the parameters are shown in Table 3. The roundness variations were measured in the plane of the cut z = 12 mm from the upper plane. For each scan sensor, the cycle is repeated three times and from the results an arithmetic mean was made. Repeating the measurement was done to verify the accuracy of the results.

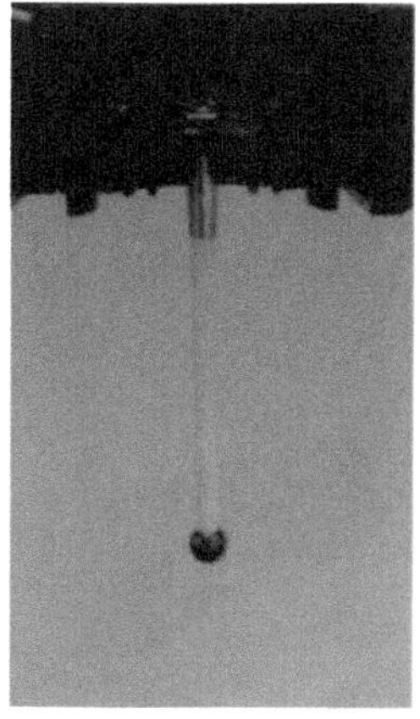

Fig. 4. Measuring probe 3001

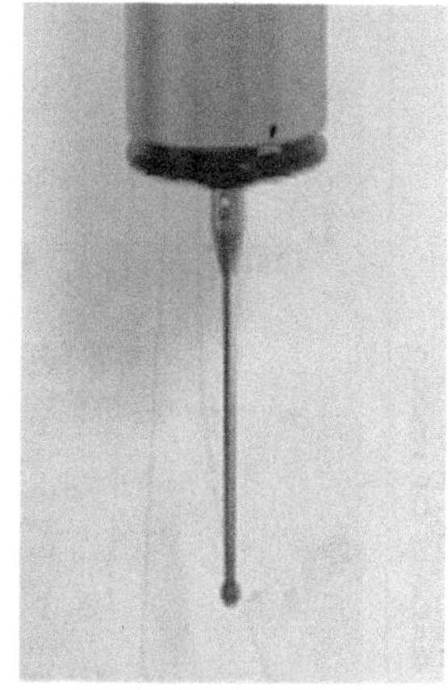

Fig. 3. Measuring probe 1001

Table 3. Parameters and types of the stylus tip [7].

Probe designation [-]	Probe diameter balls [mm]	Stylus length [mm]	Material of stem stylus [mm]	Variance probe [mm]
1001	2	35	Stainless steel	0,0013
3001	5	50	Ceramics	0,0016

In the experiment, 172 results were measured, of which 64 results were evaluated. The tolerance field was set to 50 µm. The experimental part was focused on the evaluation of roundness deviation. For measurement there were used four samples shown in Fig. 5. The process of measurement was made on a WENZEL LH 65 X3M Premium, the three-coordinate measuring machine shown in Fig. 6.

Fig. 5. Samples examined.

Fig. 6. CMM - WENZEL LH65 X3M – premium.

4   Findings

From the measured values of deviations of roundness Graphs 7, 8, 9 and Tables 4, 5, 6 were plotted. The results indicated that measuring using the filter had a smaller roundness deviation than the non-filter method. The MZC method used showed a slightly smaller variation in circularity than the LSC method, but the difference is insignificant. A similar trend of values was measured using both methods, without a filter and using the continuous method. None of the result values exceeded a criterion of 50 µm. The lowest measurement deviation was 1,66 µm in the case of a roughness Ra = 0,8 mm using the MZC method with a filter and a stylus tip 1001 (ø d = 2 mm). Whereas the maximum deviation of circularity (18,7 µm) was measured using the LSC method without a filter on a workpiece with a roughness of Ra = 6,3 mm and a stylus tip of 3001 (ø d = 5 mm) (Fig. 7).

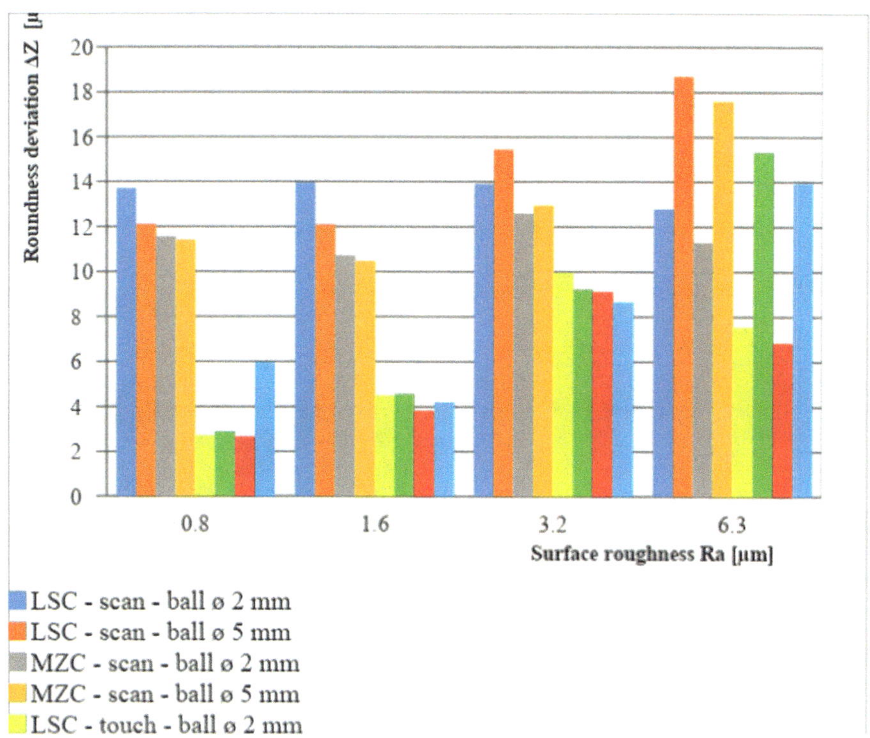

Fig. 7. Comparison of LSC and LSC methods, the touch and scan method, used without a filter.

Table 4. Measurement results without a filter [μm].

Roughness Ra	Scanning methods - LSC		Scanning methods - MZC		Touch methods - LSC		Touch methods - MZC	
	ø 2 mm	ø 5 mm	ø 2 mm	ø 5 mm	ø 2 mm	ø 5 mm	ø 2 mm	ø 5 mm
0,8	13,7	12,13	11,56	11,43	2,73	2,9	2,66	5,96
1,6	13,96	12,1	10,73	10,5	4,5	4,56	3,83	4,2
3,2	13,9	15,46	12,6	12,96	9,96	9,23	9,1	8,66
6,3	12,8	18,7	11,3	17,6	7,53	15,33	6,83	13,93

The figure above shows the lowest deviation of roundness measured without a filter in the case of roughness Ra = 0,8 μm by the MZC, using the touch method and a stylus tip ø 2 mm. On the contrary, the highest roundness deviation was measured without a filter on devices whose roughness was Ra = 6,3 μm. This deviation was determined by a LSC scan and a stylus tip ø 5 mm. Generally, the scan method results show higher circularity variations than in case of the touch method. LSC and MZC show similar variations, the difference occurred with the type of stylus tip used or the chosen method, whether it be the scan or touch method (Fig. 8).

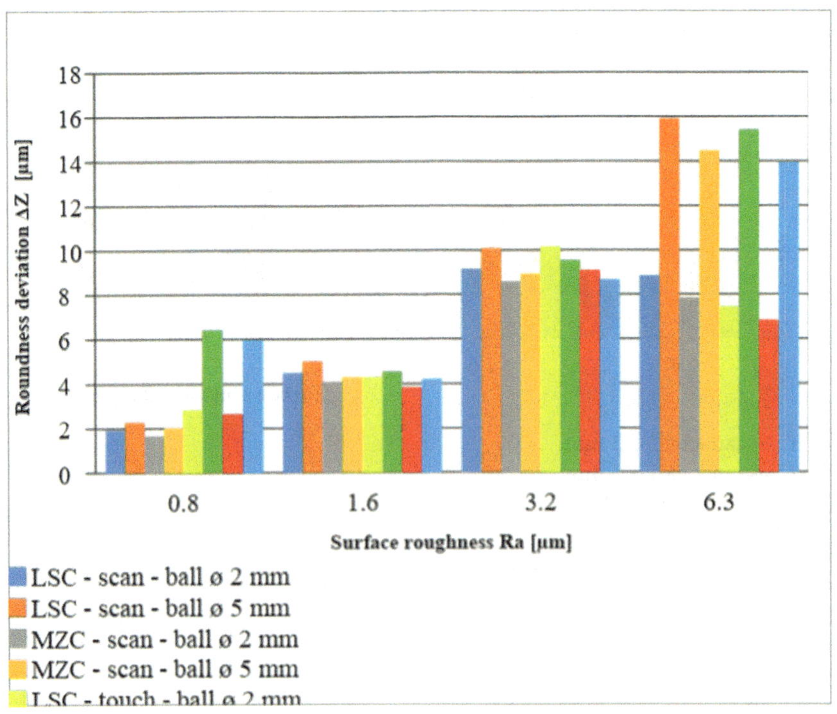

Fig. 8. Comparison of LSC and MZC methods, the touch and scan method, used with the filter

Table 5. Measurement results, used filter [μm].

Roughness Ra	Scan - LSC method		Scan –MZC method		Touch – LSC method		Touch – MZC method	
	ø 2 mm	ø 5 mm	ø 2 mm	ø 5 mm	ø 2 mm	ø 5 mm	ø 2 mm	ø 5 mm
0,8	1,93	2,3	1,66	2,03	2,86	6,43	2,66	5,96
1,6	4,5	5,03	4,1	4,3	4,3	4,56	3,83	4,2
3,2	9,16	10,1	8,6	8,93	10,16	9,56	9,1	8,66
6,3	8,86	15,93	7,83	14,5	7,46	15,43	6,83	13,93

The lowest measured roundness deviation was discovered in case of samples of roughness Ra = 0,8 μm by the MZC, the scan method and a stylus tip size of ø 2 mm.

On the contrary, the highest deviation in the case of using a filter was measured on a sample with a roughness Ra = 6,3 μm. This deviation was determined by the LSC method, the scan method and a stylus tip size of ø 5 mm. Overall, the lowest deviations were shown on workpieces with a roughness Ra = 0.8 μm and Ra = 1.6 μm. The samples Ra = 3,2 μm and Ra = 6,3 μm showed about two times higher deviations than the previous samples.

A detailed analysis of the MZC method and the influence of using a filter or not, and the scan or touch method, show us these results (see Fig. 9). In comparing these results there is used one size of stylus tip, at a size of ø 2 mm. The figure on the left compares the influence of using a filter or not to the touch method and subsequently the deviation of roundness. In this case the deviations have a similar trend. The visible difference is that each measured value using a filter is slightly higher than the value without a filter.

Regarding the scan method, seen in the right figure, the difference of deviations is noticeable. The deviations without the filter move at a relatively equal amount. Using filters affected the deviations which gradually increased with higher roughness. The only exception is the roughness variation Ra = 6,3 μm. For this roughness, the deviation was lower than the deviation of the previous Ra = 3,2 μm.

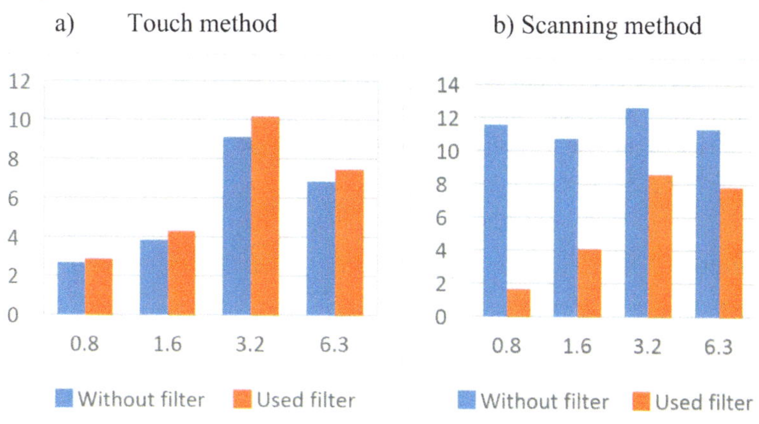

Fig. 9. MZC method - comparing the scanning and touch method with the same ball size.

Table 6. Measurement results [μm].

Roughness Ra [mm]	Scanning method		Touch method	
	Without filter	Used filter	Without filter	Used filter
0,8	11,56	1,66	2,66	2,86
1,6	10,73	4,1	3,83	4,3
3,2	12,6	8,6	9,1	10,16
6,3	11,3	7,83	6,83	7,46

5   Conclusions

Based on the results it could be said that the values measured by using a filter showed a generally lower deviation of roundness than the values measured without a filter. Furthermore, the MZC method has slightly lower circular deviations than the LSC method. In the case of values measured without filter and by both methods, the resultant roundness deviation was almost the same for each type of roughness. When using the filter and scan method and a stylus tip 2 mm in diameter, it caused lower deviation of roundness than using a stylus tip 5 mm in diameter. These results prove the dependence of roughness and size of the stylus tip on roundness deviation.

The experiment done using the touch method consisted of 38 scanned points. The other measured method used were the scan method. In this case the measured points were set 0,1 mm apart.

Further possible research could be in comparing different types of stylus tips due to their different measurement properties. For example, a carbon stylus tip is lighter and stiffer. Another area of interesting research could be an analysis of different types using the touch method for different roughnesses. Therefore, it creates the possibility of determining the optimal touch method suitable for measuring parts with a certain roughness. Nevertheless, the higher number of measured points in the case of the touch method could create a greater accuracy. In addition, the scan method could be measured using various speeds.

Acknowledgments. The research work reported here was made possible by conjunction with projects Student Grant Competition SP2018/136 Specific Research in Machining and Engineering Metrology Areas and SP2018/150 Specific research of modern manufacturing technologies, financed by the Ministry Education, Youth and Sport Faculty of Mechanical Engineering VŠB - TU Ostrava.

References

1. Krolczyk J, Krolczyk G, Legutko S, Napiorkowski J, Hloch S, Foltys J, Tama E (2015) Material flow optimization – a case study in automotive industry. Tehnicki Vjesnik – Technical Gazette 22(6): 1447–1456
2. EN ISO 12181 – 1:2011 (2011) Geometrical product specifications (GPS) - Roundness - Part 1: Vocabulary and parameters of roundness
3. Novák Z (2007) Optimization of surface roughness measurement. Mmspektrum 6:14

4. Gapinski B, Rucki M (2008) The roundness deviation measurement with CMM. In: IEEE international workshop on advanced methods for uncertainty estimation in measurement, pp 108–111
5. Ali SHR (2017) Automotive engine metrology. Pan Stanford Publishing, p 280
6. EN ISO 1101:2017 (2017) Geometrical product specifications (GPS) - Geometrical tolerancing - Tolerances of form, orientation, location and run-out. Czech Office for Standards, Metrology and Testing
7. Zelinka J (2016) The dependence of the measured value from the size stylus tip. Technical university of Ostrava, p 47
8. Metrosoft QUARTIS R12 – User's manual (2015) Wenzel Mertomec AG
9. EN ISO 4291:1985. Methods for the assessment of departure from roundness - Measurement of variations in radius
10. Otrusinová L, Čepová L, Petřkovská L (2013) Measurement of deviation of roundness. Technol Eng 10(1):21–24
11. Novák Z (2015) 3D analysis and filtration of the surface profile. Mmspektrum (5)
12. Hobson T (ed) (2011) Exploring roundness, a fundamental guide to the measurement of cylindrical form. 3rd edn, Leicester, LE4 9JQ, England
13. Holub N (2015) Influence of methodology measuring of selected deviations of shape and position in the company. University of West Bohemia, p 52
14. Metrosoft QUARTIS R6 – User's manual (2011) Wenzel Mertomec AG
15. Drastík F (1996) Precision of machine parts according to international standards. Tolerance of dimensions and geometric properties, Montanex

The Problems of Measuring Selected Geometric Deviations on a CMM After Machining

Ondřej Mizera, Lenka Čepová, Marek Sadílek, Robert Čep[(✉)],
Radek Hrubý, and Jan Zelinka

Faculty of Mechanical Engineering, VŠB - Technical University of Ostrava,
17. listopadu 15/2172, 708 33 Ostrava-Poruba, Czech Republic
robert.cep@vsb.cz

Abstract. The article is concerned with the measuring of the chosen geometrical deviations as follows: flatness, parallelism, cohesiveness and perpendicular. All the surfaces of the samples are milled in two milling centers which are DMG Mori NLX 2500 with C axis & DMG Mori NLX 2500 with a Y axis. After their creation, the samples were measured on the Coordinate Measuring Machine (CMM) ZEISS Prismo 7. The end of this article contains an evaluation of the impact of the method of milling on geometrical deviations while measuring on a 3D CMM.

Keywords: Coordinate measuring machine · Geometrical deviation · Flatness · Parallelism · Cohesiveness

1 Introduction

Geometrical tolerances and shape deviations are an inseparable part of every technical document because of the determination of an accuracy of the shapes and dimension values [1]. It is possible to consider it as an attribute of the product from a geometrical point of view. These attributes are called GPS – Geometrical Product Specifications. They define the deviations of the shapes and positions towards real elements. Their determination ensures the proper function of a part. It is possible to determinate more geometrical deviations if it is necessary for the right characteristics of the part, and it ensures that the right process is used for creation. It is standard EN ISO 1101 [2].

2 Creation of Geometrical Deviations

In regards to the ideal geometrical shape of an element, it is possible to determine all possible causes of influencing factors, for example a mechanical machine, a tool, a workpiece, and an environment.

The biggest and the most frequent cause of the problem of a geometrical deviation arising is the milling machine itself and its location [3]. During the running of the machine, the machine shakes, a flexible deformation arises, and there is a deviation in

© Springer Nature Switzerland AG 2019
M. Diering et al. (Eds.): *Advances in Manufacturing II* - Volume 5, LNME, pp. 158–171, 2019.
https://doi.org/10.1007/978-3-030-18682-1_13

the position of the rotation axis and the conductive surface of the milling machine. The shaking of a machine is produced not only by the function of the machine itself, but it can occur as a result of a nearby railway or a press machine [4, 5].

If it is not correctly positioned in its position, with its feed secured, a workpiece tool changes its position before or during operations making monitoring tool wear a priority, changing instrument geometry, creating gains, or other unwanted influences adversely affecting geometric deviation [6].

2.1 Flatness

Flatness it is defined as the minimum distance between two planes within which all the points on a surface lie (Fig. 1). A surface along which all the points lie along single plane is called as a perfectly flat surface [7].

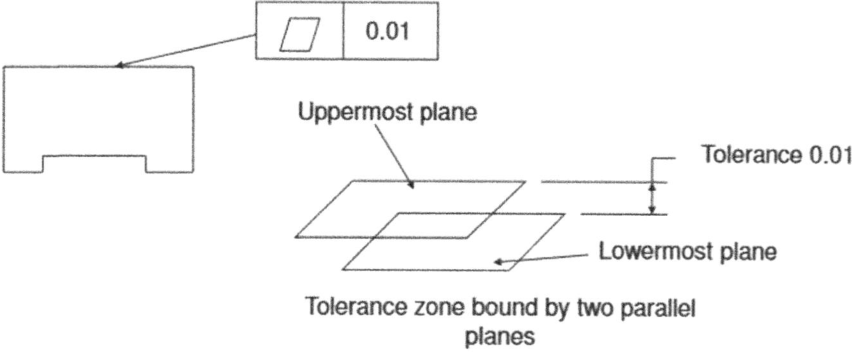

Fig. 1. Schematic definition of flatness [7].

2.2 Parallelism

Parallelism is defined for a feature (like a surface or line) with reference to another feature called a reference (Fig. 2). It defines the distance between two lines or surfaces that are parallel to each other and parallel to the datum surface and encompass the line or surface in question [7].

2.3 Concentricity

Defines the position of the axis relative to the other reference axis (Fig. 3). It defines a cylinder that matches the reference axis and the diameter given by the geometric tolerance [7].

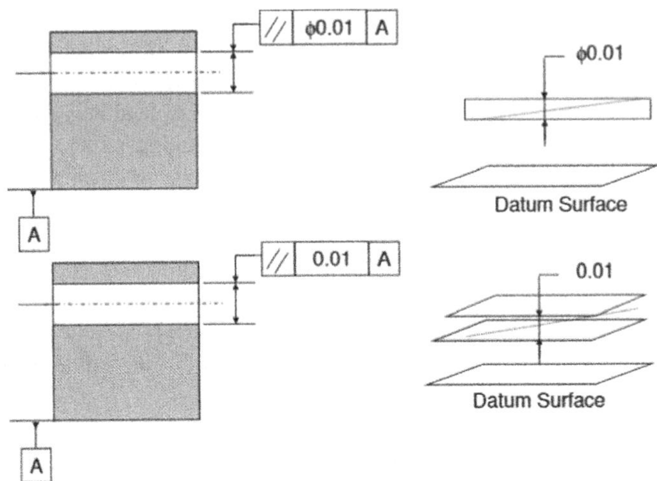

Fig. 2. Schematic definition of parallelism [7].

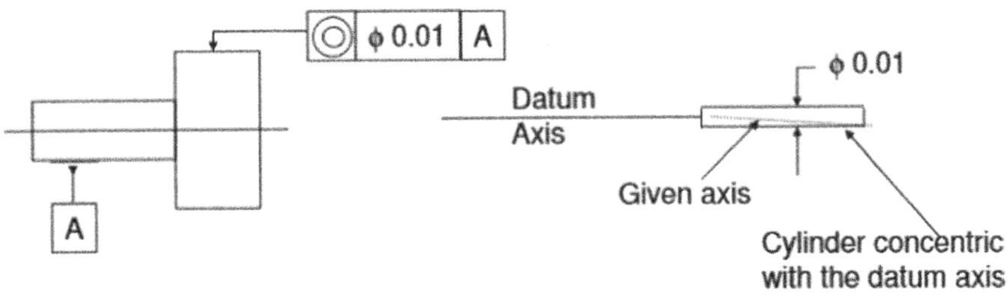

Fig. 3. Schematic definition of concentricity [7].

2.4   Perpendicularity

There are two parallel planes or lines which are oriented perpendicular to the datum feature or surface (Fig. 4). These planes are held perpendicular to the datum, but only ensure that the entire feature falls into the tolerance zone [8].

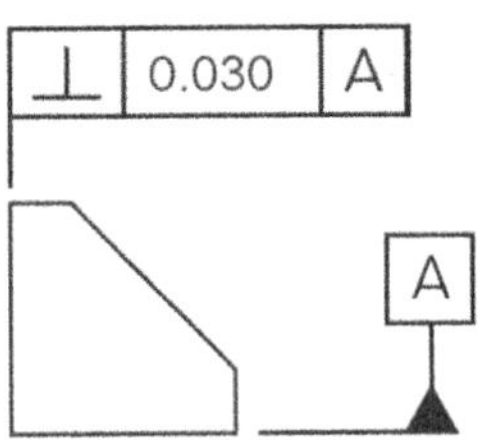

Fig. 4. Schematic definition of perpendicularity [8].

When measuring a geometric element, make sure that as many points as possible are scanned evenly throughout the element. The number of scoring points depends on the size, the quality of the surface and the prescribed tolerance of the geometric element [7]. The higher the number of scanned points, the better the geometric element is measured on the component, however, the measurement period is prolonged. In most cases, it is better to increase the number of measurement points and use the same distribution to cover the entire area of the measured workpiece [9, 10].

According to the British Standard [12], a minimum number of points is recommended for every other geometric element. There are no other binding regulations [11, 12].

3   Measuring Machines for the Measurement of Geometrical Deviations

Currently, there is a trend towards automatization, that is the digitalization of processes, the trend is called the Fourth Industry (the fourth industrial revolution) for which it is typical that changes in the workplaces occur. Workplaces are becoming automatized, all data are being archived and processed. Metrology doesn't stay behind. We can meet calipers every day, which automatically send the obtained results to a file set in a computer beforehand. The age of verifying geometrics using a dial indicator has been slowly disappearing and the age of digitalization has arisen. Moreover, measuring machines operated using PC software are much more accurate, quicker and reliable.

The most common and widespread measuring machine in the industry is the3D CMM. We classify this device in the category of universal measuring machines, where the accuracy of measuring reaches up to 1 μm value for a geometrical deviations [13]. The accuracy of the device depends on a type of the design and on the type of attachments (contacts, modules, the types of heads) (Fig. 5).

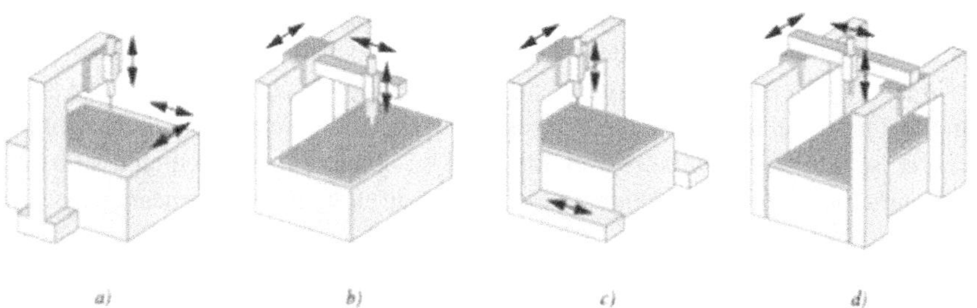

Fig. 5. Type CMM (a) horizontal arm (b) cantilever (c) gantry (d) bridge [2].

Besides CMM, optical non-contact measuring instruments are on the rise, it doesn't matter if they are based on the laser method, the shade method or the reflective method, as well as camera systems. Perfect shade heads work with a differentiation of 0,01 μm. Moreover, the system is calibrated many times a second. Nowadays they are used mainly for a measuring the diameter of shape complicated shaft parts and they are absolutely irreplaceable for devices determined for measuring solid elements (ceramics, hard steel, coating surfaces) and components, whose surface mustn't be damaged (highly accurate surfaces with a low roughness) [3].

Measuring microscopes belong to the most frequently used devices for laboratories measuring small components, and if need be for the calibration of manual measuring tools. Due to their mechanical performance and optical components it is possible to measure in a truly accurate way. They are equipped with classical bar measuring scales and there are spiral microscopes for deducting the values obtained. New types of microscopes are represented by contactless measuring systems with optical projection technology combined with a video system. Contemporary machines make use of a two-axis optical and video optical system, which is equipped with a colorful CCD camera, a link to a PC, software and a touch display for smart operating software [3].

4 Measurement Strategy

In the process of measuring geometric elements, it is necessary to consider that the scanned points are evenly spaced across a geometric element, and measured with a maximum number of scattered points. This is because the higher the number of scanned points, the more accurate the real surface and its true deviation will appear. There must always be a direct proportionality between the measurement time and the number of points, the more points used in the measurement process, the longer the total sample time will be. Each geometric element has a recommended minimum number of points to capture the accuracy of the results, whereby it produces the most appropriate result using software and mathematical calculations. By using the minimum number of points to determine distances or diameters, the actual shape of the measured element cannot be read according to the minimum number of points. As described in the British Standard [14], it clearly defines the strategy and the minimum number of scanned points on elements such as surface, cylinder, and circle. All surfaces were measured with 9 evenly spaced points and a sample cylinder was measured using four points in three planes. The initial setting of the sample base depends on the measurement of the individual characteristics. By subsequently aligning the space on the bottom axis of the cylindrical base and towards it there was measured the middle and upper element of the square cross section (Fig. 6a). It was followed by other elements and an evaluation of their geometrical deviations, namely deviations of straightness, flatness, perpendicularity (Fig. 6b, c).

The evaluation of the measured results was done using the Gauss Pensetting Method, the least squares method, because we are interested in oddities from the ideal shape and not the actual surface, including the outlying points. We can say that in using this method, we arithmetically averaged all the measured values in the plane with the same mathematical formula (Fig. 7), so all the resulting data are mathematically evaluated uniformly. For the smallest squares method, we required that the sum of the squares (second powers) of the difference values of the measured values and the real values from the CAD model for the same value be as small as possible.

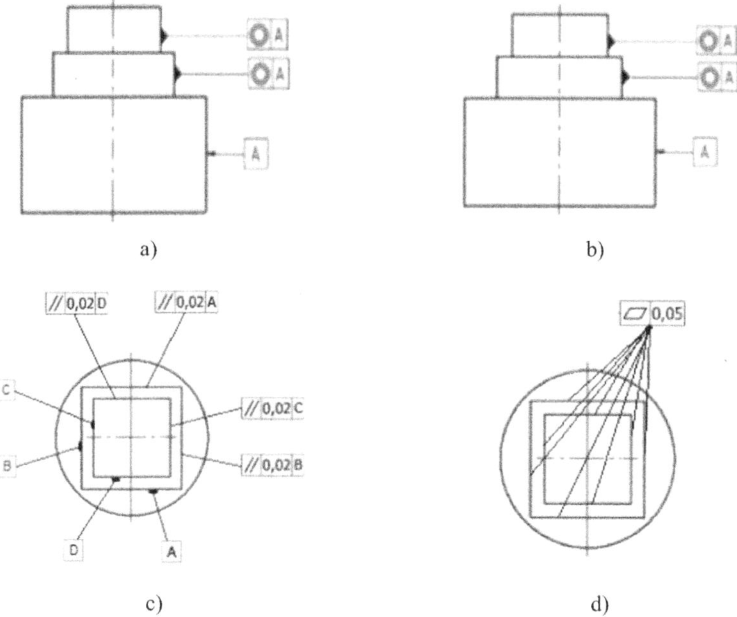

Fig. 6. Base and measurement strategy.

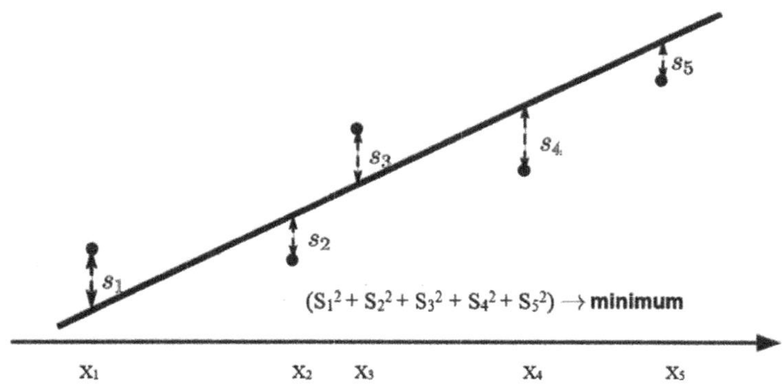

Fig. 7. Straight through the smallest squares method – Gauss.

5 Experimental Work

The creation of 18 samples were made on a two CNC turning centers, where the different factor was a shift in milling and it is marked in Table 1. In the first case it was C-axis and Z axis, at this point another division was made in milling before and after calibration.

The other case was the Y axial and Y radial axis. All the following comparisons and calculations are shown (Fig. 8). The machined samples were measured with the

Table 1. Distribution of produced pieces on way of Milling

CNC machine				DMG MORI NLX				
Division		2500MC/700		2500Y/700		Division		
		Before calibration	After calibration					
Axis	C axial	Milling a larger square	1PVC	4POVC	9VC	14VC	Milling a larger square	C axial created on CNC 2500MC/700	Axis
				5POVC	10VC	15VC			
			2PVC	6POVC	11VC	16VC			
				7POVC	12VC	17VC			
			3PVC	8POVC	13VC	18VC			
	Z axial	Milling a smaller square	1PMZ	4POMZ	9MY	14MY	Milling a smaller square	Y 9–13 axial 14–18 radial	
				5POMZ	10MY	15MY			
			2PMZ	6POMZ	11MY	16MY			
				7POMZ	12MY	17MY			
			3PMZ	8POMZ	13MY	18MY			
Number of samples		3	5	5	5	–		
Summarization		18						

coordinate measuring machine Zeiss Prismo 7 (Fig. 9) and the Calypso software-equipped computer station. The maximum permissible error of the applied measurement equipment is equal to:

$$\text{MPEE} = (2,5 + L/300) \, [\mu m] \tag{1}$$

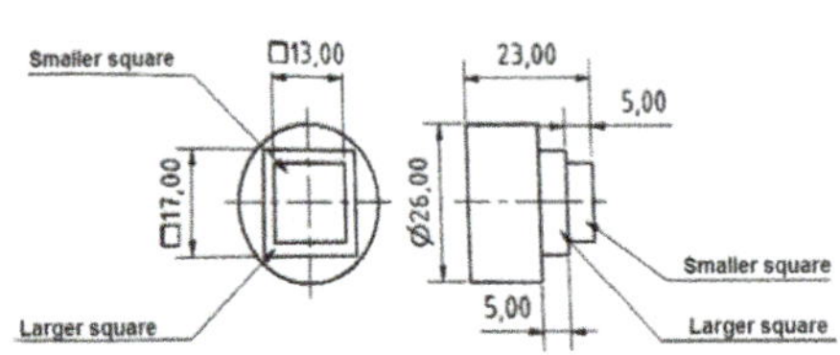

Fig. 8. Sketch with the marking of the scanning surfaces.

Fig. 9. Coordinate measuring machine - Zeiss Prismo 7.

For the measurement there was used the solid head VAST and with a diameter tip 2 mm (Fig. 10).

Abbreviation	Description abbreviation
P	Measured before calibration
PO	Measured after calibration
V	Measured on a larger square
M	Measured on a smaller square
C	Axis milling
Y	Axis milling
Z	Axis milling

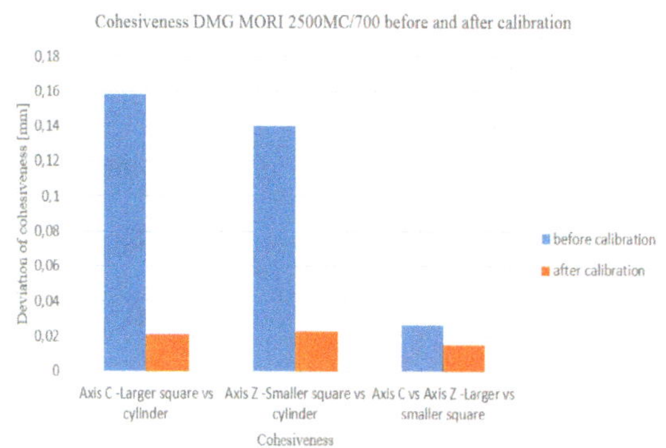

CONCRETE IMPROVEMENTS:

-o 84,9 % (131,6µm) axis C, -o 81,4 % (113,5µm) axis Z, -o 51,5 % (13,7µm) axis C vs. Z.

Fig. 10. The calibration of the rotating part is very noticeable here, especially in its alignment with the axis of the cylinder, which is also manifested in the plane of squareness reciprocity.

The results of measuring were statistically processed. Seven average values had to be excluded, because they were subordinated to the calculation of a gross mistake Hn. The results of the gained values are marked and evaluated in Fig. 11, 12, 13 and 14.

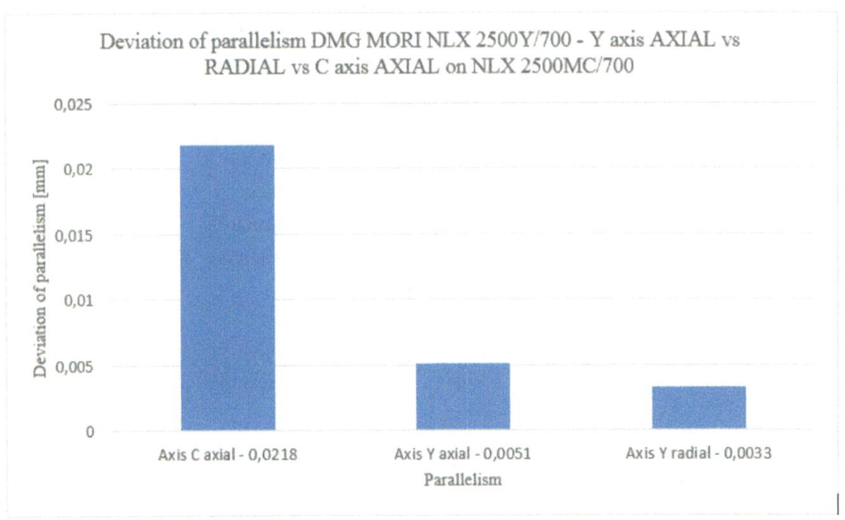

Valorization

-about 76,6 % (16,7μm) a better result when machining the Y axis axially than the rotary C axis,
-about 84,9 % (18,5μm) better result when machining the Y axis radially than the rotary C axis,
-about 35,3 % (1,8μm) a better result when machining the Y axis axially than the Y radially axis.

Fig. 11. Deviation of parallelism shows the trend of better results in Y axis machining. The difference in axial and radial milling of the Y axis is smaller than in previous deviations.

Comparing the calibrations of the lathe-milling Machines DMG MORI NLX 2500MC/700 is marked in Fig. 8 and the measurement methods are shown in Fig. 13. Comparison of geometric accuracy deviations is marked in Table 2 (Fig. 15).

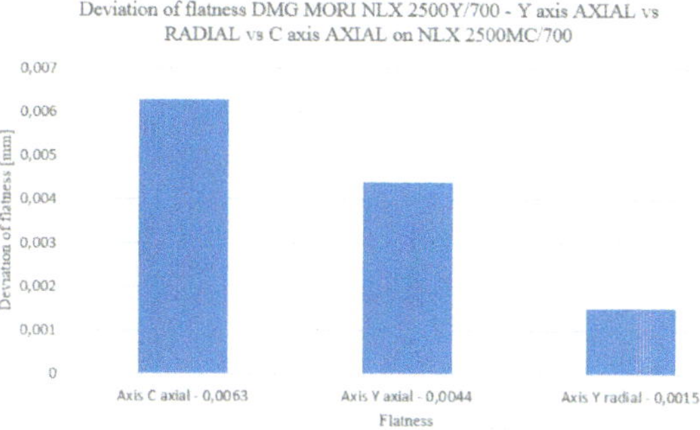

Valorization

-about 30,2 % (1,9µm) a better result when machining the Y axis axially than the rotary C axis,
-about 76,2 % (4,8µm) better result when machining the Y axis radially than the rotary C axis,
-about 65,9 % (2,9µm) a better result when machining the Y axis axially than the Y radially axis.

Fig. 12. Deviation of flatness the results are similar to the deviations of parallelism and perpendicularity. Y-axis machining achieves better results in both cases in both axial and radial milling.

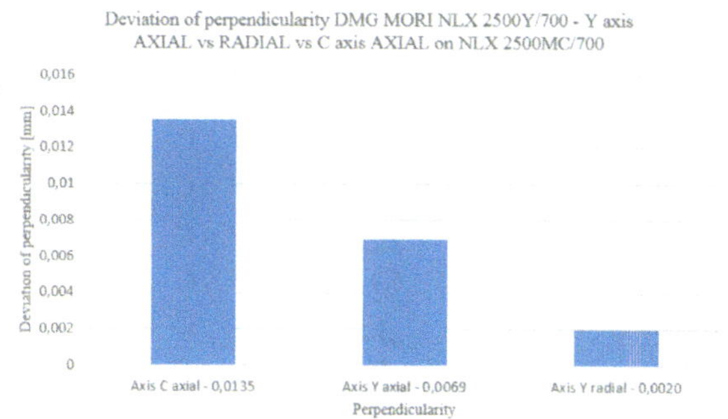

Valorization

-about 48,9 % (6,6µm) a better result when machining the Y axis axially than the rotary C axis,
-about 85,2 % (11,5µm) better result when machining the Y axis radially than the rotary C axis,
-about 71 % (4,9µm) a better result when machining the Y axis axially than the Y radially axis.

Fig. 13. Deviation of perpendicularity the results are similar to those in the case of deviations of parallelism and flatness. Y-axis machining achieves better results than C-axis machining.

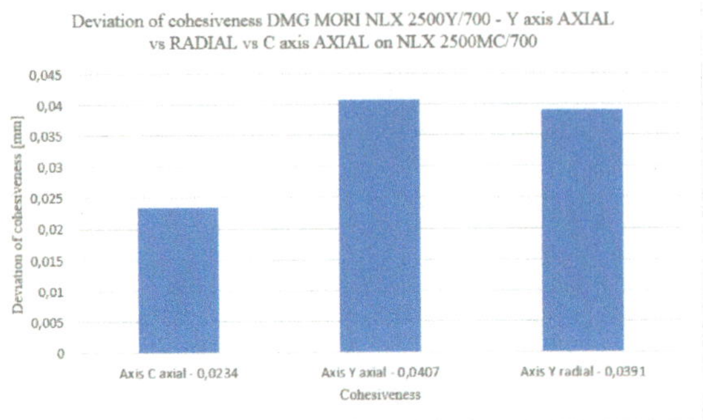

Valorization

-about 42,5 % (17,3μm) a better result when machining the Y axis axially than the rotary C axis,
-about 3,9 % (1,6μm) better result when machining the Y axis radially than the rotary C axis,
-about 40,2 % (15,7μm) a better result when machining the Y axis axially than the Y radially axis.

Fig. 14. Deviation of cohesiveness with the rotary axis C the lowest deviation of alignment was achieved. On the contrary, the worst axial Y axis was shown.

Table 2. Comparison of geometric accuracy deviations

Technology	Monitored geometric tolerances	Symbol	Axis	Difference deviation values [μm]
Radial milling	Perpendicularity	⊥	Y < C	11.5
	Parallelism	//	Y < C	18.5
	Flatness	⊿	Y < C	6.9
	Concentricity	◎	Y < C	15.7
Axial milling	Perpendicularity	⊥	Y < C	6.6
	Parallelism	//	Y < C	16.7
	Flatness	⊿	Y < C	1.9
	Concentricity	◎	C < Y	17.3

Fig. 15. Methods of measurement.

6   Conclusion

The aim of the research was the establishment and evaluation of the geometrical deviations of 2 methods of milling on 2 different machines, where the factor of defining the rate of inaccuracy using a non-calibrated milling machine was important.

Above all, the research focused on the accuracy of geometric tolerances, when the axis of rotation of the milling centers was tilted. The evaluated parameters were the calibration of the individual devices and a comparison between the results and the individual deviations in calibrations for precise production. In addition, research activities have been followed, finding the most optimized way and the most efficient way of manufacturing using one fastening. This means the most precise and fastest production depending on the accuracy of the measured geometric parameters. The selected geometric shapes are the most used areas, and research has been focused on these aspects. This is putting research into practice.

By comparing the calibration of milling centers it is evident, at first sight that calibration make sense. The gained values are mostly obvious when we talk about tenths of mm, in other cases there are differences in a hundredths of mm. All measurements and evaluations were made on the 3D CMM Zeiss Prismo 7.

During comparing the geometrical deviations of chosen parameters using 2 methods of milling, it is possible to determine which the best (the smallest geometrical deviations) is. The DMG MORI 2500 Y/700 with a radial Y axis, not just for its concentricity results was according to statistical calculations the best, but also for the complexity of the result used for this strategy of milling with a radial Y axis, if we take into account that the creation of the composition was made only on one device.

Another fact is that we have achieved the greatest geometrical precision by radial turning with the Y axis, both in millenniums and in hundreds of millimeters. The unique disadvantage of turning with this Y axis is a greater deviation of alignment than when using a rotary C axis. That is why it depends mainly on the company itself, which the turning-milling center decides to use, especially with regard to the production applied to the machine, the required precision of the production and, last but not least, the acquisition costs of the construction center.

The results of the analysis will be used to decide about purchasing a new device or determining and planning line of production and deciding about the process of production under the geometrical accuracy of each milling center and the possibilities of which axes can be used.

Acknowledgments. This paper was supported by project Students Grant Competition SP2018/136 Specific Research in Machining and Engineering Metrology Areas and SP2018/150 Specific research of modern manufacturing technologies, financed by the Ministry Education, Youth and Sports on Faculty of Mechanical Engineering, VŠB-Technical University of Ostrava.

References

1. Olszar K (2018) Deviation measuring after machining at CMM. Diploma thesis, VŠB - Technická univerzita Ostrava
2. EN ISO 1101:2017 (2017) Geometrical product specifications (GPS) - geometrical tolerancing - tolerances of form, orientation, location and run-out. Czech Office for standards, metrology and testing
3. Gapinski B, Janicki P, Marciniak-Podsadna L, Jakubowicz M (2016) Application of the computed tomography to control parts made on additive manufacturing process. Proc Eng 149:105–121
4. Kostka J (2010) Comparison of roudness deviation at conventional and coordination machines. Bachelor thesis, VŠB - Technická univerzita Ostrava
5. Stępień K (2014) In situ measurement of cylindricity - problems and solutions. Precis Eng 38 (3):697–701
6. Gapinski B, Zachwiej I, Kolodziej A (2015) Comparison of different coordinate measuring devices for part geometry control. In: Digital industrial radiology and computed tomography
7. Suhas SJ. Department of Mechanical Engineering, Indian Institute of Technology, Bombay, Powai, MUMBA– 400 076 (India)
8. https://www.gdandtbasics.com/perpendicularity/. Accessed 28 Oct 2018
9. Starczak M, Rus T, Plowucha W (2000) Documentation of measuring strategies. In: IVth international scientific conference: coordinate measuring technique. scientific bulletin of lodz technical university branch in Bielsko-Biala, vol 53
10. Janecki D, Stępień K, Adamczak S (2016) Sphericity measurements by the radial method: I. Math Fundam Meas Sci Technol 27(1):015005

11. Starczak M, Jakubiec W (2001) Optimisation of measuring strategies in coordinate measuring technique. Measur Sci Rev 1(1):191–194
12. STANDART BS 7172:1989 (1989) Guide to assessment of position, size and departure from nominal form of geometric features
13. Sadílek M, Fojtík F, Sadílková Z, Kolařík V, Petrů J (2015) A study of effects of changing the position of the tool axis to the machined surface. Trans FAMENA, Zagreb 39(2):33–46
14. Standard BS 7172:1989 (1989) Guide to assessment of position, size and departure from nominal form of geometric features, p 20

Optimal Prioritization of the Model
of Distribution of Measurement Points
on a Free-Form Surface in Effective Use
of CMMs

Marek Magdziak[1(✉)] and R. M. Chandima Ratnayake[2]

[1] Faculty of Mechanical Engineering and Aeronautics,
Rzeszów University of Technology,
al. Powstańców Warszawy 12, 35-959 Rzeszów, Poland
marekm@prz.edu.pl
[2] Department of Mechanical and Structural Engineering and Materials Science,
University of Stavanger, Stavanger, Norway

Abstract. Investigation of the optimal model of the distribution of measurement points (DoMPs) on a free-form surface (FFS) for performing coordinate measurements is vital for the effective use of coordinate measuring machines (CMMs). The selection of the optimal model is currently being made in an ad hoc manner. The selection criteria of the best distribution of measurement points in general may depend on the possibility of estimating the curvature on an FFS, the time taken for measurements, the deviations of the substitute geometry from the nominal free-form surface and the deviations calculated based on the probe radius correction process. This manuscript demonstrates the use of the analytic hierarchy process (AHP) to deal with the multi-criteria nature in the selection of the optimal model of the distribution of measurement points. It also demonstrates how to prioritize the measurement points' distribution models based on the selection criteria.

Keywords: Metrology and measurement systems ·
Coordinate measuring technique · Free-form surface ·
Distribution of measurement points · Analytic hierarchy process

1   Introduction

Contact coordinate measurements of free-form surfaces (FFSs) on different products require the selection of the appropriate measurement strategy, which may include a number of factors that have an influence on the accuracy of measurements [1]. The measurement strategy must be decided at the preparation phase of a measurement program [2]. For instance, a measurement strategy shall be based on the number and the method of the distribution of measurement points located on measured surfaces [3–6]. The allocation of too many measurement points increases the time taken for coordinate measurements. On the other hand, a small number of points may contribute to a reduction in the accuracy of coordinate measurements of objects composed of curvilinear surfaces. Although the measurement software available with coordinate measuring machines

© Springer Nature Switzerland AG 2019
M. Diering et al. (Eds.): *Advances in Manufacturing II* - Volume 5, LNME, pp. 172–182, 2019.
https://doi.org/10.1007/978-3-030-18682-1_14

(CMMs) supports users' ability to define the distribution of measurement points, the software has its own limitations. In most cases, especially in industrial conditions, measurement points are distributed uniformly [7] or based on curvatures of an investigated free-form surface [8]. When the measurements are carried out in the single-point probing mode, the uniform distribution of measurement points on free-form surfaces of workpieces makes coordinate measurements inefficient (i.e. coordinate measurements take a long time). In addition, the single-point probing mode is still widely used in CMM-based measurements. The methods for distributing measurement points are divided into three main groups: blind strategies (based on nominal models of measured free-form surfaces), adaptive strategies (using the information about already distributed measurement points), and strategies that use the information about the manufacturing process of a workpiece [9, 10].

This paper presents an approach to prioritize the strategy of the distribution of measurement points (DoMPs) using the Analytic Hierarchy Process (AHP). The AHP has been widely used in CMM-based applications, structural engineering applications, and manufacturing processes-related prioritizations [11–13]. For instance, the AHP analysis has been used when selecting the most appropriate coordinate measuring machine, taking into account the following criteria: technology, design, firm image, consultancy and training, and costs [11]. The following sections of this article present the methodology for selecting the best model of the distribution of measurement points on the selected FFS, the data collection, analysis and results.

2 Methodology

Figure 1 illustrates the FFS that has been used to demonstrate the prioritization of models of the distribution of measurement points. The AHP has been used to prioritize the measurement points' distribution models. Figure 2 illustrates the five different randomly chosen models of the distribution of measurement points located on the selected FFS. The first model of the distribution includes the largest number of points. The other four models were created by modifying the first model. The modifications consisted of changing the numbers of points and their locations on the selected FFS in relation to the first model of the distribution. Moreover, it was assumed that coordinate measurements would be performed in the single-point probing mode, which is widely used, e.g. when coordinate measurements are carried out using CNC machine tools in manufacturing industry.

The criteria used for the prioritization and the selection of the most appropriate model of the distribution of measurement points by using the AHP analysis were as follows:

- the curvature of the considered free-form surface (CoFFS),
- the time for coordinate measurements (ToM),
- the accuracy of the substitute model of the measured curvilinear surface (DoSG),
- the accuracy of the probe radius correction process (PRC).

Figure 3 illustrates the AHP hierarchical structure that has been used for the multi-criteria prioritization process, to select the best model of the distribution of measurement

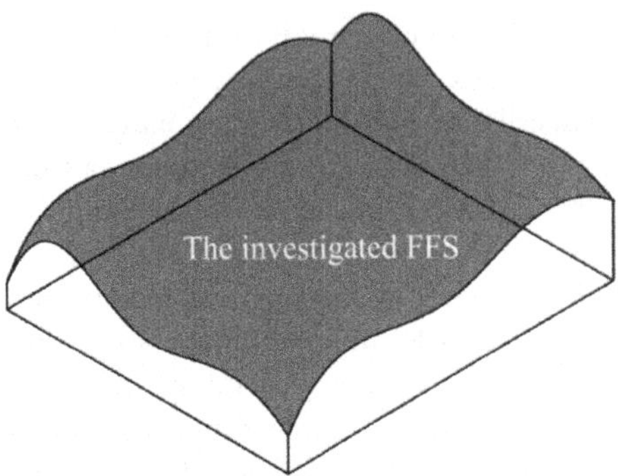

Fig. 1. The measured free-form surface used to demonstrate the application of the AHP analysis.

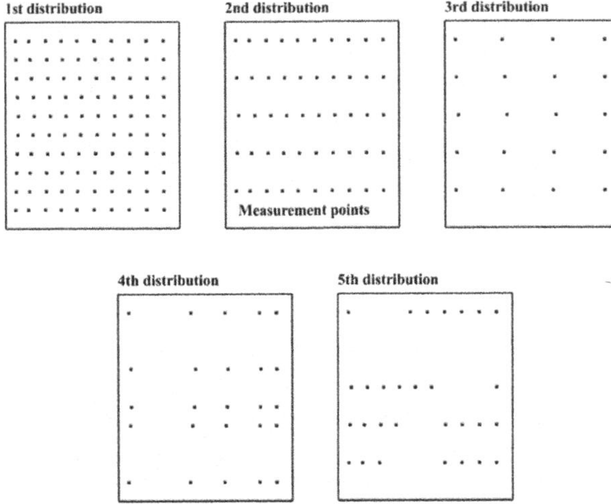

Fig. 2. The analyzed distributions of measurement points located on the selected FFS.

points. The pairwise comparisons have been carried out in relation to the hierarchical structure, focusing on the goal of the AHP analysis, the criteria and possible alternatives (i.e. models of the distribution of measurement points). The pairwise comparisons were carried out using the scale presented in Table 1 [14]. The calculations for all investigated models of the distribution of measurement points and four considered criteria were performed by using the CATIA V5-6 and Rhinoceros software packages.

2.1   1st Criterion – Curvature of a Free-Form Surface

The application of the curvature criterion results from the analysis of literature [8] and the possibilities of commercial measurement software cooperating with CMMs. The example of the measurement software is the Calypso software produced by the Carl

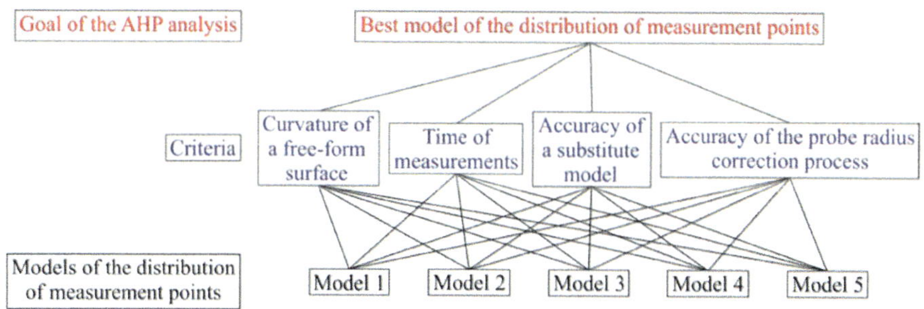

Fig. 3. The AHP structure used for the prioritization process to select the best model.

Table 1. The scale for performing comparisons (pairwise comparison of activities) [14].

Intensity of importance	Definition	Explanation
1	Equal importance	Two activities contribute equally to the objective
3	Moderate importance one over another	Experience and judgment favor one activity over another
5	Essential or strong importance	Experience and judgment strongly favor one activity over another
7	Very strong importance	An activity is strongly favored, and its dominance is demonstrated in practice
9	Extreme importance	The evidence favoring one activity over another is of the highest possible order of affirmation
2, 4, 6, 8	Intermediate values between the two adjacent judgments	When compromise is needed

Zeiss company. The Calypso software package enables the inclusion of more measurement points on FFSs that have large curvature. More measurement points should be located in the areas of surfaces with small radii of curvature, in order to identify the largest form deviations of the measured FFS.

In the case of the first criterion, the curvatures of the analyzed free-form surface were calculated at all measurement points. In the next step, the average curvature was calculated based on the curvatures measured at individual points. The calculations were made by means of the CATIA V5-6 software. The value of the average curvature depends on the number and distribution of measurement points on the selected free-form surface.

2.2   2nd Criterion – Time of Measurements

The time taken for coordinate measurements was estimated based on the distance that the stylus tip of a measuring probe of a CMM travels between individual measurement points located on the selected curvilinear surface. Figure 4 presents the way of moving a measuring probe between points in the case of the last model of the distribution. The same way of moving of a probe was used for all other models. It has been assumed that the measuring probe moves at a constant velocity along the investigated surface. There should not be too many measurement points distributed on the FFS because of the long time necessary for coordinate measurements. Lengthy measurement time makes coordinate measurements inefficient.

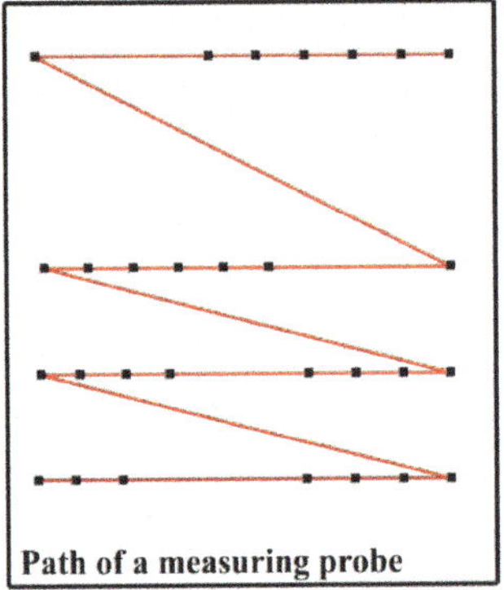

Fig. 4. The applied path of a measuring probe moving between the selected measurement points.

Analogously to the first criterion, the distance traveled by a stylus tip of a measuring probe was also computed by using the CATIA V5-6 software. It was assumed that the considered distance was calculated as the length of the polyline connecting all measurement points. Moreover, it was assumed that there are no auxiliary movements of a measuring probe of a coordinate measuring machine.

2.3   3rd Criterion – Accuracy of a Substitute Model

The substitute model of the measured workpiece helps to assess whether the applied number of measurement points is sufficient, considering the geometrical complexity of the investigated curvilinear surface. Many researchers determined locations of measurement points on measured workpieces, based on the comparison of the substitute model to the nominal model of a surface [3, 4, 15].

The substitute models of the considered surface for all investigated models of the distribution of points were created by using the Multi-Section Surface command available in the CATIA V5-6 software. The substitute models were prepared, based on curves interpolating measurement points and representing the cross-sections of the measured free-form surface. Moreover, guide curves were applied, in order to create surfaces representing the substitute models. The accuracy of the substitute models of the considered free-form surface was checked by using the Deviation Analysis command of the CATIA V5-6 software. In the case of the investigated surface, deviations between the substitute models and the nominal one were calculated, in order to assess the accuracy of the substitute models.

2.4 4th Criterion – Accuracy of the Probe Radius Correction Process

The last criterion, used to select the best model, relates to the probe radius correction process, which has a significant influence on the accuracy of contact coordinate measurements. The compensation of the radius is applied to calculate the coordinates of corrected measurement points on the basis of the coordinates of indicated measurement points. Indicated points represent the center of a stylus tip of a measuring probe during its contact with a measured workpiece. Corrected measurement points represent the real contact points of a stylus tip with a measured surface [16]. The accuracy of the probe radius correction process depends on the applied method of compensation, the number of measurement points, their distribution and the shape of considered free-form surfaces of products. Therefore, the adopted distribution of points may lead to the minimalization of measurement errors resulting from the applied method of the probe radius compensation process.

In the case of the last criterion, the method based on the second degree Bézier curves was used to calculate the coordinates of corrected measurement points. The calculations of the deviations resulting from the applied method of the probe radius compensation and the used model of the distribution of measurement points were conducted by using the CATIA V5-6 and Rhinoceros software packages. For each of the five models of the distribution of points, three cross-sections of the measured surface, along which measurement points were distributed, were selected. Two of them were extreme cross-sections and one was located approximately in the middle of the investigated free-form surface. For all analyzed cross-sections, indicated measurement points were exported from the CATIA software to the Rhinoceros software package. In the next stage, coordinates of corrected measurement points were calculated on the basis of indicated measurement points by using the probe radius compensation, based on the second degree Bézier curves. The applied method of the compensation was presented in detail by Kawalec and Magdziak [16]. Measurement points representing the contact points of the stylus tip of a measuring probe with the investigated surface were compared to the offset curves calculated by means of the Rhinoceros software. The offset curves were generated based on the curves interpolating indicated measurement points and the radius of the stylus tip of a measuring probe cooperating with a coordinate measuring machine. The radius of the stylus tip results from the qualification process of a probing system, which should be performed before conducting real coordinate measurements. The radius of 1 mm was assumed for the calculations of corrected measurement points. On the basis of the deviations calculated for three cross-sections of the measured surface, in the case of each model of the distribution of measurement points, the average deviation was computed. The average deviation indicates the accuracy of the probe radius correction process for the selected model of the distribution of measurement points.

3 Selection of the Best Strategy for the DoMPs

The AHP approach (i.e. using the Expert Choice software) has been used to prioritize the possible alternative models and to investigate the most appropriate model that shall be used to distribute measurement points on the selected FFS. First, the pairwise comparisons were carried out (ref. Table 1) at the highest level of the AHP hierarchy (ref. Fig. 3) by using the nine-point scale. The comparisons of criteria in relation to the goal (i.e. the selection of the optimal DoMPs) were made. Figure 5 illustrates a matrix that presents the results of the comparisons in relation to Table 1.

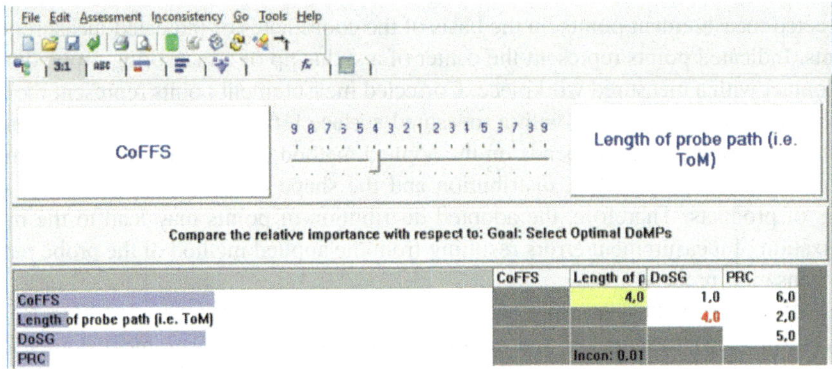

Fig. 5. Comparisons of criteria in relation to the goal (i.e. to select the optimal DoMPs).

It has been revealed that comparisons do not demonstrate inconsistency (i.e. inconsistency index shall be equal to or less than 0.1), as the inconsistency index is 0.01.

Then, comparisons of the alternatives (i.e. the possible models that shall be used to distribute the measurement points) with respect to each criterion were made. Figures 6, 7, 8 and 9 present the results of the comparisons, as well as the resulting inconsistency indexes for sets of comparisons. In all cases, the inconsistency indexes were less than 0.1. In order to compare the models of the distribution of measurement points, the normalization of data calculated by using the CATIA V5-6 and Rhinoceros software packages to the range from 1 to 9 was applied.

	Model-1	Model-2	Model-3	Model-4	Model-5
Model-1		1,0	9,0	2,0	1,0
Model-2			9,0	2,0	1,0
Model-3				7,0	9,0
Model-4					2,0
Model-5		Incon: 0,01			

Fig. 6. Comparisons of alternatives (models 1–5) with respect to criterion 1 (i.e. CoFFS).

Compare the relative preference with respect to: Length of probe path (i.e. ToM)				
	Model-1	Model-2	Model-3	Model-4	Model-5
Model-1		6,0	6,0	6,0	9,0
Model-2			1,0	1,0	2,0
Model-3				1,0	2,0
Model-4					2,0
Model-5		Incon: 0,00			

Fig. 7. Comparisons of alternatives (models 1–5) with respect to criterion 2 (i.e. the length of probe path, which is a measure of ToM).

Compare the relative preference with respect to: DoSG				
	Model-1	Model-2	Model-3	Model-4	Model-5
Model-1		4,0	4,0	9,0	5,0
Model-2			1,0	3,0	1,0
Model-3				3,0	1,0
Model-4					2,0
Model-5		Incon: 0,01			

Fig. 8. Comparisons of alternatives (models 1–5) with respect to criterion 3 (i.e. DoSG).

Compare the relative preference with respect to: PRC				
	Model-1	Model-2	Model-3	Model-4	Model-5
Model-1		1,0	9,0	3,0	2,0
Model-2			9,0	3,0	2,0
Model-3				1,0	3,0
Model-4					1,0
Model-5		Incon: 0,04			

Fig. 9. Comparisons of alternatives (models 1–5) with respect to criterion 4 (i.e. PRC).

Moreover, Figs. 10, 11, 12 and 13 present the sequence of alternatives (i.e. models 1, 2, 3, 4, and 5) after performing AHP analysis (i.e. based on the weights calculated with respect to each criterion).

Fig. 10. Priorities with respect to the criterion CoFFS.

Figure 14 presents the overall synthesized priority weights that have been derived by taking the criteria and alternatives-based pairwise comparisons into consideration.

The results revealed that model 1 is the number one priority, whilst models 2, 5, 4, and 3 have subsequent priorities. It is worth noting that model 1 has the least priority in

Fig. 11. Priorities with respect to the criterion ToM.

Fig. 12. Priorities with respect to the criterion DoSG.

Fig. 13. Priorities with respect to the criterion PRC.

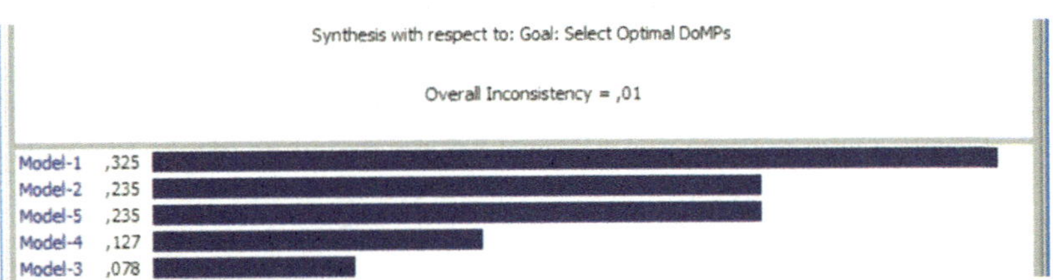

Fig. 14. Overall priorities.

relation to the second criterion (i.e. time taken for coordinate measurements). However, in relation to all other criteria (i.e. CoFFS, DoSG, and PRC), model 1 has the highest priority (i.e. the best distribution of measurement points located on the selected FFS). Hence, model 1 shall be used in practical coordinate measurements, if accuracy is the most important aspect of measurements.

4   Conclusions

This manuscript demonstrated the use of theoretical foundation in AHP for the multi-criteria prioritization of possible models of the distribution of points on an FFS within the context of coordinate measurement related applications. The findings revealed that the AHP approach and its theoretical foundation have the potential to prioritize and select the optimal model of the distribution of points to perform coordinate measurements of the selected FFS. The AHP enabled the optimal model (i.e. the alternative) to be selected, based on several criteria with significant transparency and it also enabled digitalization of the prioritization process. The AHP-based digitalization of the prioritization process enables enhancing the collaborative manufacturing and operations' management as intended in Industry 4.0 paradigm whilst aggregating expert knowledge utilization. Moreover, the findings revealed that the AHP-based prioritization enables users of CMMs to be supported at the measurement program preparation and planning phase with especial focus on having potential in increasing collaborative prioritization. Hence, it is possible to conclude that the suggested approach enhances the performance of efficient coordinate measurements in an automated manufacturing environment whilst having opportunity for digitalized prioritizations.

Future research shall be carried out to investigate the potential of using different multi-criteria prioritization approaches such as fuzzy AHP to make the analysis more realistic.

Acknowledgments. The work was supported by The National Science Centre (Poland) within the Miniatura 1 programme and the project entitled 'Evaluation of the new method for the distribution of measurement points on curvilinear surfaces' (DEC-2017/01/X/ST8/02020).

References

1. Magdziak M, Wdowik R (2017) Comparison of selected methods of probe radius correction based on measurements of ceramic workpieces. Procedia CIRP 62:391–395
2. Magdziak M, Ratnayake CRM (2018) Investigation of best parameters' combinations for coordinate measuring technique. Procedia CIRP 78:213–218
3. ElKott D (2011) Coordinate metrology of freeform surfaces. A CAD-based, practical approach. Saarbrücken, (VDM Verlag Dr. Müller GmbH & Co. KG)
4. ElKott DF, Veldhuis SC (2005) Isoparametric line sampling for the inspection planning of sculptured surfaces. Comput Aided Des 37(2):189–200
5. Raghunandan R, Venkateswara RP (2008) Selection of sampling points for accurate evaluation of flatness error using coordinate measuring machine. J Mater Process Technol 202(1–3):240–245
6. Yadong L, Peihua G (2004) Free-form surface inspection techniques state of the art review. Comput Aided Des 36(13):1395–1417
7. Ren M, Kong L, Sun L, Cheung C (2017) A curve network sampling strategy for measurement of freeform surfaces on coordinate measuring machines. IEEE Trans Instrum Meas 66(11):3032–3043
8. Rui-song J, Wen-hu W, Ding-hua Z, Zeng-qiang W (2016) A practical sampling method for profile measurement of complex blades. Measurement 81:57–65

9. Moroni G, Petrò S (2014) Optimal inspection strategy planning for geometric tolerance verification. Precis Eng 38(1):71–81
10. Zahmati J, Amirabadi H, Mehrad V (2018) A hybrid measurement sampling method for accurate inspection of geometric errors on freeform surfaces. Measurement 122:155–167
11. Kumru M, Kumru PY (2015) A fuzzy ANP model for the selection of 3D coordinate-measuring machine. J Intell Manuf 26(5):999–1010
12. Samarakoon SMK, Ratnayake RMC (2015) Strengthening, modification and repair techniques' prioritization for structural integrity control of ageing offshore structures. Reliab Eng Syst Saf 135:15–26
13. Antosz K, Ratnayake RMC (2016) Machinery classification and prioritization: Empirical models and AHP based approach for effective preventive maintenance. In: IEEE international conference on industrial engineering and engineering management (IEEM), pp 1380–1386
14. Saaty TL (1990) How to make a decision: the analytical hierarchy process. Eur J Oper Res 48(1):9–26
15. Rajamohan G, Shunmugam MS, Samuel GL (2011) Practical measurement strategies for verification of freeform surfaces using coordinate measuring machines. Metrol Measur Syst 18(2):209–222
16. Kawalec A, Magdziak M (2017) The selection of radius correction method in the case of coordinate measurements applicable for turbine blades. Precis Eng 49:243–252

The Geometric Surface Structure
of X5CrNiCuNb16-4 Stainless Steel in Wet
and Dry Finish Turning Conditions

Kamil Leksycki[✉] and Eugene Feldshtein

University of Zielona Góra, Zielona Góra, Poland
k.leksycki@ibem.uz.zgora.pl

Abstract. The paper presents the research results on forming the texture of the X5CrNiCuNb16-4 stainless steel surface when finish turning. Surface texture parameters were measured using the Sensofar S Neox optical profilometer. The tests were carried out in wet and dry conditions for varying cutting speeds and feeds with the constant depth of cut. The Parameter Space Investigation (PSI) method was used for planning the research, as this allows the research to be realized with a minimum number of experimental points. The results obtained were subjected to statistical analysis using Statistica 13 software. It was found that the feed influences greatly the values of surface roughness parameters, while the cutting speed affects insignificantly. For both cooling conditions, a reduction in the feed rate reduced surface roughness parameters. Lower values for surface roughness parameters were obtained when cutting with cooling in comparison with dry cutting.

Keywords: Surface texture · Finish turning ·
X5CrNiCuNb16-4 stainless steel · Cooling conditions

1 Introduction

The development of medicine is largely dependent on the production of increasingly attractive materials for medical devices. Actually, many biomedical materials are used; their features should ensure, inter alia, an enhanced corrosion and wear resistance, high mechanical strength, fatigue strength and resistance to cracking [1, 2].

Today, many metals and alloys are used for biomedical purposes, among which are titanium alloys, nickel-based alloys, cobalt-chrome alloys, stainless steels, etc., with the most commonly used materials being 316 L stainless steel, Ti-6Al-4V titanium alloy or Co-20Cr-15W-10Ni cobalt-based alloy [3, 4]. Additionally, as it was stated in [5, 6], X5CrNiCuNb16-4 stainless steel is becoming an increasingly popular material, as a result of its favourable properties.

X5CrNiCuNb16-4 steel is a chromium-nickel, hardened stainless steel that has the martensite structure with a small amount of austenite, containing a low carbon percentage and approximately 3% of copper, further enhanced by the precipitation of copper-rich particles [7, 8].

© Springer Nature Switzerland AG 2019
M. Diering et al. (Eds.): *Advances in Manufacturing II* - Volume 5, LNME, pp. 183–194, 2019.
https://doi.org/10.1007/978-3-030-18682-1_15

X5CrNiCuNb16-4 stainless steel is distinguished by its high strength and excellent resistance to corrosion compared to the stainless steels AISI 304 or AISI 316 L [6]. Therefore, it is widely used in medicine, particularly in surgery [5].

In the process of production, final processing has an important role because it provides the required quality of the treated surface [9]. At the same time, an increase in machining productivity can be achieved by increasing cutting conditions, which also affect the quality of the machined surface [10, 11].

An evaluation of the product quality in industry is particularly carried out by measuring the surface roughness, on the basis of which the functionality and life of the product in a hostile environment can be determined. It is also a factor that significantly affects the time and cost of production [12]. The surface roughness makes it possible to evaluate the mechanical and chemical properties of the element machined, while ensuring a low roughness of the surface proceeded improves fatigue strength, creep resistance, tribological characteristics and corrosion resistance of this surface. During operation cycle, it affects sliding surface friction, ability to hold lubricant substances, an also a contact resistance of mating area. As a result, specific machining parameters should be chosen to achieve the level of finish required. Many effects may influence the surface roughness when processing, such as parameters of cutting, hardness of workpiece material, cutting tool features and its geometry. The cutting parameters (cutting speed, feed rate and depth of cut) as well as coolant/lubricant conditions should be selected carefully to ensure optimum surface finishing [13]. On the other hand, the surface roughness formation depends on numerous uncontrolled effects that influence this process and make it difficult to operate. Therefore, the most common method of treatment conditions selecting is to use traditional conditions that cannot provide the desired surface roughness and do not guarantee the desired manufacturing performance [14, 15].

In [16], the surface quality of X5CrNiCuNb16-4 stainless steel was tested after finish turning in wet conditions. The tests were carried out in the cutting speed range of 75–200 m/min, the feed range of 100–200 mm/rev and the cutting depth of 0.25–0.8 mm. Lower surface roughness occurred when machining at higher cutting speeds and medium feed values. As the feed increased, the surface roughness increased. No significant changes in surface roughness were observed when the depth of the cut was changed.

In [17], the surface roughness and topography of X5CrNiCuNb16-4 stainless steel obtained in the range of higher cutting speeds of finish turning were analysed. The tests were carried out in the cutting speed range of 350–400 m/min, feed rates of 0.1–0.2 mm/rev and cutting depths of 0.25–0.35 mm. Lower surface roughness occurred with lower feed rates. No interaction between the wedge and the workpiece materials was observed. Traces of the cutting edge were noticed on the machined surface, the pitch between which enlarges with the increase of the feed.

In industry, the qualitative requirement regarding the geometrical characteristics of products are the characteristic parameters of the surface roughness, i.e. Ra or Rz.

However, application of these parameters does not guarantee a proper comparison of the performance properties of surfaces for various applications and different cutting parameters [18]. Therefore, special attention should be focussed on the remaining parameters of surface roughness, such as Rp, Rv, Rq, which determine the surface texture of the workpiece.

It should be noted the high efficiency of near dry cutting in forming the low roughness of the machined surface [19–21]. However, these processing methods were not considered in this study.

The aim of the research conducted was to determine the impact of cutting parameters on the shaping of the geometric structure of the X5CrNiCuNb16-4 stainless steel surface under the wet and dry finish turning process.

2 Experimental Assumptions

2.1 Material and Method

X5CrNiCuNb16-4 stainless steel, the chemical composition of which is shown in Table 1 [22], was machined. The length of a single sample was 15 mm, diameter 50 mm.

Table 1. Chemical composition of the processed material, %.

Cr	Ni	Cu	Mn	Si	C	P	S	Tb + Na	Fe
15.0–17.0	3.0–5.0	3.0–5.0	≤ 1.0	≤ 1.0	≤ 0.07	≤ 0.04	≤ 0.03	0.15–0.45	73.0

The machining was carried out on a CTX 510 machining centre, produced by DMG MORI. A turning tool with a CoroTurn SDJCR 2525 M 11 holder and a CoroTurn DCMX 11 T3 04-WM 1115 insert with (Ti, Al)N + (Al, Cr)$_2$O$_3$ PVD coating were used. Angles of the wedge: tool cutting edge angle $|_r = 93°$, rake angle © = 18, 〈 clearance angle = 7°, nose radius $r_{\searrow} = 0.4$ mm, shear width $b_{n©} = 0.1$ mm [23].

The tests were carried out in the range of cutting speed of 150–500 m/min and feed rates of 0.05–0.4 mm/rev and a constant cutting depth of 0.5 mm, corresponding to the conditions of the finishing processes. Turning was carried out both under dry conditions and with the use of an aqueous emulsion, based on Castrol Alusol SL 51 XBB emulsifying oil with a working concentration of 7% (wet). The selected profile amplitude parameters of surface roughness were analysed according to the PN-EN ISO 4287:1999 standard: Rp − maximum peak height of the roughness profile, Rv − maximum valley depth of the roughness profile, Rz − maximum height of the roughness profile, Ra − medium arithmetical deviation of the roughness profile, Rq − mean square deviation of the roughness profile. The measurements of the surface texture parameters were made using the Sensofar S Neox optical profilometer. The results of the measurements were obtained using Mountains Maps Premium 7.4 software.

2.2 The Research Method

The research plan was based on the Parameter Space Investigation (PSI) method [24], which allows the experiments to be planned with a minimum number of experimental points. For this purpose, the points were arranged sequentially, in strictly defined coordinates of a multi-dimensional space, in such a way that their projections on the axes of coordinates X_1 and X_2 were located at equal distances to each other (Fig. 1).

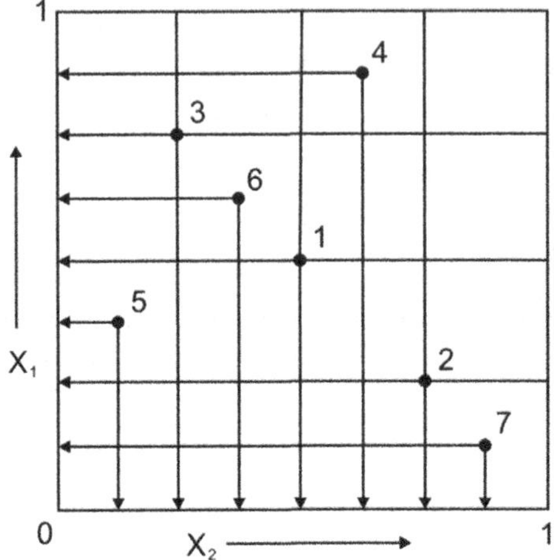

Fig. 1. The distribution of the test points on X_1 and X_2 axes in accordance with the PSI method.

The coordinates of the test points are introduced in Table 2, where $X_{min} = 0$ and $X_{max} = 1$. The 7 test points on each axis allow statistical calculations to be made.

Table 2. Coordinates of PSI method test points.

Variables	1	2	3	4	5	6	7
X_1	0.5000	0.2500	0.7500	0.8750	0.3750	0.6250	0.1250
X_2	0.5000	0.7500	0.2500	0.6250	0.1250	0.3750	0.8750

The statistical analysis of the results obtained was made using the Statistics 13.1 programme.

3 Research Results and Their Analysis

Figure 2 presents images of X5CrNiCuNb16-4 stainless steel surface obtained for the 7 test points when wet and dry finish turning. During dry machining, smaller *Ra* parameter values were obtained in points 3, 5 and 6, while larger values were obtained in point 7. On the other hand, during wet turning, lower *Ra* values occurred at points 3 and 5, while the *Ra* values were higher at point 7.

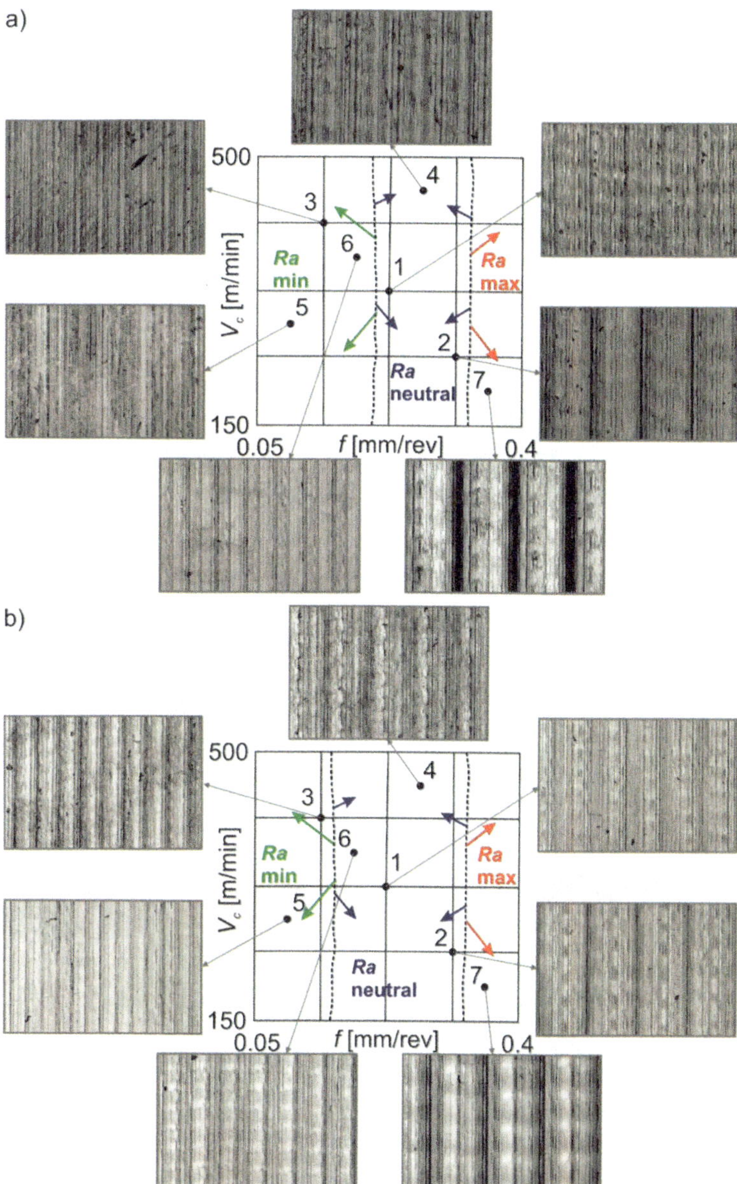

Fig. 2. Layout of 7 PSI method points tested and images obtained in the following conditions: a) dry, b) wet; (the area of low surface roughness is marked as green, the area of high roughness is marked as red and the area of middle roughness is marked as violet).

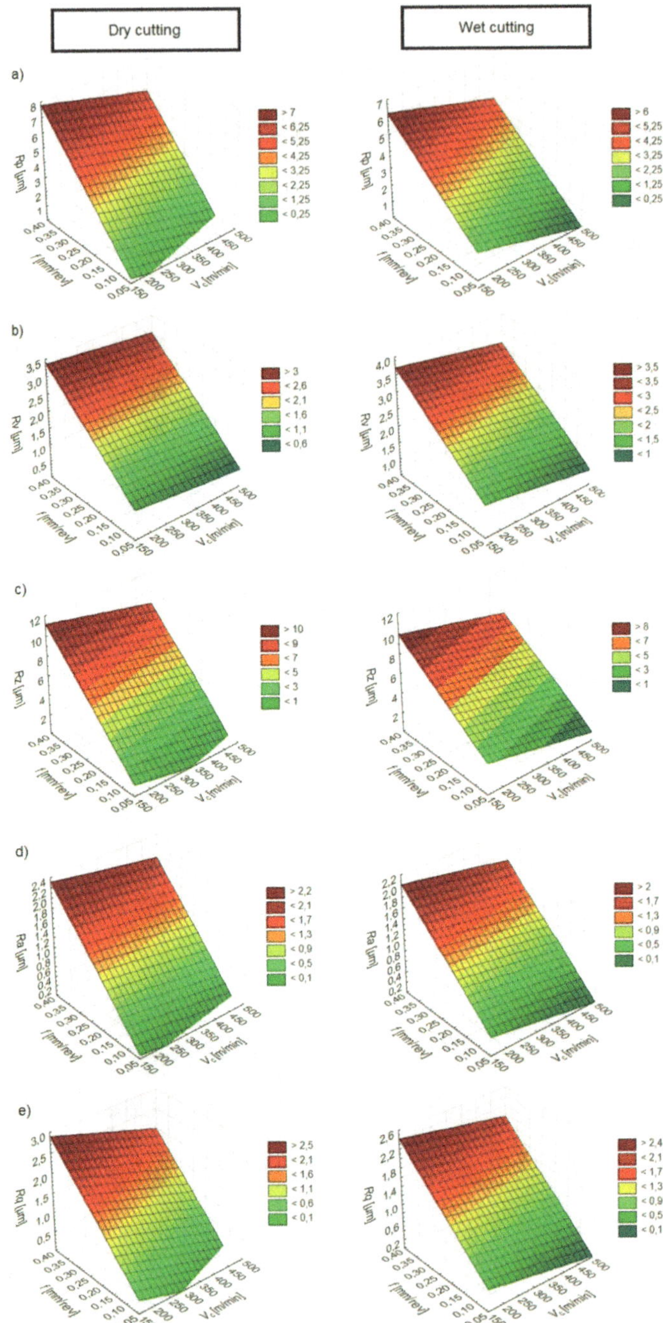

Fig. 3. Changes in roughness parameters depending on the cutting speed and feed under wet and dry conditions for: (a) *Rp*, (b) *Rv*, (c) *Rz*, (d) *Ra*, (e) *Rq*.

For both cooling conditions, three clear surface roughness limits were distinguished (the black dotted line at Fig. 2). Under dry machining lower Ra parameter values were obtained in the feed range ~ 0.05–0.15 mm/rev, while larger ones were obtained for 0.35–0.4 mm/rev. On the other hand, under wet cutting, smaller Ra values were achieved in the feed range of ~ 0.05–0.1 mm/rev, while this was higher for ~ 0.35–0.4 mm/rev. The divisions of the borders are also confirmed by the images registered. The black vertical lines mark the elevation of the profile, with their intense colours indicating the height of the rise, the distance between them and also the value of the feed. In the range of smaller surface roughness, there are smaller elevations; their widths and the distances between them are small. In the range of higher surface roughness, the elevations are clear; their widths and the distances between them are higher too.

The results of surface roughness measurements were subjected to statistical analysis and are presented in Fig. 3. In addition, statistical analysis made it possible to generate regression equations for the impact of the cutting speed and feed rate on the values of the selected parameters of surface roughness; these equations have been presented in Table 3.

Table 3. Results of the statistical processing for selected surface roughness parameters when finish turning of X5CrNiCuNb16-4 stainless steel.

Parameter	Dry machining	Wet machining
Rp, µm	$-1.122V_c + 21.363f$	$-1.785V_c + 12.977f$
Rv, µm	$-0.700V_c + 7.084f$	$-1.427V_c + 6.297f$
Rz, µm	$-0.823V_c + 28.446f$	$-3.214V_c + 19.286f$
Ra, µm	$-0.072V_c + 6.382f$	$-0.273V_c + 4.776f$
Rq, µm	$-0.375V_c + 7.285f$	$-0.407V_c + 5.522f$

Tests show that when wet machining, both surface roughness parameters Rp, Rv, Rz, Ra, Rq and the intensity of the impact of the cutting speed and feed rate decrease. It was noticed when turning, both dry and wet, that smaller feed rates cause smaller values of surface roughness parameters. The impact of the cutting speed on the surface roughness parameters Rp, Rv, Rz, Ra, Rq was insignificant.

The surface roughness parameters obtained at individual test points in accordance with the PSI method were also analysed in relation to variable feed rates and cutting speeds. Figure 4 shows the average percentage changes in Rp, Rv, Rz, Ra, Rq roughness parameters for X5CrNiCuNb16-4 stainless steel surface under wet machining when compared with dry machining.

A reduction in Rp, Rv, Rz, Ra, Rq surface roughness parameters analysed was observed when wet machining in the range from 2% to 25% for different PSI points. However, the increase when dry machining was in the range from 2% to 240%. The principal boundary of changes for the roughness parameters was observed for $f \sim 0.2$ mm/rev; in Fig. 4 it is marked with a black dotted line.

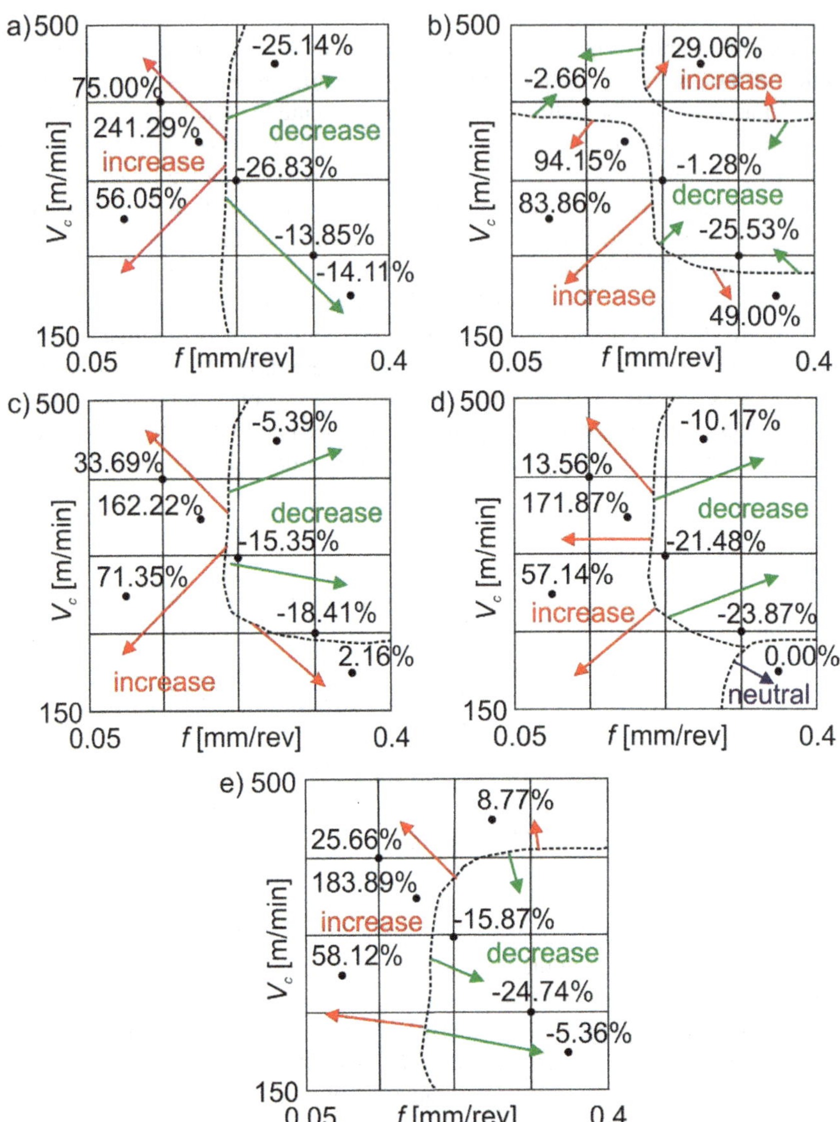

Fig. 4. Average percentage reduction of the selected roughness parameters in 7 test points when wet machining compared with dry machining for: (a) *Rp*, (b) *Rv*, (c) *Rz*, (d) *Ra*, (e) *Rq*.

The surface roughness profiles, obtained at individual test points according to the PSI method, were also analysed. Figure 5 presents the profiles of surface roughness, obtained when dry machining while Fig. 6 presents the profiles when wet machining. In Figs. 5 and 6, lower values of surface roughness were highlighted in green, the neutral values in violet and the higher values in red colors.

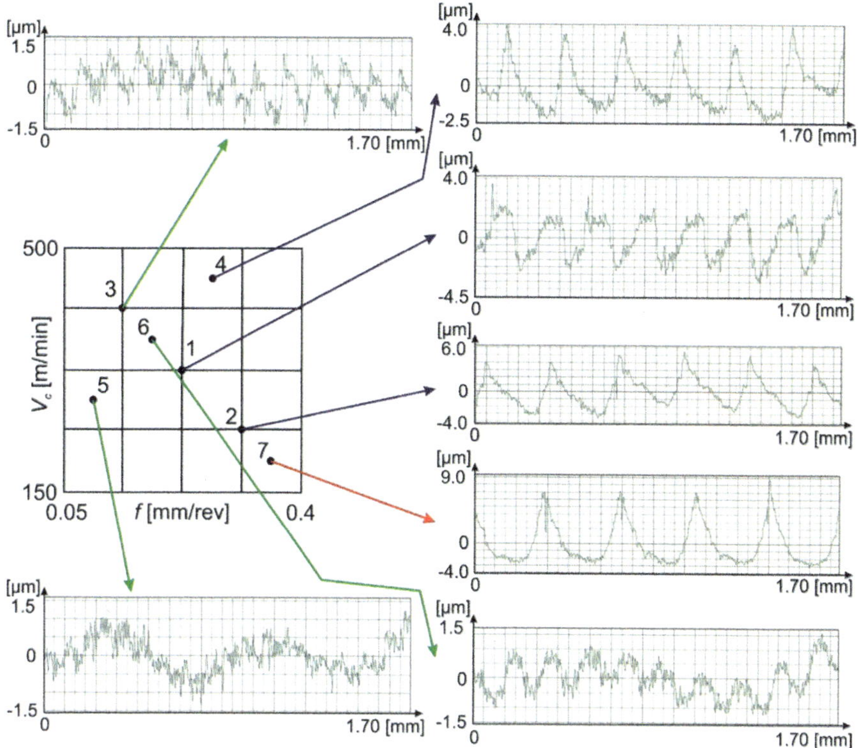

Fig. 5. Surface roughness profiles obtained at the 7 test points when finish turning of X5CrNiCuNb16-4 stainless steel under dry conditions.

For dry machining, the smallest Ra values ranged from 0.259 μm to 0.472 μm; neutral values were in the range from 1.150 μm to 1.550 μm. For wet machining, the smallest values of the Ra parameter were in the range from 0.407 μm to 0.536 μm with the neutral values being from 0.903 μm up to 1.180 μm. Higher Ra values for both cooling conditions were 2.060 μm. For both cooling conditions, point 5, at which the machining was carried out, with $V_c = 281.25$ m/min and $f = 0.093$ mm/rev, deserves particular attention, since the lowest roughness values were obtained there. The similar conclusions could be made for point 7, where the machining was carried out with $V_c = 193.75$ m/min and $f = 0.356$ mm/rev and where the highest values of roughness parameters were obtained.

For wet and dry machining, lower values of surface roughness parameters were obtained in the range of lower feeds and higher cutting speeds. On the other hand, higher values were obtained in the range of higher feeds and lower cutting speeds.

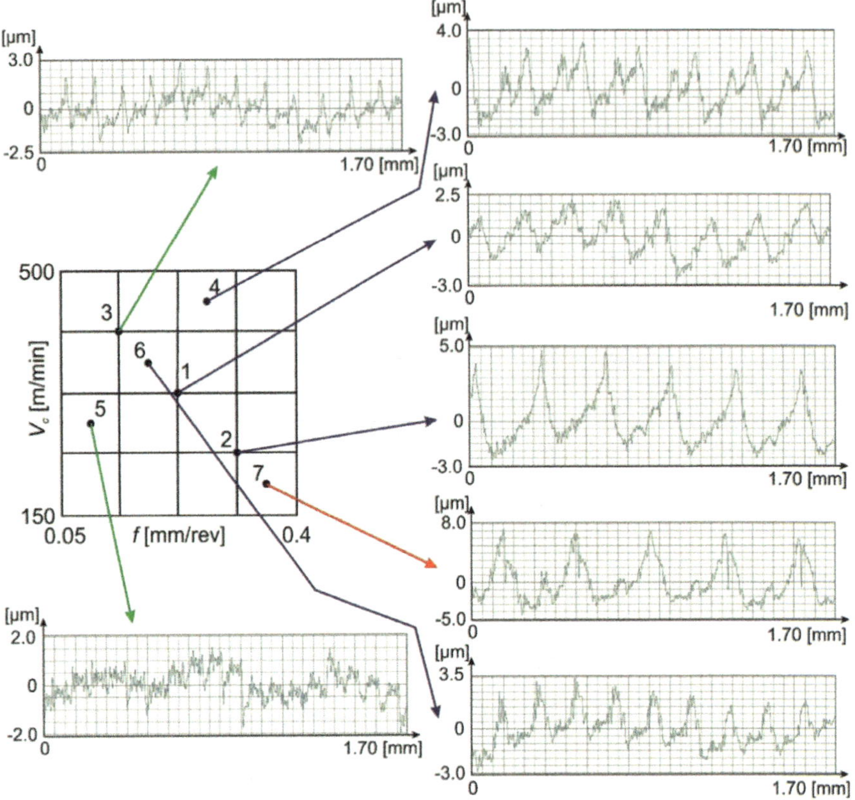

Fig. 6. Surface roughness profiles obtained at the 7 test points when finish turning of X5CrNiCuNb16-4 stainless steel under wet conditions.

4   Conclusions

In this article, the forming of the texture of X5CrNiCuNb16-4 stainless steel surface when finish turning, under wet and dry cooling conditions was investigated. Selected parameters of the amplitude of the surface roughness profile were analysed.

Based on the results obtained, the following conclusions can be formulated:

- For both cooling conditions, there were three clear areas dividing smaller, neutral and higher surface roughness values.
- Research shows that when wet machining, both the surface roughness parameters Rp, Rv, Rz, Ra, Rq and the intensity of the impact of the cutting speed and feed rate decrease.
- The feed rate significantly influences the surface roughness. In both cooling conditions, reduction of the feed rate results in a reduction in Rp, Rv, Rz, Ra, Rq parameters of surface roughness. However, the impact of the cutting speed was insignificant.

- During processing under wet conditions, as compared to dry machining, a reduction in the surface roughness parameters of 20–25% was observed.
- For both cooling conditions, smaller feeds and higher cutting speeds within the range tested reduce values of the surface roughness parameters.

Acknowledgments. The authors would like to thank Mariusz Władowski of OPTOTOM for sharing the Sensofar S Neox optical profilometer and for his support in carrying out the measurements.

References

1. Chen Q, Thouas GA (2015) Metallic implant biomaterials. Mater Sci Eng 87:1–57
2. Kulinets I (2015) Biomaterials and their applications in medicine. In: Amato SF, Ezzell Jr RM (eds) Regulatory affairs for biomaterials and medical devices, pp 1–10. Elsevier
3. Hamidi MFFA, Harun WSW, Samykano M, Ghani SAC, Ghazalli Z, Ahmad F, Sulong AB (2017) A review of biocompatible metal injection moulding process parameters for biomedical applications. Mater Sci Eng 78:1263–1276
4. Niinomi M, Nakai M, Hieda J (2012) Development of new metallic alloys for biomedical applications. Acta Biomater 8:3888–3903
5. Ahlhelm M, Günther P, Scheithauer U, Schwarzer E, Günther A, Slawik T, Moritz T, Michaelis A (2016) Innovative and novel manufacturing methods of ceramics and metal-ceramic composites for biomedical applications. J Eur Ceram Soc 36:2883–2888
6. Mutlu I, Oktay E (2013) Characterization of 17-4 PH stainless steel foam for biomedical applications in simulated body fluid and artificial saliva environments. Mater Sci Eng 33:1125–1131
7. Guoliang L, Huang C, Zou B, Wang X, Liu Z (2016) Surface integrity and fatigue performance of 17-4PH stainless steel after cutting operations. Surf Coat Technol 307:182–189
8. Kochmanski P, Nowacki J (2006) Activated gas nitriding of 17-4 PH stainless steel. Surf Coat Technol 200:6558–6562
9. Krolczyk GM, Maruda RW, Krolczyk JB, Nieslony P, Wojciechowski S, Legutko S (2018) Parametric and nonparametric description of the surface topography in the dry and MQCL cutting conditions. Measurement 121:225–239
10. Maruda RW, Krolczyk GM, Wojciechowski S, Zak K, Habrat W, Nieslony P (2018) Effects of extreme pressure and anti-wear additives on surface topography and tool wear during MQCL turning of AISI 1045 steel. J Mech Sci Technol 32:1585–1591
11. Maruda RW, Krolczyk GM, Michalski M, Nieslony P, Wojciechowski S (2017) Structural and microhardness changes after turning of the AISI 1045 steel for minimum quantity cooling lubrication. J Mater Eng Perform 26:431–438
12. Bin RW, Goel S, Davim JP, Joshi SN (2015) Parametric design optimization of hard turning of AISI 4340 steel (69 HRC). Int J Adv Manuf Technol 82:451–462
13. Abu-Mahfouz I, Rahman AHME, Banerjee A (2018) Surface roughness prediction in turning using three artificial intelligence techniques; a comparative study. Procedia Comput Sci 140:258–267
14. Benardos PG, Vosniakos GC (2003) Predicting surface roughness in machining: a review. Int J Mach Tools Manuf 43:833–844

15. Chen S, Feng F, Zhang R, Zhang Y (2018) Surface roughness measurement method based on multi-parameter modeling learning. Measurement 129:664–676
16. Kiran PC, Clement S (2013) Surface quality investigation of turbine blade steels for turning process. Measurement 46:1875–1895
17. Zou B, Zhou H, Huang C, Xu K, Wang J (2015) Tool damage and machined-surface quality using hot-pressed sintering Ti(C7N3)/WC/TaC cermet cutting inserts for high-speed turning stainless steels. Int J Adv Manuf Technol 79:197–210
18. Krolczyk GM, Krolczyk JB, Maruda RW, Legutko S, Tomaszewski M (2016) Metrological changes in surface morphology of high-strength steels in manufacturing processes. Measurement 88:176–185
19. Mia M, Singh G, Gupta MK, Sharma VS (2018) Influence of Ranque-Hilsch vortex tube and nitrogen gas assisted MQL in precision turning of Al 6061-T6. Precis Eng 53:289–299
20. Gupta MK, Mia M, Singh G, Pimenov DY, Sarikaya M, Sharma VS (2018) Hybrid cooling-lubrication strategies to improve surface topography and tool wear in sustainable turning of Al 7075-T6 alloy. Int J Adv Manuf Technol 101:1–15
21. Mia M, Gupta MK, Singh G, Krolczyk GM, Pimenov DY (2018) An approach to cleaner production for machining hardened steel using different cooling-lubrication conditions. J Cleaner Prod 187:1069–1081
22. http://www.matweb.com
23. Sandvik Coromant (2017) Turning tools, Catalogue
24. Statnikov RB, Matusov JB (2002) Multicriteria analysis in engineering. Springer

Analysis of the Application of Gypsum Moulds for Casting Strength Samples of Aluminium Alloys

Lukasz Bernat^(✉)

Institute of Materials Technology, Division of Foundry,
Poznan University of Technology, Poznań, Poland
lukasz.bernat@put.poznan.pl

Abstract. The paper presents tooling for simultaneous casting of a set of strength samples made of aluminium alloy in gypsum moulds. Using the NovaFlow&Solid simulation program, the filling and solidification process of variants used for casting strength-testing samples made with the traditional investment casting method and the casting concepts proposed by the author were analyzed. Simulations and experimental tests were performed on proportional and disproportional samples designed in accordance with the PN-EN ISO 6892-1 standard. Application of the disproportional casting samples set, proposed by the author, has been confirmed by measuring the microstructure of selected samples expressed by the distance between the Dendrite Arms Spacing (DAS).

Keywords: Precise casting · Aluminium alloys · Virtual prototyping · Gypsum mould · Casting process simulation

1 Introduction

Nowadays, application of 3D printing in the foundry industry is becoming increasingly common. Several years ago, due to high costs of equipment and materials, usage of additive technologies was limited to prototyping or producing small series of castings, mostly of alloys difficult to machine and precious metals. Modern 3D printing technologies facilitate the manufacturing of, e.g., precise models applied in casting with the use of Replicast FM method, silicon moulds for wax models, and ready-made sets of moulds and complex cores of silica sand [1–5].

One of the precise casting methods where 3D printing technology is increasingly used for larger elements is a variant of the investment casting process [6, 7], in which, as opposed to the classical method, the printed model is not burnt out, but instead becomes degasified during the mould annealing process. Apart from unquestionable advantages of this method, there are also disadvantages, such as high time-consumption of the mould production, production of castings mainly of expensive and difficult to machine alloys, and uneconomical use for single-piece production.

At the other end, as regards the size of the manufactured elements, is the jewelry and dental industry, where mainly small models are printed using technologies such as: MJP (Multijet Printing), SLA (Stereolithography), DLP (Digital Light Processing),

© Springer Nature Switzerland AG 2019
M. Diering et al. (Eds.): *Advances in Manufacturing II* - Volume 5, LNME, pp. 195–211, 2019.
https://doi.org/10.1007/978-3-030-18682-1_16

DPP (Daylight Polymer Printing). Considering small and complex shapes of models, moulds are made of a special casting gypsum, what has a significant impact on the casting quality [8–12]. The downside to the use of gypsum as a mould material is its lifetime and preparation process, as it must be degassed before pouring into a sleeve, in order to remove the air bubbles inside.

Despite the fact that gypsum moulds are mainly used for small castings in serial production, it can be an alternative to making single precise castings, without the need of a time-consuming preparation of traditional shell moulds. Their unquestionable advantage is the possibility of use with virtually any casting alloy and metal composites, as well as the possibility of pouring them at the temperature of about 100 °C without damaging the mould (this concerns alloys such as aluminium, zinc, etc.), what has a significant impact on the cooling rate of a casting and thus, refinement of its structure, inclusions of intermetallic phases, volume fraction of porosity and mechanical properties of obtained castings [13–18].

Regardless of the casting method, strength tests of samples are some of the basic tests determining metallurgical quality of casting alloys. A very important factor affecting the reproducibility of strength test results and their subsequent analysis related to the cooling rate, for instance, is the design of a set of moulds in such a way that each of the mould cavity reproducing a sample is poured simultaneously and that the same grain size, e.g. α phase dendrites, occurs over the entire length of the L_0 sample.

According to the PN-EN ISO 6892-1 standard, strength-testing samples may be proportional, what is recommended, or disproportional, which may be used after consultation with a client. Examples of recommended L_0 sizes of proportional samples with a circular cross-section are shown in Table 1.

Table 1. Recommended sizes of samples with round cross-section according to PN-EN ISO 6892-1.

Proportionality coefficient k	Diameter d [mm]	Initial measuring length L_0 [mm]	Minimum length of the parallel part L_c [mm]
5,65	20	100	110
	14	70	77
	10	50	55
	5	25	28

For disproportional samples, the recommended minimum initial measuring length of the L_0 sample is 15 mm. When conducting research, such as examination of influence of mould thermal parameters, chemical composition of a casting alloy and heat treatment of a casting on its mechanical properties and connecting the obtained results with the casting cooling rate, the sample breaking point corresponding to the place of the thermocouple insertion in the thermal axis of the sample is very important (this applies to a set of samples).

It is obviously very rare for a sample to break precisely in the position where the thermocouple is placed, so it is important that the microstructure of the casting be the same over the entire length of the L_0 sample. Unfortunately, with proportional samples, there are big differences in the solidification times between the central part of the sample and its upper and lower parts. In this case, it is impossible to accurately determine effects of a sample cooling rate on its mechanical properties.

Since gypsum moulds are used mainly for small precision castings, it is very difficult to find guidelines for manufacturing strength testing samples in gypsum moulds in the literature. Example drawings of moulds for casting aluminium alloy samples by traditional methods of melting models can be found in [19].

One of the methods, developed by Brown, features two systems: direct and indirect (Fig. 1a, b). Another method was developed by Wood and Ludwig (Fig. 1c).

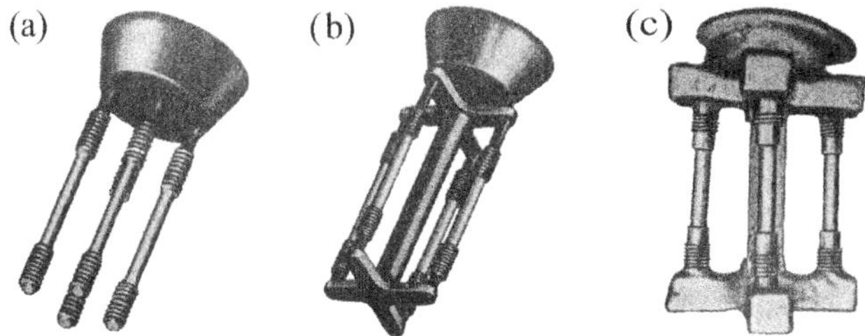

Fig. 1. Distribution variants of strength samples used for aluminium alloys casting: (a) according to Brown with the direct system, (b) according to Brown with the indirect system, (c) according to Wood and Ludwig [19].

Considering the lack of information in the literature on the subject of preparing samples for strength tests of aluminium alloys in gypsum moulds, the author proposes an original approach to the topic.

2 Research Methodology

The design of tooling for simultaneous gypsum casting of a set of strength test samples of an aluminium alloy was based on the following main assumptions: a minimum number of 4 samples in a mould of 4 pieces (3 pieces for strength tests, 1 piece for the cooling curve, porosity and microstructure tests), reproducibility of solidification conditions in each piece, low cost of tooling, and the possibility of selecting the number of samples in the mould (to a maximum of 10 samples).

The dimensions of samples for strength tests were assumed in accordance with the PN EN ISO 6892-1 standard. However, two variants were proposed: proportional, with a diameter of 10 mm (Fig. 2a), where the $L_0 = 5,65 \cdot \sqrt{S_0}$ value equals 50 mm (L_0 – initial measuring length of the sample, S_0 – cross-sectional area of the sample) and disproportional, with a diameter of 12 mm (Fig. 2b), where, according to the standard, the L0 value should have a minimum of 15 mm.

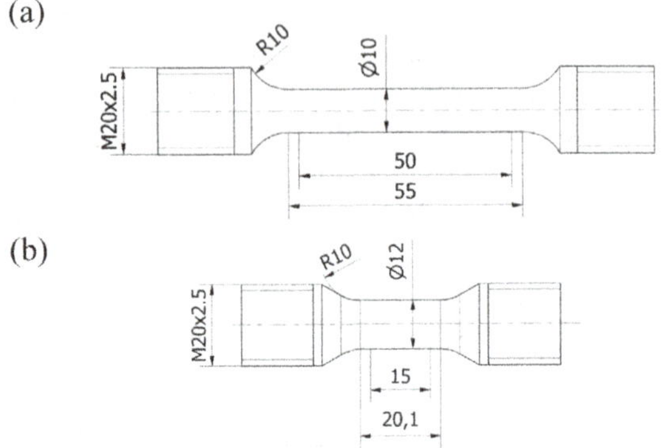

Fig. 2. Shape and dimensions of the mechanical properties test sample: (a) proportional, (b) disproportional.

A need of using disproportional samples in strength tests arises from seeking to obtain the smallest differences in solidification time in the entire initial length of the sample L_0 and thus obtaining constancy of the DAS microstructure parameter. The minimum recommended initial sample length of 15 mm, according to the PN–EN ISO 6892-1 standard, allows simultaneous solidification and ensures homogeneous structure throughout the entire measurement section, which is crucial for the spot where a sample breaks during the tensile testing process. In addition, it has a significant impact on the future connection between the cooling rate, the microstructure and mechanical properties of a casting, because regardless of the spot where a sample breaks during the strength tests, the microstructure parameters on the entire sample length remain unchanged.

With proportional samples, depending on the gating system construction and the mould, there are samples that break in the thread area or at the thread itself, due to large differences in values of the DAS microstructure parameter in the measurement part of the sample.

Due to a lack of data on the production of tooling for strength samples in gypsum moulds, it was decided to analyze the existing solutions for casting of aluminium samples in typical moulds used for the investment casting method and proposed by the author. 3D geometries of the analyzed variants are shown in Fig. 3.

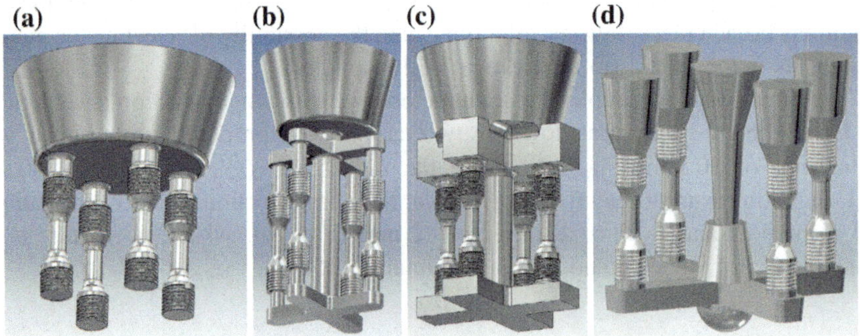

Fig. 3. Designed arrangement of samples in the mould according to: (a) the direct system by Brown, (b) the indirect system by Brown, (c) the Wood and Ludwig system, (d) author's own concept.

For all the variants, the velocity of a liquid alloy during pouring of mould cavity, temperature distribution inside the mould and casting during solidification, solidification times of the initial measuring length of the L_0 sample and proportion of shrinkage porosity in the samples were analyzed. The analysis was based on the results of simulations executed in the NovaFlow&Solid CV6.0r3 program.

The selected initial parameters of the virtual filling and solidification process were as follows:

- alloy material: EN AC-42100 (AlSi7Mg0.3),
- temperature of liquid alloy during pouring: 760 °C,
- mould material: gypsum,
- initial temperature of the mould: 200, 300, 400 °C,
- flow rate during the pouring: 0.6 kg/s,
- size of the discretization grid: ≈ 2 mm.

After an analysis of the simulation results, a set of models consisting of 10 samples was built. The samples were produced using the FDM (Fused Deposition Modeling) process, on the Zortrax M200 3D printer. Selected 3D printing parameters are presented in Table 2.

Table 2. Selected 3D printing parameters of models set for casting strength samples.

Material	Z-HIPS
Layer thickness	0.14 mm
Filling amount	10%
Printing time of the models set	ca. 35 h
Used material weight	316 g
Cost of the material	ca. EUR 10

A set consisting of 10 samples and elements of the gating system printed and assembled with wax is shown in Fig. 4.

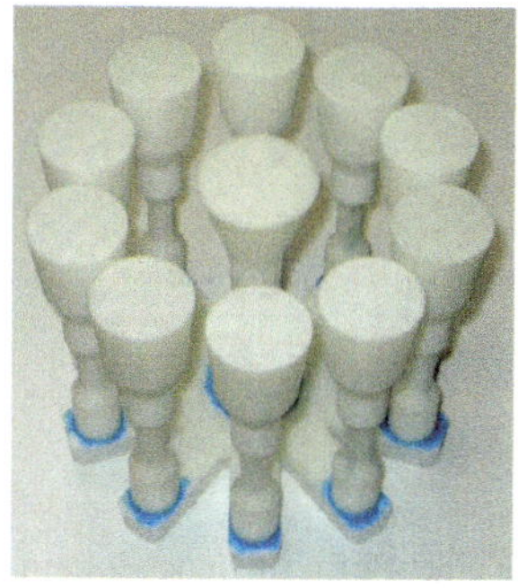

Fig. 4. A set of printed and assembled models.

The mould was made of R&R Argentum foundry gypsum, which was vacuum degassed before being poured into the sleeve. Author's own research shows that with a gypsum mould of such dimensions (wall thickness of the mould is max. 50 mm) the drying time should be at least 24 h in a temperature of 50 °C, which prevents mould cracking during the annealing process.

The time and temperature of individual stops of the annealing cycle of the gypsum mould are shown in Fig. 5.

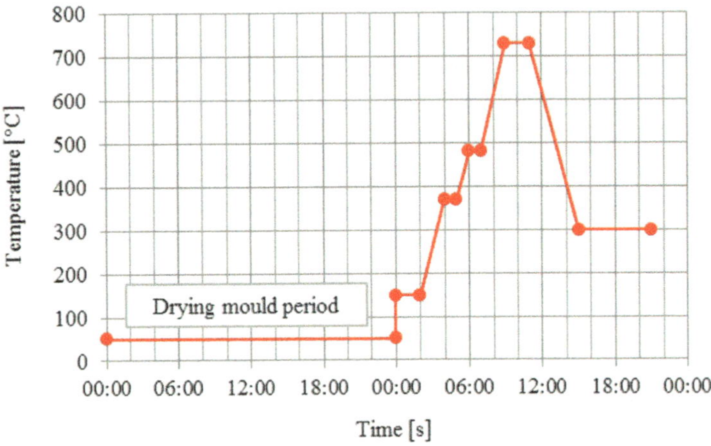

Fig. 5. The annealing cycle of the gypsum mould.

The mould at a temperature of 400 ± 5 °C was poured with a liquid AlSi7Mg0.3 alloy at a temperature of 750 ± 5 °C (Fig. 6a). The temperature measurement in the thermal axis of a sample was executed using a Cr-Ni-Cr thermocouple (Fig. 6b)

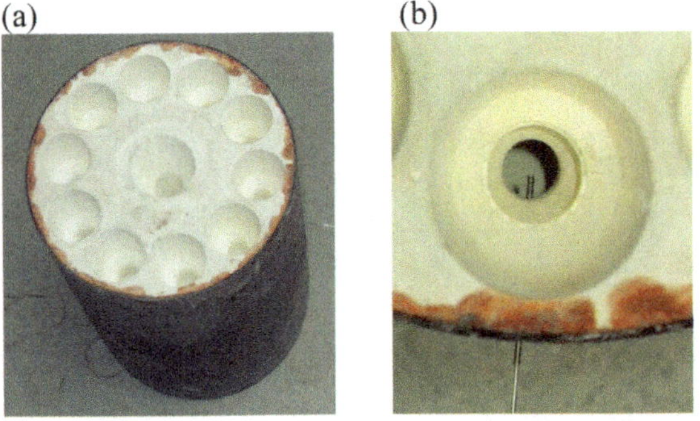

Fig. 6. (a) Gypsum mould before pouring, (b) Placement of a thermocouple in sample's axis.

In order to verify reproducibility of the casted proportional and disproportional strength samples, values of the distance parameter between arms of DAS dendrites were measured in five pieces taken from each set. The microstructure measurement points and sample marking method are shown in Fig. 7.

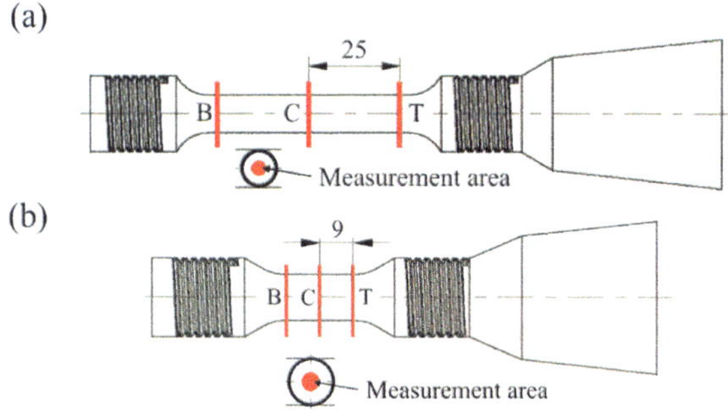

Fig. 7. Measurement points of the DAS microstructure parameter and their marks (B – bottom, C – centre, T – top) for samples: (a) proportional, (b) disproportional.

The distance between the dendrite arms of the α_{Al} DAS phase was calculated according to the formula:

$$DAS = L/(n-1)\ [\mu m],\qquad(1)$$

where: L – length of the dendrite measuring section [μm], n – number of visible arms of dendrites.

For the calculation of the DAS parameter, dendrites that had a minimum of 4 distinct arms were used, the number of dendrites measured in one analyzed area of the sample was 10.

3 Study Results

Figures 8, 9 and 10 show results of the simulation of the mould cavity pouring process for the proportional (Ø10 × 50 mm) and disproportional (Ø12 × 15 mm) samples according to the concepts of Brown as well as Wood and Ludwig.

The most unfavorable way of pouring the mould cavity is the variant proposed by Brown, where the samples are filled directly, omitting the runner and gates (Fig. 8). Regardless of the type of the samples, separation of the stream into smaller, single drops occurs in the initial phase of the pouring process, what causes strong turbulences of a liquid alloy at the bottom of the sample, and thus its oxidation. In addition, with such a construction of a pouring cup, depending on the place of "hitting" the base of the pouring cup by a stream of liquid alloy, it is highly probable that the cavities reproducing the samples will not be poured at the same time.

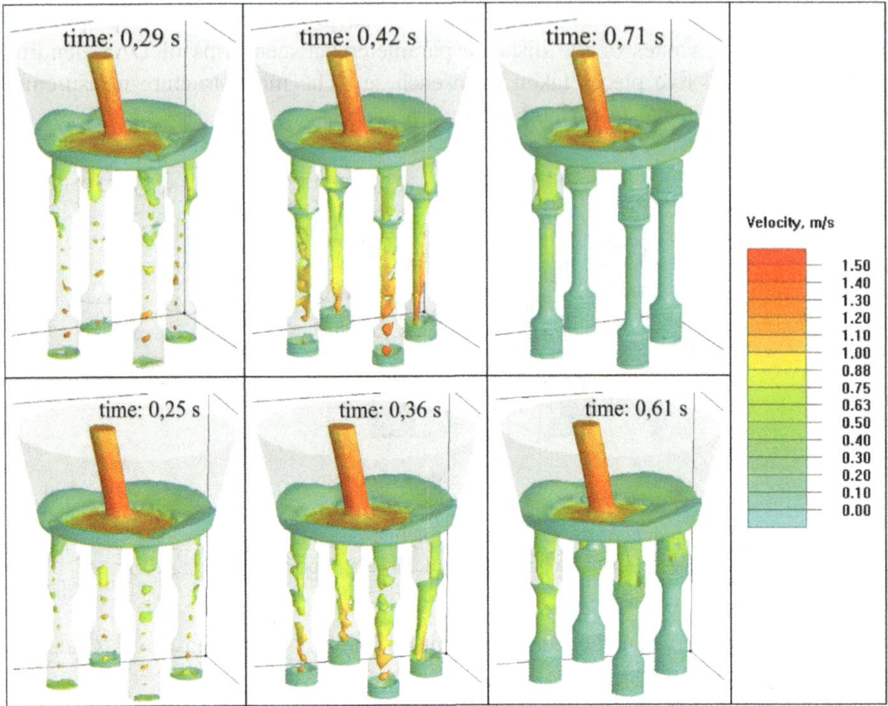

Fig. 8. Pouring level of the mould cavity according to the Brown concept (direct pouring): proportional samples Ø10 × 50 mm (upper), disproportionate samples Ø12 × 15 mm (bottom).

The disadvantages of this solution have been partially eliminated by using the sprue (Figs. 9 and 10). However, as in the previous case, due to contact of the pouring cup with the risers, a liquid alloy may directly enter the sample reproducing cavity, which may cause a defect called the "cold drop" and discontinuity of the casting structure (Fig. 9 and 10, locations marked with a red circle).

Table 3 presents predicted solidification times for the analyzed concepts of sample casting in the upper, middle and lower parts of the initial measurement length of the L_0 sample.

Figure 11 shows percentage differences in solidification times for the proportional samples Ø10 × 50 mm in relation to their central part, including pouring temperature of the gypsum mould.

For samples poured directly, according to Brown's concept, difference in solidification times in the upper part of a sample decreases along with the decrease of the mould temperature during pouring, from approx. 55% at mould temperature of 400 °C to 10% at 200 °C. A much more regular distribution of percentage differences in

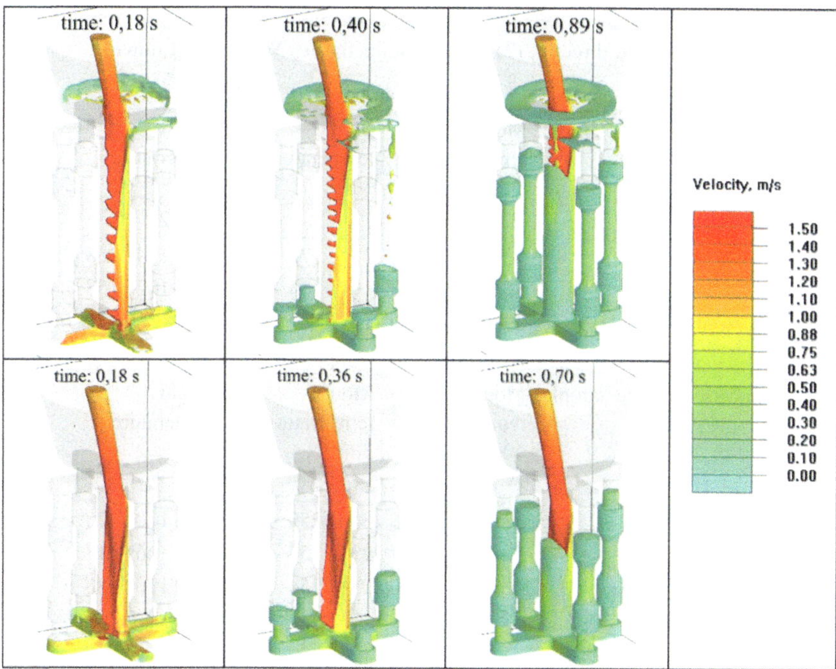

Fig. 9. Pouring level of the mould cavity according to the Brown concept (indirect pouring): proportional samples Ø10 × 50 mm (upper), disproportional samples Ø12 × 15 mm (bottom).

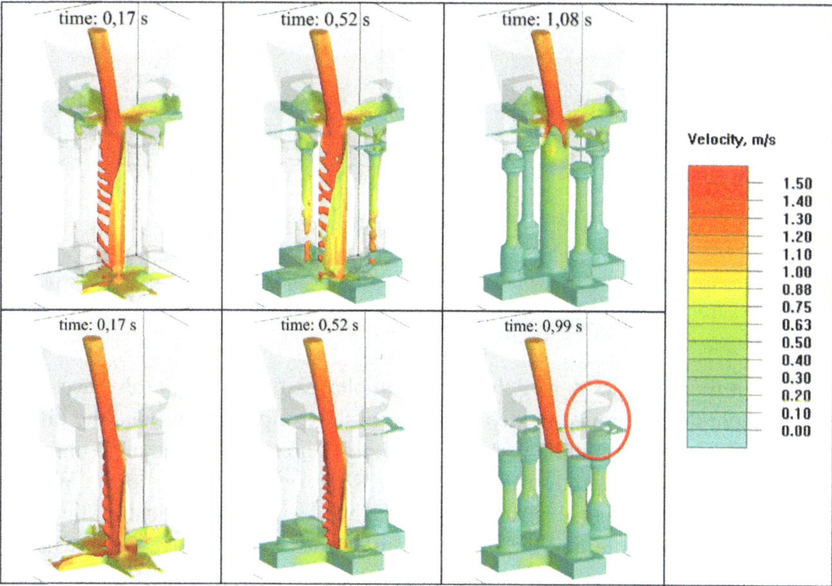

Fig. 10. Pouring level of the mould cavity according to the Wood and Ludwig concept: proportional samples Ø10 × 50 mm (upper), disproportional samples Ø12 × 15 mm (bottom).

Table 3. Solidification times in seconds, obtained from simulation of sample casting according to the concepts: (1) Brown directly, (2) Brown indirectly, (c) Wood and Ludwig.

Location on the sample	Proportional sample Ø10 × 50 mm								
	Mould temperature 400			Mould temperature 300			Mould temperature 200		
	1	2	3	1	2	3	1	2	3
Top	531	318	869	279	178	539	112	114	268
Centre	350	286	665	191	163	323	102	102	158
Bottom	290	301	610	172	180	333	115	115	179

Location on the sample	Disproportional sample Ø12 × 15 mm								
	Mould temperature 400			Mould temperature 300			Mould temperature 200		
	1	2	3	1	2	3	1	2	3
Top	579	412	911	346	259	677	206	170	475
Centre	508	402	857	291	254	618	180	166	418
Bottom	450	395	810	258	254	571	160	170	388

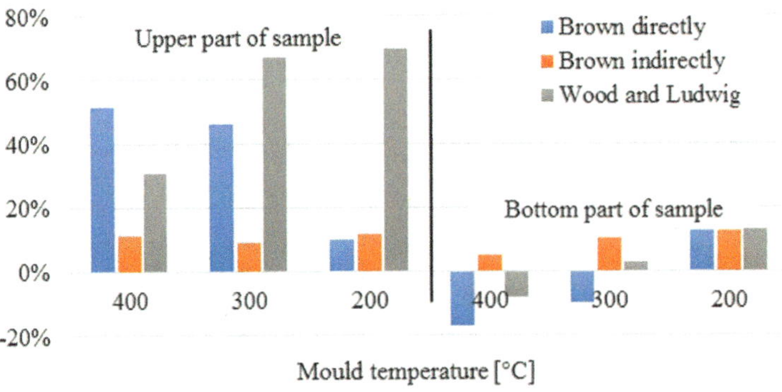

Fig. 11. Percentage difference of solidification times in the upper and lower part of proportional samples in relation to their central part.

solidification times occurs during the direct samples casting concept (also according to Brown), where the difference in the upper part of the sample is 10%, and the difference between the bottom and middle part of the sample is up to 12%. The most unfavorable method of producing strength-testing samples is the one according to the Wood and Ludwig casting concept. In this case, difference between the solidification times in the central part of the sample and the upper part of the sample is above 70%, what has a significant impact on microstructure of a casting and will contribute to future fracture of a sample at its weakest point. Therefore, the proportional samples of Ø10 × 50 mm, according to the above-mentioned concept, will not meet a requirement of maintaining the same microstructure on the entire initial measuring length of the L_0 sample.

The application of disproportional samples Ø12 × 15 mm significantly reduced differences between solidification times in the middle part of the sample and its upper and lower parts, as shown in Fig. 12. In this comparison, similarly as in the case of proportional samples, the Brown's indirect pouring concept is the most favorable one, because differences of the solidification times are at a maximum level of 3%.

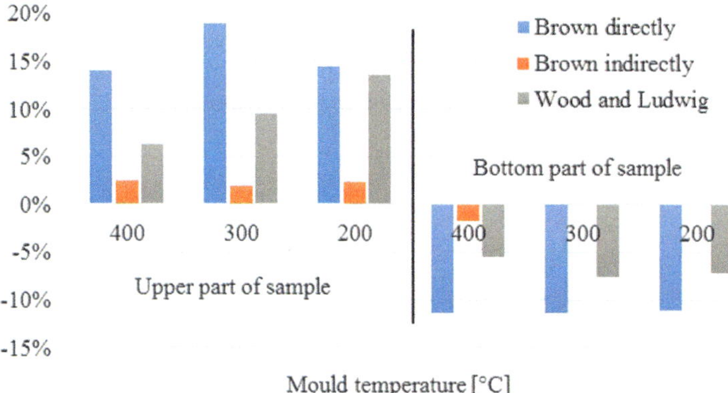

Fig. 12. Percentage difference in the solidification times in the upper and lower parts of the sample disproportional to the central part.

Unfortunately, despite small differences in the solidification times, shrinkage porosity may occur in the samples, regardless of the mould temperature during the pouring (Fig. 13).

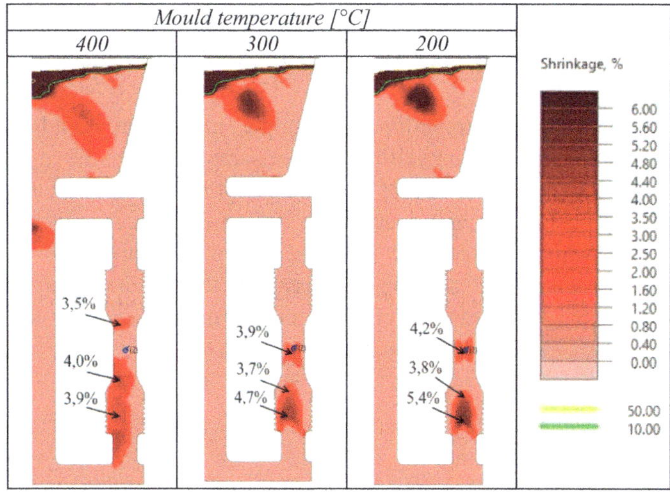

Fig. 13. Prediction of the shrinkage porosity for a Ø12 × 15 mm sample according to the Brown's indirectly poured concept.

Although the shrinkage porosity occurring in the axis of the sample is at a low level of approx. 4%, the porosity in the thread itself can cause a break in a location of the weakest link.

Due to the fact that the concepts presented in the literature concerning casting of strength-testing samples in gypsum moulds did not meet the assumed requirements, the author proposed own design of tooling for casting of such samples. Several variants of the gating system, calculated on the basis of the literature have been analyzed [20], especially regarding possibility of simultaneous pouring of all the mould cavities reproducing the strength-testing sample and maintaining the same solidification time throughout the initial L_0 length of a sample. In order to ensure the same, laminar flow of a liquid alloy in the runner, narrowing of sprue at its base and use of base of the sprue was proposed – concepts that were not applied in the previously analyzed casting concepts (Fig. 14).

(a)   **(b)**

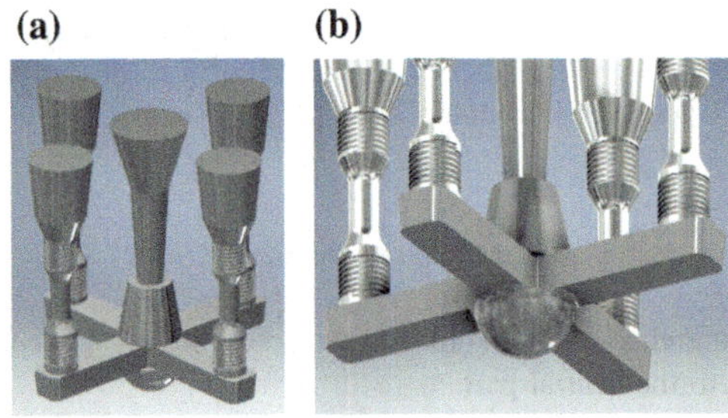

Fig. 14. Proposed elements of the gating system: a) sprue narrowing, b) base of the sprue.

Application of these solutions contributed to regular and simultaneous pouring of the whole mould cavity, as shown in Fig. 15.

Solidification times and percentage differences between the solidification times in the middle part of a sample and its upper and lower parts are shown in Table 4 and Fig. 16.

In the concept of casting strength-testing samples in gypsum moulds, proposed by the author, usage of proportional $\varnothing10 \times 50$ mm samples causes large differences in solidification times between the central part of a sample and its upper and lower parts, just as in the previously analyzed cases. These differences range from approx. 10% to 25%. Usage of disproportional, $\varnothing12 \times 15$ mm samples significantly reduces these differences (up to 2.5%), what brings a possibility of obtaining good correlation between cooling rate of a sample and its microstructure, as well as mechanical properties.

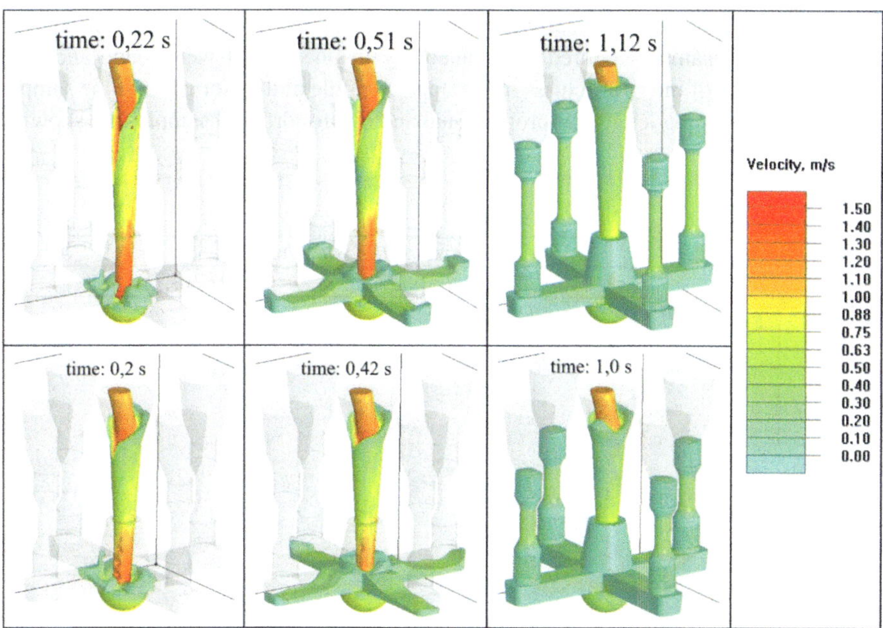

Fig. 15. Pouring level of the mould cavity according to the author's concept: proportional samples Ø10 × 50 mm (upper), disproportional samples Ø12 × 15 mm (lower).

Table 4. Prognosis of solidification times of samples in seconds, according to the author's concept.

Location on the sample	Mould temperature 400		Mould temperature 300		Mould temperature 200	
	Ø10 × 50 mm	Ø12 × 15 mm	Ø10 × 50 mm	Ø12 × 15 mm	Ø10 × 50 mm	Ø12 × 15 mm
Top	452	571	254	370	152	278
Centre	413	572	221	365	121	270
Bottom	450	580	248	370	144	278

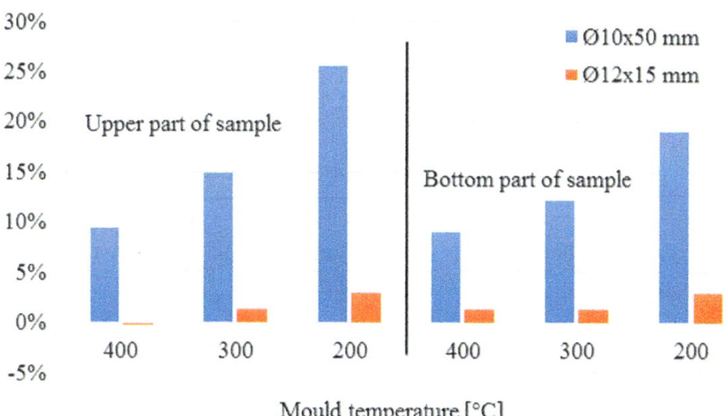

Fig. 16. Percentage differences in solidification times in the upper and lower parts of a sample in relation to its central part, including temperature of the mould and type of the sample.

In order to confirm the assumption of disproportional samples casting in gypsum moulds, 10 real samples poured simultaneously in one mould were made and compared. Examples of microstructures in the upper, middle and lower parts of the samples over the L_0 measuring length for proportional and disproportional samples are shown in Fig. 17.

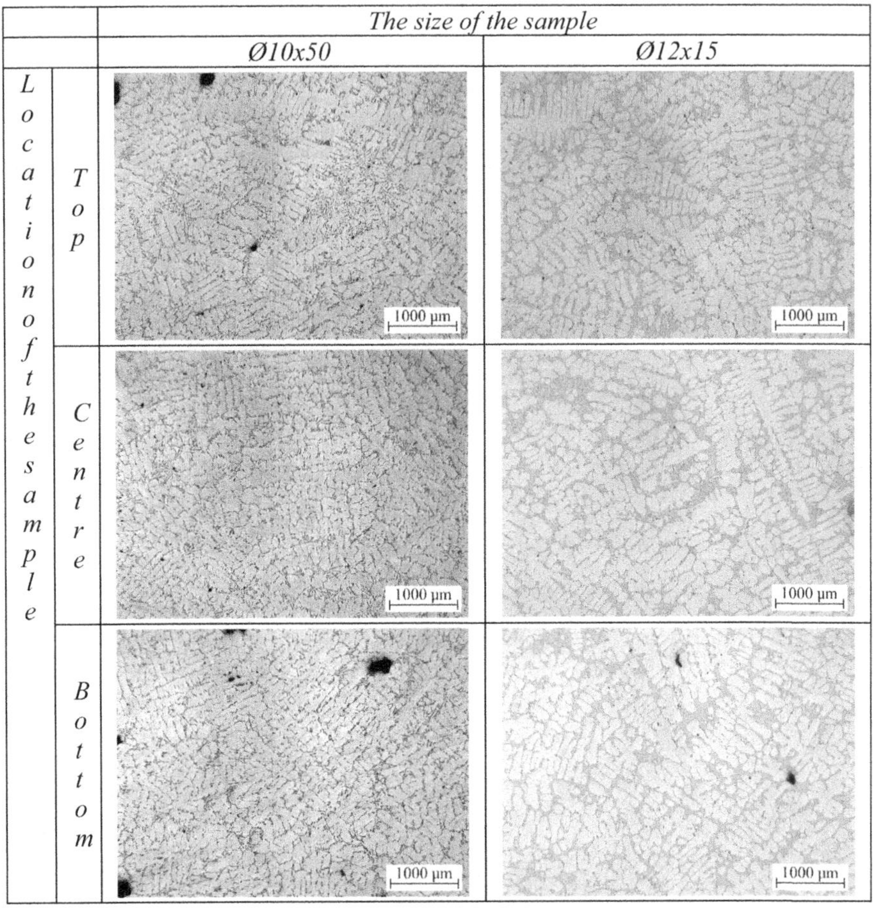

	The size of the sample	
	Ø10x50	Ø12x15
Location of the sample — Top		
Location of the sample — Centre		
Location of the sample — Bottom		

Fig. 17. Microstructure of analyses samples.

Measured distances between the DAS dendrite arms for both types of analyzed variants of the samples are shown in Figs. 18 and 19.

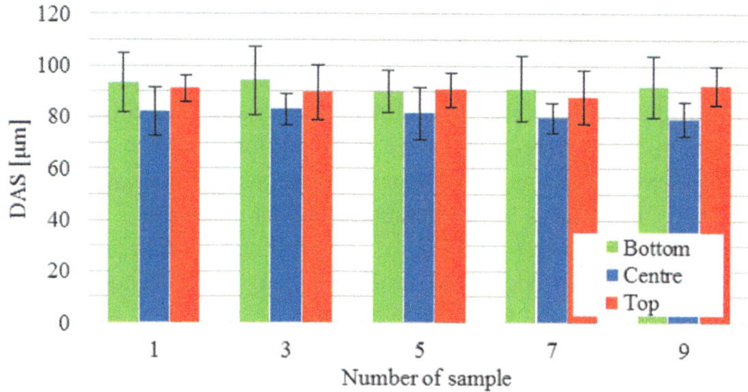

Fig. 18. Values of the DAS parameter obtained from proportional samples Ø10 × 50 mm.

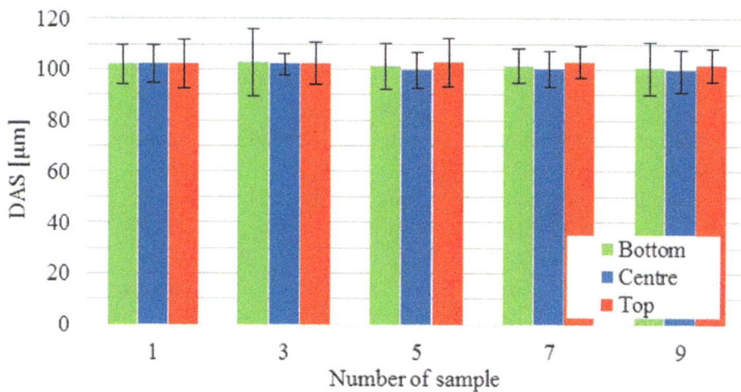

Fig. 19. Values of the DAS parameter obtained from disproportional samples Ø12 × 15 mm.

The measurements show that the proposed concept of casting samples in gypsum moulds is correct, because regardless of the samples casting variant, there is very high repeatability of microstructure measurements in each sample. Comparing the size of the DAS parameter in both casting variants, it is much more advantageous to use the disproportional Ø12 × 15 mm samples, due to very small differences in the determined DAS parameter values in the upper, middle and lower parts of the initial L_0 sample measurement section.

4   Conclusions

The conducted research allowed to formulate the following conclusions:

1. The application of virtual mould cavity pouring simulation and its solidification process facilitates proper design of the tooling for casting strength samples of aluminium alloys in gypsum moulds.

2. The application of additive technologies for tooling production allows easy change of a number of simultaneously cast samples, in the range of 4 to 10 samples.
3. The designed tooling allows to maintain very high repeatability, from one pouring to 10 samples for strength tests and in the case of casting strength-testing samples in gypsum moulds, it is preferable to use disproportional Ø12 × 15 mm samples, in order to maintain the same microstructure parameters expressed by the distance between arms of dendrites in the initial length of the sample.
4. The designed tooling and the concept of sample casting can be successfully used in further research to determine the possibility of using gypsum moulds for castings of aluminium alloys.

References

1. https://www.materialise.com/en – producer's information. Accessed 21 Nov 2018
2. https://www.sculpteo.com/en/materials/metal-casting-material/ – producer's information. Accessed 21 Nov 2018
3. https://www.3dsystems.com/applications/casting – producer's information. Accessed 21 Nov 2018
4. https://ultimaker.com/en/explore/how-is-3d-printing-used/molds-casting – producer's information. Acceseds 21 Nov 2018
5. https://zortrax.pl/ – producer's information. Accessed 21 Nov 2018
6. Budzik G, Sobolak M, Kozdęba D (2006) Rapid prototyping method of investment pattern of blade in silicone mould. Arch Foundry 18(2/2):201–206
7. Budzik G, Sobolak M, Kozdęba D (2006) Rapid prototyping technology using for investment casting process. Arch Foundry 18(2/2):207–212
8. Guzera J (2010) Making casts with the method of melting models in autoclaved gypsum molds. Arch Foundry Eng 10(3):307–310 (in Polish)
9. Guler KA, Mustafa C (2012) Casting quality of gypsum bonded block investment casting moulds. Adv Mater Res 445:349–354
10. Phetrattanarangsia T, Puncreobutra C, Khamkongkaeoa A, Thongchaia C, Sakkomolsrib B, Kuimaleec S, Kidkhunthodd P, Chanlekd N, Lohwongwatanaa B (2017) The behavior of gypsum-bonded investment in the gold jewelry casting process. Thermochim Acta 657:144–150
11. Nor SZM, Ismail R, Isa MIN (2015) Porosity and strength properties of gypsum bonded investment using terengganu local silica for copper alloys casting. J Eng Sci Technol 10(7):921–931
12. Bobby SS, Singamneni S (2014) Influence of moisture in the gypsum moulds made by 3D printing. Procedia Eng 97:1618–1625
13. Pietrowski S, Rapiejko C (2011) Temperature and microstructure characteristics of silumin casting AlSi9 made with investment casting method. Arch Foundry Eng 11(3):177–186
14. Taylor JA (2004) The effect of iron in Al-Si casting alloys. In: 35th Australian foundry institute national conference, Adelaide–South Australia, 31 October–3 November 2004, pp 148–157
15. Maa Z, Samuel AM, Samuel FH, Doty HW, Valtierra S (2008) A study of tensile properties in Al–Si–Cu and Al–Si–Mg alloys: effect of β-iron intermetallics and porosity. Mater Sci Eng, A 490:36–51

16. Fang QT, Granger DA (1989) Porosity formation in modified and unmodified A356 alloy castings. AFS Trans 97:989–1000
17. Espinoza-Cuadra J, Gallegos-Acevedo P, Mancha-Molinar H, Pikado A (2010) Effect of Sr and solidification conditions on characteristics of intermetallic in Al–Si 319 industrial alloys. Mater Des 31:343–356
18. Hajkowski J, Popielarski P, Sika R (2018) Prediction of HPDC casting properties made of AlSi9Cu3 alloy. In: Manufacturing 2017. Springer, pp 621–631
19. Allendorf H (1969) Precision casting using melted models, Warszawa (in Polish)
20. Perzyk M, Błaszkowski K, Haratym R, Waszkiewicz S (1990) Materials for the design of foundry processes. Państwowe Wydawnictwo Naukowe, Warszawa (in Polish)

A Study of Raters' Agreement in Quality Inspection with the Participation of Hearing Disabled Employees – Continuation

Beata Starzyńska, Karolina Szajkowska[✉], and Magdalena Diering

Poznan University of Technology, Piotrowo 3, 61-138 Poznan, Poland
karolina.strykowska@interia.pl

Abstract. This paper looks at the results of a repeated study of the level of agreement between raters in a sensory quality inspection, with the participation of hearing-impaired individuals. The study was conducted in a manufacturing company supplying to the automotive sector. Hearing-impaired employees perform quality inspection which includes visual inspection of products. The level of agreement between raters, including hearing-impaired ones, was assessed using the MSA procedure for non-measurable characteristics, but with the application of Gwet's AC_1 coefficient. The study outcomes show that hearing-impaired employees perform at a quality level at least equal to that achieved by able-bodied employees. The results are valid for both a study conducted in a laboratory environment and one conducted at the manufacturing site.

Keywords: Quality inspection · Sensory inspection · Hearing impairment · Measurement and control system analysis (MSA) · Gwet's AC_1

1 Introduction

Despite fast technological progress and machine-oriented production solutions, human presence in production (especially in assembly and quality control stages of the production process) is still evident and needed. In quality inspection, human uses own senses to decide about products quality. Thus, key factors influencing on inspection performance are, among others, employment support, motivation, procedures and work ergonomics [1, 2].

Employment support is the mainstream of activities aimed at improving the life quality of people with disabilities. The importance of this issue is also reflected in the actions undertaken by the EU member states at the legislative and executive levels [3].

According to various sources, there are ca. 4.5 m of disabled people in Poland [4], what accounts for 12.2% of the population. Ca. 2 m of them are at the working age. Among them, ca. 100,000 are people with hearing impairment, 44% of whom are at the working age [5].

The issues of people with hearing disabilities in literature focuses mainly on the medical [6, 7], pedagogical and educational [8–10] or sociological and cultural aspects [11]. On the other hand, the work context of people with hearing impairments is rarely discussed [12–14].

© Springer Nature Switzerland AG 2019
M. Diering et al. (Eds.): *Advances in Manufacturing II* - Volume 5, LNME, pp. 212–220, 2019.
https://doi.org/10.1007/978-3-030-18682-1_17

Experiences resulting from cooperation with people with hearing impairments led to the conclusion that the functioning of employees representing this type of disability is difficult due to communication problems and mutual understanding between deaf and hearing persons [14–16].

Publications on the subject of work of people with disabilities are most often an analysis of the labour market [17–19]. Thus, there is a gap in the literature related to the studies in the field of quality assessment of work performed by people with hearing disabilities, because the subject matter is narrowed to the elimination of barriers [20, 21] and the qualifications to work [22].

Employment opportunities and life quality of this social group is the research focus of the authors.

Knowledge of the subject is used in raising social awareness of limitations experienced by the disabled and opportunities for minimising or eradicating them through the implementation of various facilitations. Companies employing the disabled create jobs and plan and organise the scope of responsibilities taking into consideration their impairments. The same applies to groups of employees with hearing deficiency.

Such actions are undertaken in the enterprise where the second stage of the study under analysis was conducted. Further in the article, results of the continued study of the level of agreement between quality raters in a sensory inspection are presented. The study was aimed to examine whether hearing-impaired employees perform at the same quality level as able-bodied employees, and whether altered working conditions affect the study results.

The study was conducted in a manufacturing company supplying to the automotive industry, which has the status of a supported employment enterprise. Employees with hearing impairment perform product quality inspection which includes visual inspection. In order to assess whether the level of performance of disabled employees is acceptable, the study included able-bodied inspectors. The level of agreement between the decisions made by disabled and able-bodied raters was assessed by means of the measurement and control systems analysis (in short MSA) procedure for non-measurable characteristics or more precisely – characteristics assessed organoleptically [23], but with the use of Gwet's AC_1 coefficient.

2 Research Problem

The study under analysis is the continuation of a previous study conducted in the same company on the same group of employees [24]. The only difference between them was the working environment in which the quality inspection was performed.

The previous study was conducted in an isolated room where components for the automotive industry were verified for quality. The workstations in that room are referred to as offline, since they are located away from the production line and quality inspection is performed some time after the manufacturing process is completed.

The study under analysis was performed in a quality inspection room located by the production line, referred to as on-line, where quality inspection is performed directly after the manufacturing of components is completed.

Compared to the offline work station, the on-line workstation is characterised by a higher noise level. Due to the necessity to air-dry the components before inspection, the noise may temporarily rise to 75 dB. Moreover, several inspectors work at the same time side-by-side, and the production line operators may pass by the room.

The offline workstation, on the other hand, is a separate room, isolated from the noise coming from the production line and not frequented by other employees who might hinder proper performance of the inspection process.

The study was conducted in normal working conditions at the production line, at the final quality inspection workstation. Employees perform a 100% quality inspection of products made of plastic with a galvanised coating obtained by the electrochemical method. During the inspection, raters use templates of acceptable and non-acceptable defects, component inspection instructions, as well as catalogues of faults and defects. Using their sense of sight, raters classify products as "complying with the requirements" or "non-complying with the requirements"; "complying" or "non-complying"; "good" or "bad"; and "OK" or "NOK". In the case under analysis, the inspection was performed by both able-bodied and hearing-impaired employees.

Similarly to the previous study conducted on the same group of raters [24], the MSA procedure for alternative characteristics is used to estimate the raters' accuracy and compare their performance, taking into account characteristics of the evaluated products.

The measurement and control systems analysis is applied in the process of quality inspection improvement in manufacturing companies. The MSA methods and procedures make it possible to establish whether the inspection system under analysis is appropriate for the assigned tasks and objectives.

In previously conducted empirical studies, the analysis was typically aimed to assess the agreement of decisions made during a visual inspection conducted by able-bodied employees [24].

In this analysis, agreement between decisions made by quality raters (both able-bodied and disabled) was assessed with the use of Gwet's AC_1 coefficient.

The coefficient is applied to determine inter-rater agreement within a quality inspection and measurement system.

Inter-rater agreement provides information whether study participants rate the same products in the same way in consecutive rating series (a study of reproducibility), without any reference to the rating requirements (e.g., the criteria specified in the catalogue of faults and defects). Agreement between raters, i.e., reliability of their ratings, facilitates the determination of reproducibility of the classification (ratings) [25].

Gwet's AC_1 coefficient is determined with the following formula [26]:

$$AC_1 = \frac{p_a - p_e}{1 - p_e},\tag{1}$$

where:

$$p_a = \frac{1}{n'} \sum_{i=1}^{n'} \sum_{k=1}^{q} \frac{r_{ik}(r_{ik} - 1)}{r_i(r_i - 1)}$$

$$p_e = \frac{1}{q - 1} \sum_{k=1}^{q} \pi_k(1 - \pi_k)$$

$$\pi_k = \frac{1}{n} \sum_{i=1}^{n} \frac{r_{ik}}{r_i}$$

where:

- AC_1—Gwet's decision agreement coefficient,
- r—number of raters,
- q—number of categories,
- k—category,
- n—total number of objects in the study,
- n'—number of objects in the study rated by at least two raters,
- p_a—the share of the same decisions in the total number of possible decisions made by the team of raters,
- p_e—probability of making the same decisions by chance,
- r_{ik}—number of raters who classified object i into category k,
- r_i—number of raters who rated object i,
- π_k—probability of classifying into category k by any rater, for any component.

The value of the AC_1 coefficient falls within the range of $0 \div 1$.

The study should cover a sample of 30–50 objects representing all the non-conformities listed in the catalogue of faults and defects. The sample should include at least 30% of components explicitly faulty, 30% of components explicitly free of defects, and at least 30% of components difficult to rate, with defects at the threshold of acceptability.

Before the study, each component is assigned a unique number. The numbers are known to the coordinator of the study but unknown to the raters.

The study is conducted in conditions most closely resembling real working conditions, taking into consideration the lighting, noise level, ambient temperature, and rate of work.

After each series, the study coordinator alters the order of samples, so that the rater is not able to remember their previous rating. Breaks between consecutive rating series are aimed to minimise the effects of fatigue and loss of efficiency.

Raters use the $0 \div 1$ nominal dichotomous scale (0—bad, 1—good). Before each series, expert's decision is made to determine the reference value for the objects under analysis.

Before commencement of the rating, each rater familiarises themselves with the catalogue of acceptable and non-acceptable defects. Next, the study is conducted in three series.

The final outcome of the study is a report on the level of agreement (the value of the AC_1 coefficient) (jointly for all the raters, jointly for all the raters with the reference value, inter-rater level of agreement—for each rater separately, for particular raters with the reference value). The report data is based on the acceptance criteria for the level of agreement between raters (Table 1).

Table 1. Criteria of acceptance of the level of agreement between raters [6, 9].

AC_1 value	Level of agreement	Recommendations
$0 < AC_1 < 0.4$	Low (negligible)	Corrective actions are necessary
$0.4 \leq AC_1 < 0.75$	Good	Preventive and improvement actions should be taken
$0.75 \leq AC_1 \leq 1$	Very good	Improvement actions are recommended

According to the guidelines for the AC_1 coefficient, if the level of agreement is below 0.4, the inspection process requires corrective actions. At higher levels of the AC_1 coefficient, preventive and/or improvement actions are recommended. However, considering the objective of the study, the authors find it extremely useful to compare the AC_1 coefficient value for selected groups of raters.

3 Results

The study, aimed at assessment of a measurement system for non-measurable characteristics, covered three raters with hearing impairment, marked A (ON), B (ON), and C (ON), three able-bodied raters, marked D, E, and F, and the quality department manager (expert), marked REF.

The raters assessed the quality of a product made of plastic with a galvanised coating. In order to minimise the impact of such factors as job seniority, experience and age, raters with a similar level of competence were appointed (with their consent). The hearing-impaired study participants suffered from a similar level of hearing loss. Additionally, a sign-language interpreter assisted in the preparation of hearing-impaired raters for the MSA.

Before commencement of the study, the raters familiarised themselves with the catalogue of defects, i.e., a list of all possible defects which may occur in the process, rated (OK or NOK).

A sample of 45 knobs was used in the study. The same sample had been used in the previous study conducted according to the methodology presented above [24]. The rating was performed in three series.

The results were first used to estimate the inter level of agreement between the expert and particular raters (Table 2).

The calculations based on the criteria listed in Table 1 yielded very good inter agreement for the disabled raters A (ON), B (ON) and C (ON)—from 0.78 to 0.91. The level of inter agreement for the able-bodied raters D, E, and F was good (from 0.58 to 0.74), although lower than between the disabled raters.

The next step consisted in the estimation of the level of agreement between decisions of raters and the expert. The calculations yielded very good and good level of

Table 2. Inter agreement between raters.

Rater	Inter agreement	Joint agreement	Correspondence with the reference value	Joint correspondence with the reference value
A (ON)	0.78	0.72	0.71	0.69
B (ON)	0.91		0.74	
C (ON)	0.81		0.76	
D	0.74		0.66	
E	0.58		0.58	
F	0.71		0.70	
REF	1.00	–	–	–

agreement with the reference value for the disabled raters A (ON), B (ON), and C (ON) (0.71 for A (ON)-Expert, 0.74 for (ON)-Expert, and 0.76 for C (ON)-Expert). The level of agreement with the reference value was lower for the able-bodied raters and amounted to 0.66 for D-Expert, 0.58 for E-Expert, and 0.70 for F-Expert. Again, the level of agreement with the reference value was lower for the able-bodied raters than for the disabled ones.

4 Discussion

Both the study discussed in the previous paper [24] and the one presented here were aimed at verifying whether employees with hearing impairment deliver work of the same quality as able-bodied workers. A comparison of the outcomes of both studies for the groups of raters under analysis shows a good level of correspondence between decisions made by hearing-impaired raters, higher than between decisions made by able-bodied raters.

The better overall results obtained by the raters in the second study may reflect a lower level of stress they were under, having been familiar with the procedure.

Moreover, it must be noted here that the quality inspection is performed by a production line which generates a noise of up to 75 dB. Able-bodied quality raters pointed to noise as the key stressor in the working environment [27]. Hearing-impaired raters, on the other hand, failed to mention noise as an impeding factor.

It should be noted that the tests were carried out in a sheltered workshop. Similarly to the previous study [24], the results may be affected by the system of training dedicated to quality raters. For this reason, the two groups of employees undergo different types of training. The main difference is the duration of training. The same applies to the training of disabled workers for the performance of quality inspection in the company under analysis. The training of hearing-impaired quality raters takes longer, as types of acceptable and non-acceptable defects need to be explained and interpreted into the sign language so that the disabled employees can fully understand them. Samples (approved templates) are used, and the explanation is interpreted into the sign language, as hearing-impaired employees have problems with understanding the contents of presentations, procedures or training instructions.

5 Conclusions

This paper presents results of a study of the level of agreement between quality raters in a sensory inspection, conducted with the participation of hearing-impaired individuals. Workers with hearing impairment perform quality inspection which includes visual inspection of products. The level of agreement between raters was assessed using the MSA procedure for non-measurable characteristics, with the application of Gwet's AC_1 coefficient.

The outcomes of this repeated study show that hearing-impaired employees perform at least at the same quality level as able-bodied employees. This applies to both tests carried out previously in laboratory conditions as well as in production conditions.

The study results validate the development of methods and tools of quality engineering and organisation of work processes aimed at creating the working environment for the disabled. Focus should be put on dedicated training schemes and job aptitude tests evaluating candidates' ability to interpret results and make conclusions, perceptiveness, ability to focus, and personality profile, to enhance employment-readiness of individuals with disabilities. Introducing new solutions for these inspectors, it is worth bearing in mind the support through solutions in the cloud and on other Industry 4.0 oriented principles and technologies [28].

Such an approach is crucial, considering that impeded communication strongly fuels isolation of hearing-impaired individuals and may constrain their ability to deliver performance at an expected quality level [29, 30]. An employee with hearing impairment must be familiar with their job requirements to be able to meet the expectations [14]. Moreover, such an approach promotes equal opportunities for employees with hearing impairment, and helps them perform at the same or even better level than able-bodied employees.

Acknowledgments. The results presented in the paper come from a scientific statutory research conducted at the Chair of Management and Production Engineering, Faculty of Mechanical Engineering and Management, Poznan University of Technology, Poland, supported by the Polish Ministry of Science and Higher Education from the financial means in 2018: 02/23/DSPB/7716.

References

1. Kujawińska A, Vogt K, Hamrol A (2016) The role of human motivation in quality inspection of production processes. Advances in intelligent systems and computing, HAAMAHA, vol 490, pp 569–579
2. Suszyński M, Butlewski M, Stempowska R (2017) Ergonomic solutions to support forced static positions at work. In: MATEC web of conferences, vol 137, p 01015
3. Strykowska K, Starzyńska B (2016) Onboarding of a hearing-impaired quality inspector. Polish Association for Production Management (PTZP) Publishing Company, Opole, vol 2, pp 286–299 (in Polish)
4. The National Census of Population 2011, Population of the disabled. Retrieved from the website of the Office of the Government Plenipotentiary for Disabled People

5. Piekot T (2012) Focused research report – occupational identity of the deaf and their problems on the labour market (in Polish)
6. Kierzek A (2016) Treatment of deafness by the Cléret method in the second half of the 19th century. Otorynolarhinology Clin. Rev. 15(1):28–32 (in Polish)
7. Durko T, Jurkiewicz D, Kantor I, Klatka J (2012) Consensus on the treatment of hearing loss using bone-anchored implants. Pol Otorhinolaryngological Rev 1(1):47–50 (in Polish)
8. Bartnikowska U (2010) Ways of shaping self-esteem in children and young people with hearing impairments. In: Wójcik M (ed) Education and rehabilitation of people with hearing impairments - contemporary challenges. Educational Publisher Akapit (in Polish)
9. Dunaj M (2014) Selected aspects of professional education for the deaf. In: Sak M (ed) Deaf education, Office of the Ombudsman (in Polish)
10. Kowalski P, Nowak-Adamczyk B (2012) Education for deaf and hard of hearing persons - challenges for the education system in Poland. In: Torciuk S (ed) Equal opportunities in access to education for people with disabilities. Analysis and recommendations, Office of the Ombudsman (in Polish)
11. Teper-Solarz Z (2016) Deaf - on the margins of the "hearing world". Univ Sociol J 14(1):37–45 (in Polish)
12. Perkins-Dock RE, Battle TR, Edgerton JM, McNeill JN (2015) A survey of barriers to employment for individuals who are deaf. JADARA 49(2):66–85
13. Scherich DL (1996) Job accommodations in the workplace for persons who are deaf or hard of hearing: current practices and recommendations. J Rehabil 62(2):27–35
14. Scherich DL, Mowry RL (1997) Accommodations in the workplace for people who are hard of hearing: perceptions of employees. J Am Deafness Rehabil Assoc 31(1):31–43
15. Shuler GK, Mistler LA, Torrey K, Depukat R (2014) More than signing: communicating with deaf. Nurs Manage 45(3):20–27
16. Luft P (2000) Communication barriers for deaf employees: needs assessment and problem-solving strategies. Work 14(1):51–59
17. Dunaj M (2014) Deaf people in the labor market. In: The situation of deaf people in Poland. Report of the team for the d/Deaf, Publisher of the Office of the Ombudsman, pp 89–103 (in Polish)
18. Garbat M (2012) Employment and vocational rehabilitation of people with disabilities in Europe. The Publishing House of the University of Zielona Góra (in Polish)
19. State Fund for Rehabilitation of Disabled Persons (PFRON) (2010) Job satisfaction among individuals with disabilities, Warsaw (in Polish)
20. Bartuzi P, Bugajska J et al (2014) Adaptation of facilities, rooms and workplaces for disabled people with specific needs - good practices. CIOP Publishing, Warszawa (in Polish)
21. Zawieska WM (2014) Designing of facilities, rooms and adaptation of workplaces for disabled people with specific needs - frameworks. CIOP Publishing, Warszawa (in Polish)
22. Żołnierczyk-Zreda D, Kurkus-Rozowska B (2012) Professions recommended for people with selected disabilities. CIOP Publishing, Warszawa (in Polish)
23. AIAG-Work Group (2010) Measurement system analysis, 4th edn. Reference manual, AIAG-Work Group, Daimler Chrysler Corporation, Ford Motor Company, General Motors Corporation
24. Starzyńska B, Szajkowska K, Diering M, Rocha A, Reis LP (2018) A study of raters agreement in quality inspection with the participation of hearing disabled employees. In: Hamrol A, Ciszak O, Legutko S, Jurczyk M (eds) Advances in manufacturing. Lecture notes in mechanical engineering. Springer, Cham, pp 881–888

25. Diering M, Dyczkowski K (2016) Assessing the raters agreement in the diagnostic catheter tube connector production process using novel fuzzy similarity coefficient. In: Proceedings of the IEEE international conference on industrial engineering and engineering management, pp 228–232
26. Gwet KL (2012) Handbook of inter-rater reliability. The definitive guide to measuring the extent of agreement among multiple raters, 3rd edn. Advanced Analytics LLC, Gaithersburg
27. Szajkowska K, Kujawińska A, Lis K, Starzyńska B (2017) An ergonomic analysis of the quality inspection workstation for hearing-disabled workers. In: Ergonomics for the disabled. Łódź University of Technology Publishing House, pp 55–70 (in Polish)
28. Varela MLR, Putnik GD, Manupati VK, Rajyalakshmi G, Machado J (2018) Collaborative manufacturing based on cloud, and on other I4.0 oriented principles and technologies: a systematic literature review and reflections. Manage Prod Eng Rev 9(3):90–99. https://doi.org/10.24425/119538
29. Foster S, MacLeod J (2003) Deaf people at work: assessment of communication among deaf and hearing persons in work settings. Int J Audiol 42(Supplement 1):128–139
30. Johnson VA (1993) Factors impacting the job retention and advancement of workers who are deaf. Volta Rev 95:341–356

Electronic Nonconformities Guide as a Tool to Support Visual Inspection

Agnieszka Kujawińska[(✉)], Michał Rogalewicz,
Karolina Szajkowska, Wiktor Piotrowski, and Wojciech Parczewski

Poznan University of Technology, Piotrowo 3, 61-138 Poznan, Poland
Agnieszka.kujawinska@put.poznan.pl

Abstract. The article presents an electronic guide to the nonconformity of parts produced in an enterprise in the automotive industry. In visual inspection, it is very important to be familiar with errors that can potentially be detected at a given control position and clearly define the control criteria. In the automotive industry, in which products must conform to very high requirements, not only in terms of safety of use, but also appearance, repeatability and reproducibility of this product feature is extremely important. What appears to be helpful then is nonconformities catalogues, which contain pictures of possible defects and their descriptions. The innovatory approach of the guide consists in taking into account the needs of deaf and hearing-impaired people working in the quality control department and using Orbitvu 3D to take pictures of nonconformities.

Keywords: Electronic nonconformities guide · Visual inspection ·
Deaf and hearing-impaired people requirements

1 Introduction

Control is an activity that occurs in all phases of product life. It is defined as "a comparison of the actual state of an object with the assumed (required) state" [1, 2]. According to this definition, control is aimed at determining whether the expectations, needs and requirements of the client have been met and if so to what extent. The control process is often identified with the detection of nonconformities [3], but should be treated more broadly. The controllers' task is to detect nonconformities and the reasons for their occurrence, and often to correct deviations from the desired state [4]. It also involves securing and assuring quality by preventing products that do not comply with the arrangements from reaching the client. In order for control to bring the expected results, it should be performed by employees who possess the necessary qualifications, and the requirements should be clearly and precisely specified. Meeting these conditions increases the chance of controllers' assessment to be objective and repeatable.

Production companies use different types of quality control in assessing products' compliance with requirements [5–7]. The choice of the type of control depends on many conditions. There are external conditions, which the company has no influence on, such as regulations, standards or clients, and there are internal conditions. The latter include: the type and form of production organization as well as the specificity of the manufactured

© Springer Nature Switzerland AG 2019
M. Diering et al. (Eds.): *Advances in Manufacturing II* - Volume 5, LNME, pp. 221–238, 2019.
https://doi.org/10.1007/978-3-030-18682-1_18

product, the level of technical and organizational control currently existing in the enterprise, employees' qualifications and other [8–10]. Taking into account different criteria, many quality control classifications can be distinguished [11, 12]:

- Quality control area and object:
  - Product control.
  - Process control.
  - System control.
  - Place control.
- Control planning method:
  - Planned control, i.e. One which results from the control plan.
  - Action control, which is unannounced or random.
- Control performer:
  - Automaton.
  - Operator.
  - Specialist (e.g. Quality control department).
  - Auditor.
- Extent of the control:
  - Random control (statistical).
  - 100% control.
- Time/place of control in the process:
  - Before starting the process.
  - During the process.
  - After the process.
- The way in which data are obtained:
  - Measurement.
  - Organoleptic control (e.g. Visual).

Due to the specific character of the processes being carried out or technical and organizational conditions in the enterprise, it happens that it is mainly non-measurable features of the product that are subject to assessment, i.e. features which are impossible to be expressed using the number and unit of measurement. Then, the main role is played by organoleptic control, which is based on the assessment performed through impressions received by human senses:

- Taste – e.g. Determining tastes: sweet, bitter, sour, salty, umami (according to pn-en iso 5492: 2009 standard, the metallic taste was also considered to be the main taste).
- Hearing – e.g. Determining the noise level.
- Touch – e.g. Determining the degree of softness, hardness, brittleness and consistency.
- Smell – e.g. Determining the freshness of the product, its fragrance, as well as the presence of foreign odours.
- Sight – e.g. Determining the colour of the product, its change, shape, size, etc.

A special type of organoleptic control is visual control (inspection). It is classified among fast and cheap assessment methods - it does not usually require expensive control and measurement equipment. It is usually a non-destructive control, i.e. it does not lead to the wear-out of the item being assessed. Visual inspection uses an alternative assessment, i.e. one during which the state of the product cannot be accurately assessed. One can, however, describe a product or its properties (often several at once) using two opposite mutually exclusive states, e.g. "compliant" and "non-compliant".

Visual control, despite the extensive development of technology nowadays, is still often used in industry. In many cases human participation cannot be replaced by any other more effective form of control [13, 14]. It is valued mainly due to its flexibility during assessment in non-standard situations and its ability to recognize entirely new cases. When making decisions during the control process, it is not only the employee's professional knowledge that matters, but also an individual approach to each item controlled. A well-qualified controller has a high level of sensitivity to nonconformities and is guided by the principle of limited confidence in the manufactured product [14, 15].

Visual control has also weak sides. One of them is the fact that its result does not specify the extent to which the product meets or not the requirements - it is only possible to qualify it as one of the previously mentioned states. In addition, the same human, whose approach is mentioned as an advantage of visual control, is also seen as a disadvantage. The human is subjective in making decisions (which does not happen in the case of quantitative control), and sometimes is unreliable [16]. Considering the above, it is important to limit the influence of other factors on the quality of visual control. Visual inspection, depending on the industry and type of production, requires appropriate equipment, as well as the employee's qualifications and good eyesight. Conditions at the workplace are also important: adequate light intensity, limited noise and vibrations, proper microclimate [17, 18].

In visual assessment, it is very important to be familiar with errors that can potentially be detected at a given control position and clearly define the control criteria. In the automotive industry, in which products must conform to very high requirements, not only in terms of safety of use, but also appearance, repeatability and reproducibility of this product feature is extremely important. What appears to be helpful then is nonconformities catalogues, which contain pictures of possible defects and their descriptions. However, for the guides to fulfil their function while recognizing the defect, they must be designed and made in a neat and clear way, and the pictures of defects should be accurate and of high quality.

The article features an electronic guide to the nonconformity of parts produced in the automotive industry enterprise. The innovatory approach of the guide consists in taking into account the needs of deaf and hearing-impaired people working in the quality control department and using Orbitvu 3D to take pictures of nonconformities.

2 Characteristics of the Analyzed Process and Control Processes

The company in which the study was carried out and for which a nonconformities guide was prepared performs injection and subsequent electroplating processes, i.e. coating plastics.

Galvanic processes take place in three galvanizing machines in which baths are controlled in terms of chemical composition. All products manufactured for the needs of the automotive industry are galvanized in the same machine. In addition, auxiliary tubs are used during the process, the total capacity of which is over 44,000 m^3. The processed products are transferred between the tubs by means of an automatically controlled conveyor.

In the company, quality control is carried out after the electroplating process. This is a 100% control conducted by the controller with the unaided eye on a position specially adapted for this function.

Due to the increase in complaints from clients, assessment was made not only of the production process but also of visual control. It turned out that the effectiveness of control was at the level unacceptable for the management. This became the starting point to assess the ergonomics [19] of workstations and to conduct individual interviews with employees. As part of the conducted study, the following were assessed: methods of work of controllers, lighting of their workstations, noise, vibrations and microclimate.

It was acknowledged that the controller's working method was correct, but the work may be too monotonous (the only departure from monotony is the diversity of controlled products). There is no problem with discomfort or lack of ergonomics.

The controller's workstation (Fig. 1) is illuminated with white light, which is above the head with correct intensity. There is not a spot light source, but it is not needed. The controller's workstation is effectively isolated from the sun with a window blind.

The control process is not exposed to noise and vibrations - all noise-causing processes are located in another part of the production hall.

Microclimate was assessed through the prism of air humidity and temperature at the control station. While humidity was not considered by employees as unfavourable, temperature in the absence of air conditioning was a problem, even despite available fans.

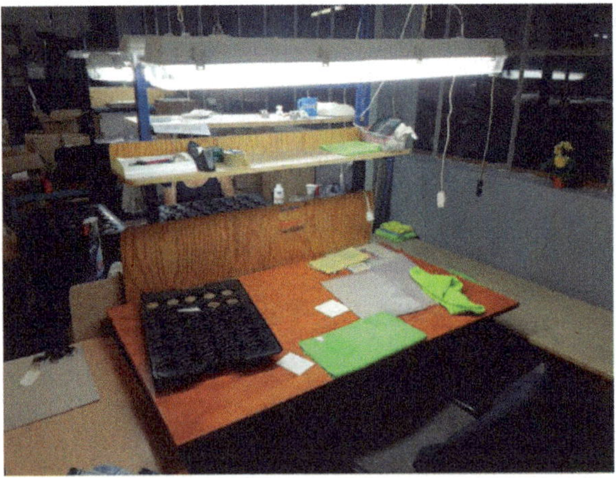

Fig. 1. Quality controller's workstation.

The results of individual interviews with quality controllers confirmed that they do not use the paper-form nonconformities guide (known as the defects catalogue). As the main reason they indicated poor quality of photos in the guide. For this reason, they relied more on their experience rather than on the existing instrument. In agreement with the management of the company, the authors of the article decided that there is a real need to improve the guide, which, as expected, was to contribute to the increased effectiveness of control after the galvanisation process.

3   Design of an Electronic Nonconformities Guide

The section presents the present nonconformities guide used in the company and indicates its weaknesses and constraints. Next, a draft electronic nonconformities guide is presented along with the requirements that were imposed on it. Because these requirements are closely related to the target group to which it is addressed - in large part to the deaf and hearing-impaired - the last part of this section discusses their specificity. It is associated with a special language used by controllers with such a dysfunction.

The last part presents the methodology of designing an electronic nonconformities guide.

3.1   Current Nonconformities Guide

The current nonconformities guide is in paper form (Fig. 2) and contains the following information:

- Type of defect (name).
- Defect no. On the reject slip.
- Verbal description of the defect.
- A photo of the defect.

Due to the frequent inability to correctly take a photo of a nonconformity (the reason is the shiny surface of the product and the size of the nonconformity), it was not possible to assign a photo to each nonconformity.

As one can see, the sizes of the images in the guide are small and unclear, which may make it difficult to use it and discourages the controller. The catalogue also contains verbal descriptions that may be useful for people who hear well, but for hearing-impaired and deaf people they are not very practical. Written language is a difficulty for them, especially for those who have been deaf since birth.

3.2   Requirements for an Electronic Nonconformities Guide

Taking into account the imperfections of the guide currently used in the company and the reluctance of employees to use it, requirements were made for an electronic non-conformities guide that is being developed. On the one hand, they are connected with functional features and simplicity as well as convenience of application, and on the

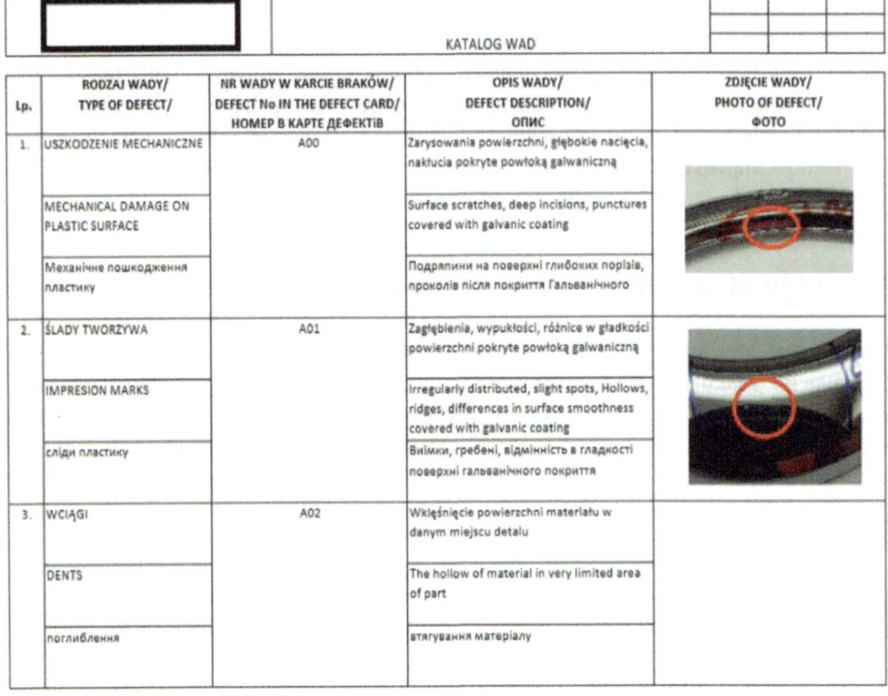

Lp.	RODZAJ WADY/ TYPE OF DEFECT/	NR WADY W KARCIE BRAKÓW/ DEFECT No IN THE DEFECT CARD/ НОМЕР В КАРТЕ ДЕФЕКТІВ	OPIS WADY/ DEFECT DESCRIPTION/ ОПИС	ZDJĘCIE WADY/ PHOTO OF DEFECT/ ФОТО
1.	USZKODZENIE MECHANICZNE MECHANICAL DAMAGE ON PLASTIC SURFACE Механічне пошкодження пластику	A00	Zarysowania powierzchni, głębokie nacięcia, nakłucia pokryte powłoką galwaniczną Surface scratches, deep incisions, punctures covered with galvanic coating Подряпини на поверхні глибоких порізів, проколів після покриття Гальванічного	
2.	ŚLADY TWORZYWA IMPRESION MARKS сліди пластику	A01	Zagłębienia, wypukłości, różnice w gładkości powierzchni pokryte powłoką galwaniczną Irregularly distributed, slight spots, Hollows, ridges, differences in surface smoothness covered with galvanic coating Виїмки, гребені, відмінність в гладкості поверхні гальванічного покриття	
3.	WCIĄGI DENTS поглиблення	A02	Wklęśnięcie powierzchni materiału w danym miejscu detalu The hollow of material in very limited area of part втягування матеріалу	

Fig. 2. Nonconformities guide in paper form currently used in the company.

other hand, they take into account the specific requirements of hearing-impaired and deaf people:

- Simple and intuitive interface.
- The use of graphic icons instead of a verbal description.
- The pictures should show typical examples for nonconformities listed.
- Any nonconformity, both acceptable and unacceptable, should have a reference to the physical standard.
- The guide should clearly distinguish zones of a given part, in situations where it consists of more than one zone.
- Each part available in the guide should have a reference to a nonconformities list containing all possible defects that may occur in the process.
- The guide must work on a computer in the room where quality controllers work.
- The guide should be available for mobile devices and for all employees of the quality control department.
- The guide should contain photographs of nonconformities, both acceptable and unacceptable.
- Descriptions of nonconformities should also be made in Polish, understandable for hearing-impaired and deaf people.

On the basis of the requirements set out above regarding an electronic nonconformities guide, an interface diagram of the future guide was drawn up (Fig. 3).

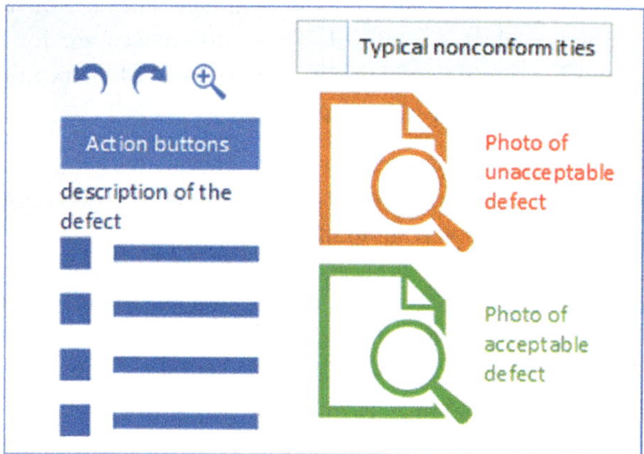

Fig. 3. Diagram of an electronic defects guide.

3.3 Specificity of Requirements for People with Impaired Hearing

There are two basic groups of people with impaired hearing:

- Deaf people - people who do not respond to acoustic stimuli at all, or who react only to very loud stimuli that have no practical significance. Such a hearing loss either cannot be compensated for at all or it can be reduced only to a small extent by using sound amplifying hearing aids. Hearing disorders of such people seriously hinder communication with other people. Their speech is usually indistinct and difficult to understand by other people. Deaf people communicate with others through sign language or lip speech.
- Hearing-impaired people - people with hearing impairments usually between 35 and 69 DB. Hearing impairment results in decreased hearing ability, but in the case of sufferers, the difficulties associated with hearing impairment can be compensated for using hearing aids. These devices do not cause much difficulty in communicating with hearing people. Social barriers for such people are much smaller than for deaf people.

The communication barrier is one of the biggest barriers faced by deaf and hard of hearing people. Unfortunately, it is often neglected by the public, management of workplaces, institutions or offices dealing with social issues.

The basic problem is the result of the fact that deaf people do not communicate in Polish. The natural language of such people is the Polish Sign Language (PSL) - a visual and spatial language used by the Polish community of the Deaf. PSL is not genetically related in any way to Polish. It is a natural language created through historical development, socially and geographically diverse. The articulator is the body of the sign language user. The grammatical status of the language is made up of factors such as facial expressions, pantomime and the position of the body.

Sign language grammar differs from the Polish language grammar. Due to the grammatical dissimilarity of the Polish phonic language and the Polish sign language, full understanding is not always obtained. These differences are for instance [1879] Fr. Józef Hollak and Fr. Teofil Jagodziński - authors of "Słownik mimiczny dla głuchoniemych i osób z nimi styczność mających".]:

- Mimic characters do not undergo any grammatical variations.
- The main mimic sign depicting an object also requires other auxiliary signs to accurately justify the expressed thought.
- Some parts of speech, such as relative pronouns or conjunctions, are not usually used in mimic speech - the deaf-mute form statements based on simple sentences.
- In mimic speech, the first person is always the doer, then the action performed, and finally the object to which the action performed refers.
- In negative sentences, negation is used at the very end.
- In complex sentences, all determiners and complements must be placed next to the words to which they refer.
- Mimics have few basic and primary signs. Sometimes one sign is determined by the combination of two or more basic signs. Deaf and mute people often, however, shorten such statements and use only the main sign.
- When creating personal nouns, the subject is often added first, e.g. "a man to do" is a worker. Non-personal nouns are expressed in the same way as the words from which they originate. It means that words such as: sitting, standing, sleeping, sewing are written as infinitives - to sit, to stand, to sleep, to sew.
- Mimic names do not know any figures of speech.

The grammatical differences mentioned above had to be taken into consideration during the creation of an electronic nonconformities guide, so that its content would be understandable also for deaf and hearing-impaired people. Initially, it was assumed to write separate descriptions to complement what is shown in the picture included in the guide. With time, however, it was decided to standardize verbal content, taking advantage of some similarities between the two languages used.

3.4 Methodology for Developing an Electronic Nonconformities Guide

The nonconformities guide was made in accordance with the following steps of the author's methodology:

STAGE 1: Indication of the part after the electroplating process subject to visual assessment.
STAGE 2: Preparation of a list of nonconformities and their description in accordance with the customer's requirements.
STAGE 3: Preparation of standards with nonconformities.
STAGE 4: Photographing in the 3D ORBITVU device.
STAGE 5: Preparation of the photo database.

STAGE 6: Execution of a guide interface that meets the expectations of people with hearing impairment.

STAGE 7: Implementation of the guide in production practice.

These stages will be developed and described in more detail in Sect. 4.

4 Electronic Nonconformities Guide for the Selected Part

The electronic nonconformities guide was designed in accordance with the previously specified requirements (Sect. 3.2), taking into account the needs of hearing-impaired and deaf people (Sect. 3.3) and according to the methodology presented in Sect. 3.4. The consecutive steps of the methodology to implement the guide in the company are detailed below.

STAGE 1: *Indication of the part after the electroplating process subject to visual assessment.*

The electronic nonconformities guide includes 8 products that are assessed visually following the electroplating process: four of them are manufactured for the needs of the automotive industry, and the remaining ones for the household appliances industry. The following is an example of a product included in the guide: a strip measuring 172 × 25.2 mm produced for passenger cars (Fig. 4).

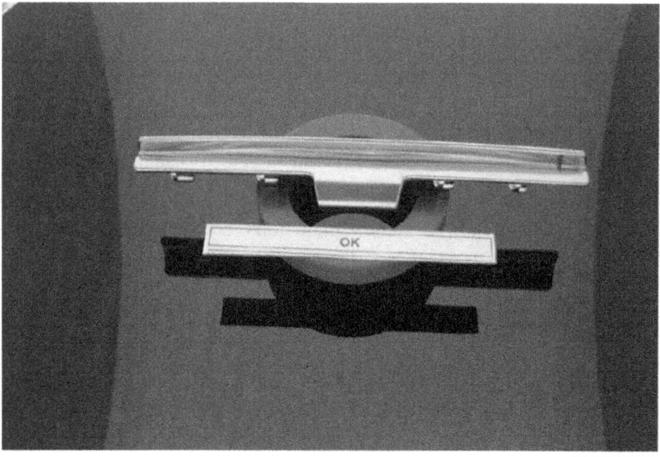

Fig. 4. Product subject to visual assessment after the electroplating process - a strip for a passenger car.

STAGE 2: *Preparation of a list of nonconformities and their description in accordance with the client's requirements.*

The defects catalogue currently used in the company defines which defects may occur both on the plastic and on the coating applied (Table 1). The defects of the plastic are marked with letter A, and the defects of the coating with letter B.

Table 1. The defects catalogue currently used in the company.

Type of defect	Defect description
Mechanical damage (A)	Surface scratches, deep incisions, punctures covered with galvanic coating
Traces of plastic (A)	Hollows, convexity, difference in the smoothness of the surface covered with galvanic coating
Collapses (A)	Material dents
Sears (B)	Dulling of the chromium-plated coating on the edges, milk/white colour, perceptible rough surface
Bubbles (B)	Formation of bubbles under the chromium-plated coating
Peeling (B)	No adhesion of the coating, uneven structure, undulations of the chromium-plated coating
Crater (B)	Holes in the coating
Inadequate chromium-plating (B)	Yellow colour of the coating
Accretions (B)	Flooded opening, protruding material
Abrasions, scratches (B)	Scratching of the coating that does not cause deep damage
Mechanical damage to the coating (B)	Deep scratch in the coating
Stains (B)	Dirty chromium-plated surface (residue of rinse water or galvanic bath)
Foreign material, inclusions (B)	Inclusions of foreign material in the chromium-plated coating, visible and palpable
Passivation (B)	White, milk streaks on the chrome-plated coating
Uneven coating (B)	Lack of chromium-plated coating, visible raw material
Gasification (B)	Small, numerous inclusions in the coating
Other (B)	Dulling of the chromium-plated coating (mps bath), silvering in the coating
Incorrect dimension/shape (B)	–
Sinks (B)	Elements that fell from the hanger during the chromium-plating process

The list of defects was extended by acceptable defects and more detailed criteria of conformity/nonconformity along with the division into product zones. This list served as a starting point for the preparation of standards with nonconformities.

STAGE 3: *Preparation of standards with nonconformities.*

At this stage, standards with nonconformities were prepared, which were then approved by clients ordering the service. Standards contain markings informing about the minimum and maximum size of nonconformity that may appear on the product.

Figure 5 shows an example of the standard that is used in the company. It presents an element that in terms of colour is acceptable (OK) and one that is rejected (NOK).

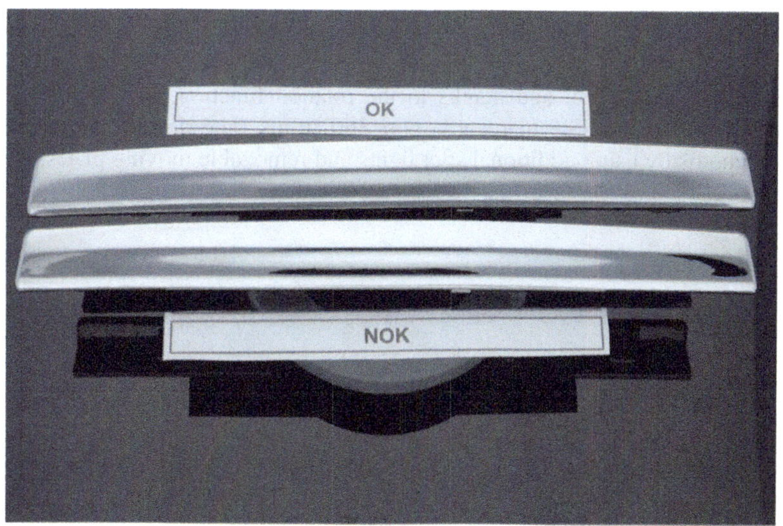

Fig. 5. Colour standard.

STAGE 4: *Taking a photograph in the 3D ORBITVU device.*

Photographs that were placed in the nonconformity database were taken using the ORBITVU 3D Alphashot Micro device shown in Fig. 6.

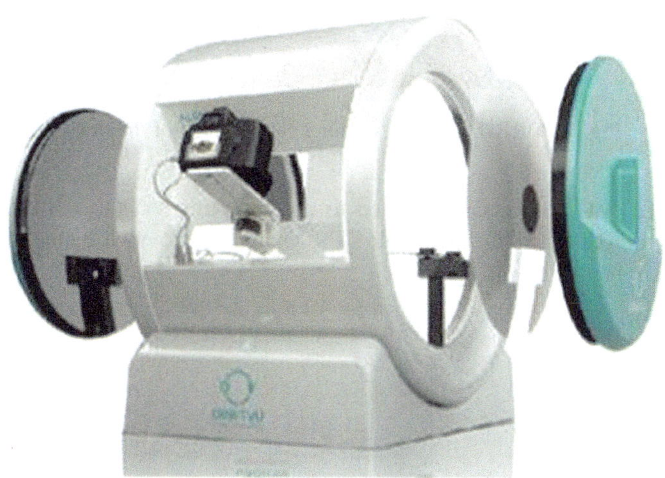

Fig. 6. ORBITVU 3D Alphashot Micro.

ORBITVU 3D is a compact 2D and 360° photo studio for creating professional photography of small objects with maximum dimensions of 18 × 16 ×15 cm. They are most often used for photos of jewellery and watches. Its use as shown in the article is an innovative solution.

The operation of the ORBITVU 3D device is based on the use of a movable lighting system moving up to 300° with the change of the camera position. Objects are placed on a special plate, and thanks to the rotation function, in addition to taking pictures, it is also possible to record short 360° films. The device allows to obtain accurate photos by using additional spot lights and removable moving plates, available in three colours (Fig. 7).

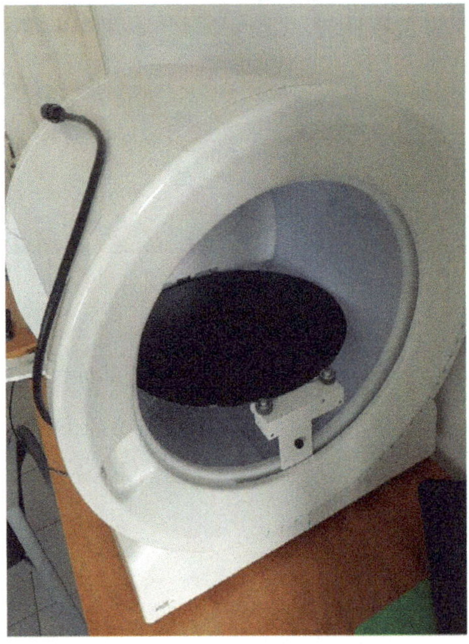

Fig. 7. ORBITVU 3D - working environment of the device.

The ORBITVU device is connected to the computer via the USB cable, and is operated in the ALPHASHOT EDITOR 2 PRO software provided by the manufacturer. Its interface is presented in Fig. 8.

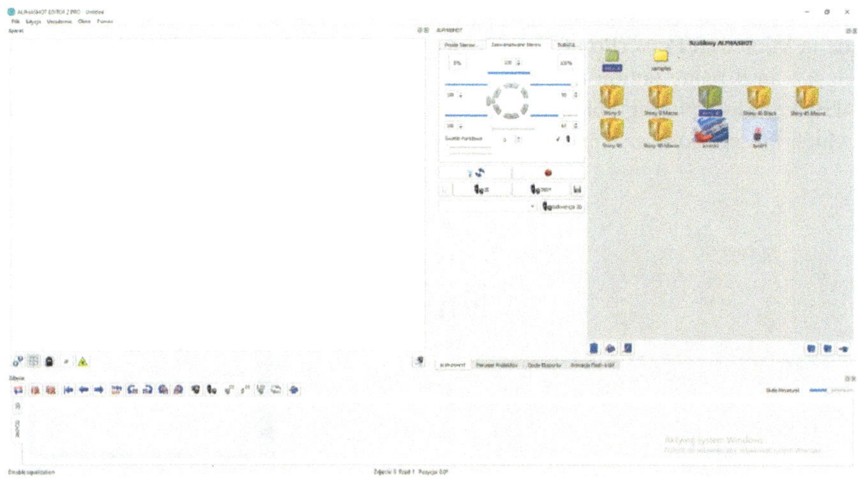

Fig. 8. Interface of the ALPHASHOT EDITOR 2 PRO software.

As previously pointed out, the serious problem that hindered taking clear and high quality images of the products was their glistening. After a series of tests, the device was properly calibrated by modulating the light and using a black-only plate.

Each standard of the defect was photographed at least twice - the first picture shows the overall view of the standard, while the second picture shows a specific defect.

STAGE 5: *Preparation of the photo database.*
STAGE 6: *Execution of a guide interface that meets the expectations of people with hearing impairment.*

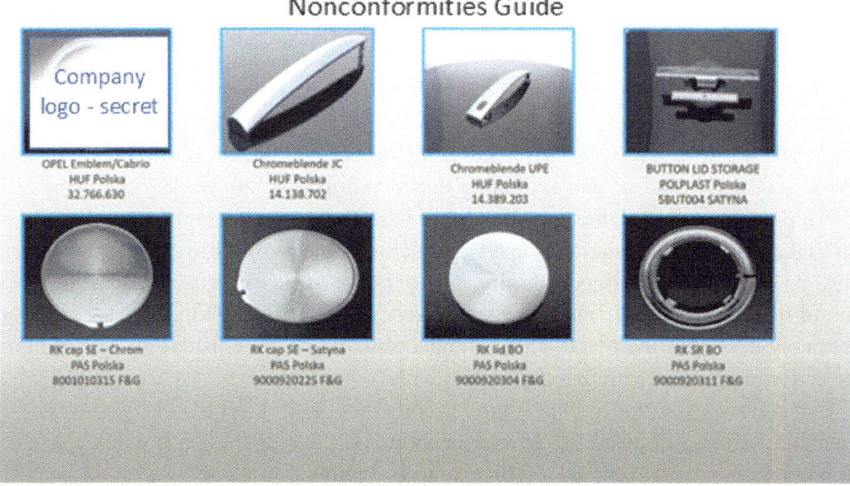

Fig. 9. Main menu of the electronic nonconformities guide.

The electronic nonconformities guide was developed in the form of a presentation made in Microsoft PowerPoint 2016. Its main menu is presented in Fig. 9, and the user navigates in it using action buttons.

Each product photograph is also an action button that takes the user to the window in which the nonconformities list applicable to it is found (Fig. 10).

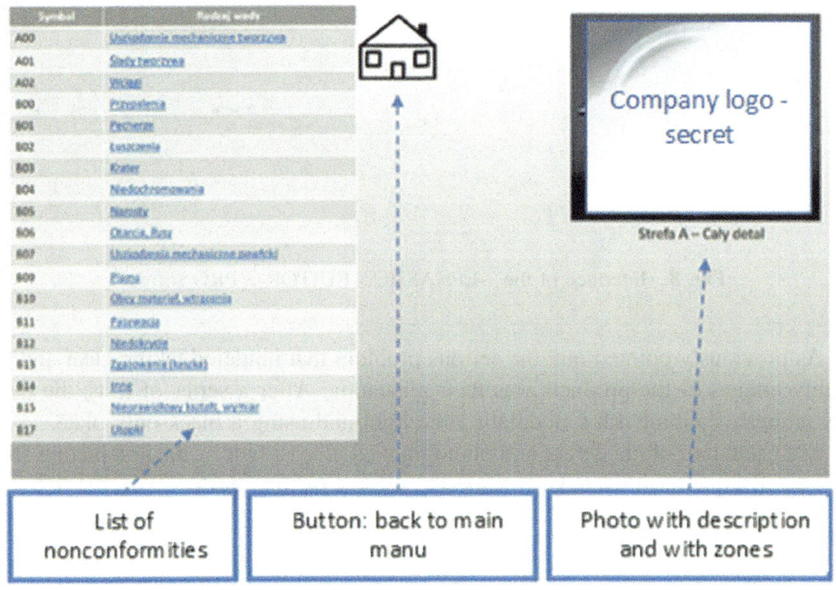

Fig. 10. List of nonconformities for the selected product.

Clicking on the selected defect from the list causes the appearance of its photo with a description (Fig. 11).

The user can return to the list of defects or to the main menu at any time. The user can also learn more about the unacceptable defect by clicking the button with that name. The software allows to view the same defect that occurs on different standards ("Next standard" button) and switching between zones within one product (Fig. 12).

All photographs of defects are displayed with standards so that in case of doubt one can find a defect on the standard itself. To facilitate the work of operators and to better illustrate the nature of nonconformity, photographs of all defects appear in three colours of frames: the black frame presents a general description (Fig. 11), the red frame accurately describes the place where an unacceptable defect exists (Fig. 12), the green frame describes an acceptable defect (Fig. 13).

For correct operation of the guide, a computer with a Windows system with the Microsoft Office suite installed, at least version 2007, is required. The guide also works on smartphones with Android - Office suite is available for free.

STAGE 7: *Implementation of the guide in production practice.*

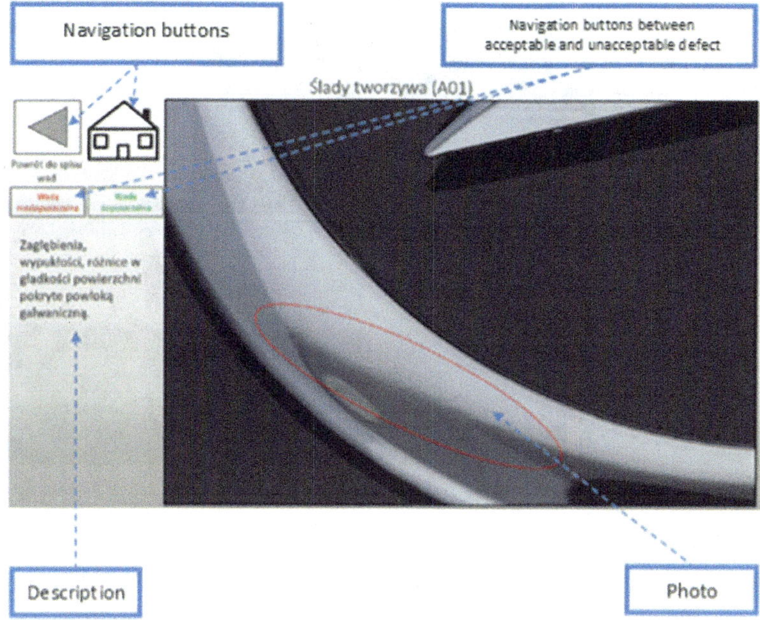

Fig. 11. An example of nonconformity for the selected product.

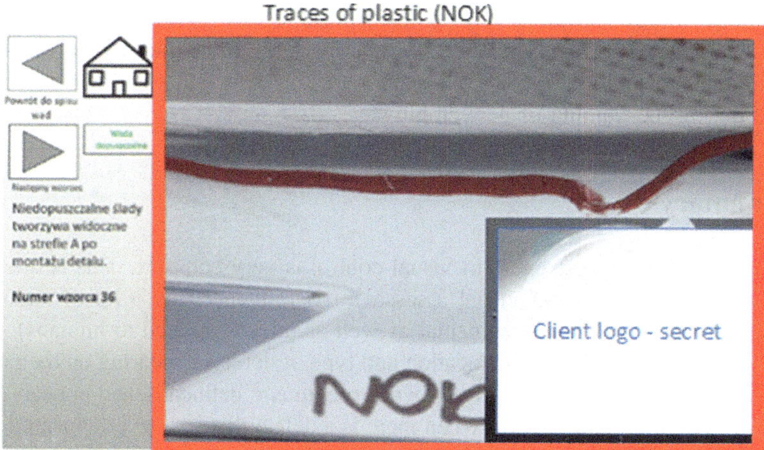

Fig. 12. Photo of an unacceptable defect.

The paper version of the guide in accordance with the employees' opinion in the company was replaced with a multimedia version. In order to be able to implement it, it was necessary to equip quality control stations with devices for reading the guide - tablets.

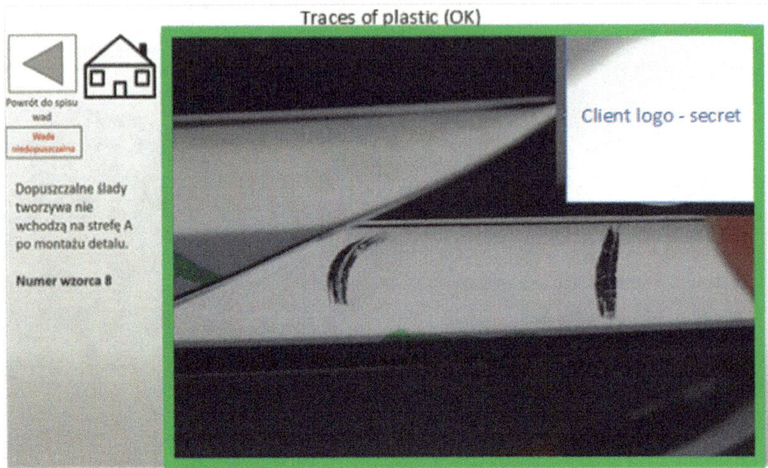

Fig. 13. Acceptable defect photo in the green frame.

The implementation of the guide for the daily work of controllers was carried out within two months. In order to find out opinions on the new nonconformities guide, it was decided to interview nineteen people who are involved in visual control on a daily basis. The selected controllers had different experience and time spent in work.

All the respondents expressed their willingness to use the guide in a new form. When asked why, the controllers responded that the new guide allows them to find information faster. It is helpful in case of doubt because in the multimedia version it is easier to distinguish defects from each other. They indicated that the new guide could be used during the training of new employees.

5   Summary

The problem of the effectiveness of visual control is very complex. This is due to the fact that the effectiveness of visual assessment is influenced by a lot of factors (ergonomic, organizational, environmental as well as directly related to humans). It also depends on the knowledge of the location and type of defects appearing on the product. The nonconformities guide is useful when a previously defined defect appears on the product. However, it is not useful when there is a defect that has never appeared on the product before. The guide may then narrow down the search area for nonconformities.

The nonconformities catalogue in the multimedia version can contribute to increased effectiveness of visual control. High-quality photos are a great convenience for controllers during training but not only for them. In the multimedia version of the catalogue, it is much easier to differentiate individual defects and identify them than in the paper version.

The multimedia version of the guide created using Orbitvu 3D device can prove useful in many production companies because of its wide range of functions and ability to take high quality photos. It may also facilitate work for people with hearing impairment due to the minimal amount of text in the catalogue.

Acknowledgments. The paper is prepared and financed by scientific statutory research conducted by Chair of Management and Production Engineering, Faculty of Mechanical Engineering and Management, Poznan University of Technology, Poland, supported by the Polish Ministry of Science and Higher Education from the financial means in 2018 (02/23/DSPB/7716).

References

1. Sagi JS (1998) The interaction between quality control and production. SSRN Electron J. https://doi.org/10.2139/ssrn.113769
2. Hamrol A (2017) Management and quality engineering. Warsaw (PWN)
3. Hamrol A (2015) Strategies and practices of efficient operation. Lean Six Sigma and other, Warsaw (PWN)
4. Benbow DW, Berger RW, Eishenawy AK, Walker HF (2002) The certified quality engineer handbook. ASQ Quality Press, Milwaukee
5. Sika R, Rogalewicz M (2017) Demerit control chart as a decision support tool in quality control of ductile cast-iron casting process. In: MATEC web of conferences, vol 121, p 05007. https://doi.org/10.1051/matecconf/201712105007
6. Włodarczyk K, Kowalczyk J, Ulbrich D, Selech J (2017) A review of non-destructive evaluation methods of elements of prototype module of drying line used to receive RDF fuel from waste recycling. Procedia Eng 192:959–964. https://doi.org/10.1016/j.proeng.2017.06.165
7. Vieira G, Reis L, Varela MLR, Machado J, Trojanowska J (2016) Integrated platform for real-time control and production and productivity monitoring and analysis. Rom Rev Precis Mech Opt Mechatron 50:119–127
8. Knop K (2017) Analiza udziału i znaczenia stosowanych metod kontroli jakości do wykrywania niezgodności profili aluminiowych. Systemy wspomagania w inżynierii produkcji. Jakość, Bezpieczeństwo, Środowisko 6(7):129–142
9. Starzyńska B, Szajkowska K, Diering M, Rocha A, Reis LP (2018) A study of raters agreement in quality inspection with the participation of hearing disabled employees. In: Hamrol A, Ciszak O, Legutko S, Jurczyk M (eds) Advances in manufacturing. Lecture notes in mechanical engineering. Springer, Cham, pp 881–888
10. Jasarevic S, Diering M, Brdarevic S (2012) Opinions of the consultants and certification houses regarding the quality factors and achieved effects of the introduced quality system. Tehnicki Vjesnik-Technical Gazette 19(2):211–220
11. Kujawińska A, Vogt K, Diering M, Rogalewicz M, Waigaonkar SD (2018) Organization of visual inspection and its impact on the effectiveness of inspection. In: Hamrol A, Ciszak O, Legutko S, Jurczyk M (eds) Advances in manufacturing. Lecture notes in mechanical engineering. Springer, Cham, pp 899–909
12. Kujawińska A, Vogt K (2015) Human factors in visual quality control. Manage Prod Eng Rev 6(2):25–31. https://doi.org/10.1515/mper-2015-0013
13. Chi CF, Drury C (2001) Limits to human optimization in inspection performance. Int J Syst Sci 32(6):689–701

14. Jiang X, Gramopadhye AK, Melloy BJ, Grimes LW (2003) Evaluation of best system performance: human, automated, and hybrid inspection systems. Hum Factors Ergon Manuf Serv Ind 13(2):137–152
15. Lee J, Ko KW, Lee S (2016) Visual inspection system of the defect of collets for wafer handling process. Int J Control Autom 9:129–138
16. See JE (2015) Visual inspection reliability for precision manufactured parts. Hum Factors 57:1427–1442
17. Kleiner BM, Drury CG (1992) Design and evaluation of an inspection training program. Appl Ergon 24:75–82
18. See J, Drury CG, Speed A, Khalandi N (2017) The role of visual inspection in the 21st century. Proc Hum Factors Ergon Soc Ann Meet 61(1):262–266
19. Suszyński M, Butlewski M, Stempowska R (2017) Ergonomic solutions to support forced static positions at work. In: MATEC web of conferences, vol 137, p 01015

Use of White Light and Laser 3D Scanners for Measurement of Mesoscale Surface Asperities

Bartosz Gapiński[1(✉)], Michał Wieczorowski[1],
Lidia Marciniak-Podsadna[1], Natalia Swojak[1], Michał Mendak[1],
Dawid Kucharski[1], Maciej Szelewski[2], and Aleksandra Krawczyk[3]

[1] Faculty of Mechanical Engineering and Management,
Institute of Mechanical Technology, Poznan University of Technology,
Poznań, Poland
bartosz.gapinski@put.poznan.pl
[2] ITA Polska Sp. z o.o. sp.k., Poznań, Poland
[3] Huta Bankowa Sp. z o.o., Dąbrowa Górnicza, Poland

Abstract. Depending on the purpose of a manufactured part, quality assessment is performed within macro-, micro-, or mesoscale. The first two apply strictly to geometry and surface texture (topography or profile) measurements, respectively. The latter however, applies for measurements of features that do not belong strictly to either of the two scales mentioned above, and its boundaries are ambiguous in their nature, often overlapping the two basic scales. In this paper the authors have undertaken the assessment of the possibility to use macro-scale dedicated scanners for surface asperities measurement – features usually considered as micro-scale. The choice of the measurement systems was inspired by the limitations of previously used roughness measurement systems, which lacked a sufficient vertical range. At the same time, due to high roughness values, large areas have to be measured, which would take significant amount of time, when performing traditional roughness measurement. Several scanners were evaluated: white and blue structural light, laser, and laser with tracking device. The collected data were compared to the results obtained from a specialized Coherent Scanning Interferometer (CSI). The final results have undergone a two-level assessment: qualitative, which compared the quality of whole surfaces; quantitative, using surface texture parameters. The results allow to state that 3D scanners can be used in some cases for surface asperities assessment, however most promising values were obtained, when using devices with higher resolution.

Keywords: Mesoscale · Surface topography · 3D scanner · CSI

1 Introduction

Selecting an appropriate measurement device to perform certain tasks is one of the primary decision when planning a measurement. The more restrictive tolerance of individual features are, the more accurate measurement systems are needed. It is always

© Springer Nature Switzerland AG 2019
M. Diering et al. (Eds.): *Advances in Manufacturing II* - Volume 5, LNME, pp. 239–256, 2019.
https://doi.org/10.1007/978-3-030-18682-1_19

a question which particular measurement system should be used, taking accuracy and cost-effectiveness ratio into account. Also nowadays, the economical aspect of measurement is still strongly determinant, even though the awareness of the importance of quality control is, in most cases, on satisfactory level. These two conditions became a motivation to conduct research on the possibility to measure topography on elements that qualify for meso scale surface measurement with systems dedicated for macro scale geometry rather.

2   Research Problem

Research was conducted on surfaces of steel element after forging, hot rolling and shot blasting manufactured by Huta Bankowa (Fig. 1).

Fig. 1. An example of analyzed element.

Evaluation of such surface might be crucial for several reasons. First of all, technological process planning is determined by information of the surface condition while the assumption of material surplus in subsequent technological operations are considered. In presented surfaces, the differences of surface points in vertical axis are significantly higher than commonly accepted and applied in mechanical engineering. Therefore, application of typical instruments dedicated to measurement of topography is limited and highly time consuming which results in its low cost-effectiveness. Typically, measurement systems are divided into those designed for measurement of geometrical elements (macro-scale), and topography and roughness (micro- and nano-scale). In this case, observed surface asperities are too high to qualify them as micro scale, whereas, its character, distribution and values do not allow to classify them as macro scale. Therefore, a new range of dimensions – the meso scale, proposed by Wieczorowski [1], was used.

In the last few years, there has been a tendency to expand the field of application of measurement systems. Among traditional macro coordinate measurement machines some solutions emerged that allow to measure geometrical features and surface topography. It transfers the border of a CMM to measure not only in macro but also in meso and micro scale. At the same time, coherence scanning interferometers with high range allow to measure not only surface parameters, but also geometrical features in a range of several dozens of millimeters.

Digitization of surface allows for its representation as a point cloud. Reproduction of surface character is strongly correlated with its nature in all scales. In this aspect, full interaction between the scales takes place, which forces the operator (who himself creates an influential factor) to take it into account, because – next to the ambient conditions – it has a crucial impact on the obtained results of measurements (Fig. 2).

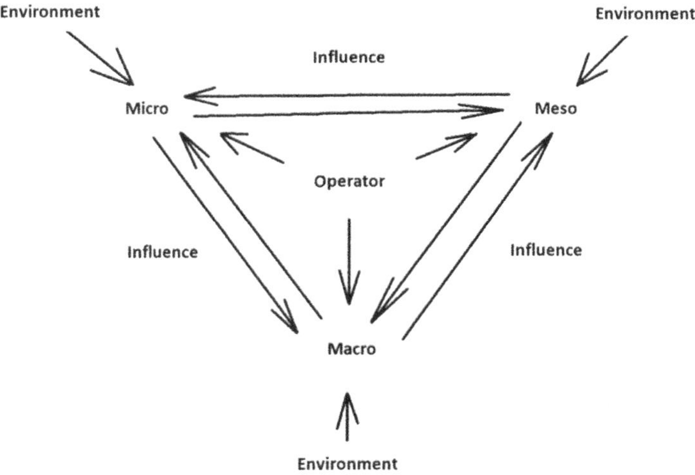

Fig. 2. Relation between scales in metrology.

In order to get comprehensive information about the measured object, it is necessary to analyze it in various scales. In this way, metrological verification covers geometrical features, shapes, and surface asperities.

Macro-scale digitization involves collecting coordinates of measuring points to determine geometrical parameters. These can be geometric figures, solids as well as free form surfaces, which are verified based on the coordinates of their points.

Micro and nano-scale digitization of surface area is under completely different conditions and purposes. Its aim is to highlight the most important features of surface asperities, rather than geometry, by calculating the reference element, functional characteristics and values of certain parameters. In this aspect the scale distinction has significant meaning. It is implemented as separation of individual components from the signal obtained from the surface.

By analyzing signal irregularities, one can focus on both large and small scale, depending on the band of frequencies that were used. The standards related to surface topography include the S filter that removes small-scale horizontal components from the original surface and the L filter that removes large-scale components. All parameters refer to a surface of limited scale, which can be considered as the S-F surface (after removing the shape) or S-L surface (after removing shape and waviness). This is the first level to extract the scale from the signal, which might not be sufficient in modern production reality. Some irregularities on the examined surface are strongly influenced by various phenomena and processes. Surface metrology allows them to be distinguished precisely by using the appropriate scale. The most common techniques are filtration, discrete wavelet transformation (DWT) and discrete modal decomposition (DMD). Using a very different frequency range, a number of its components can be separated from the surface, as shown on Fig. 3.

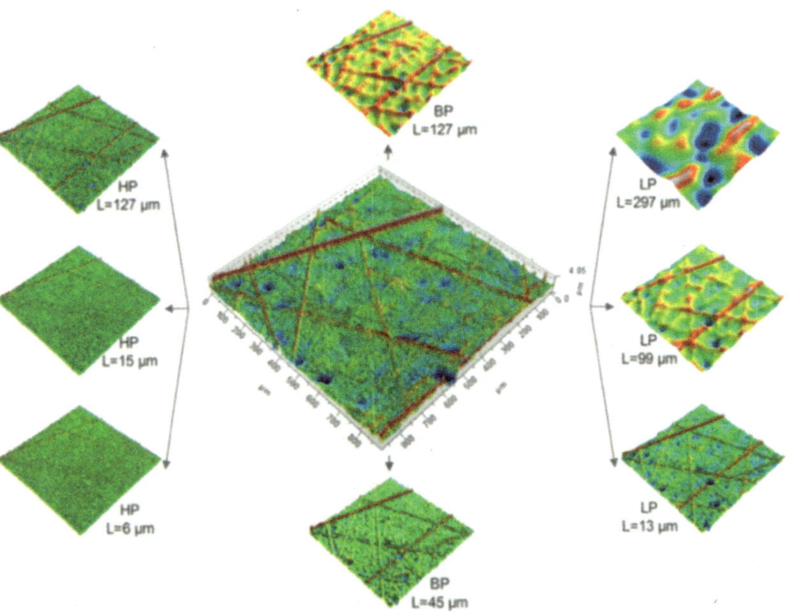

Fig. 3. Highlight of various surface features using different filters [3].

This approach to surface asperities measurement is also a part of the recently developed information-rich metrology which uses not only the measurement results, but also knowledge related to its determinants [2].

Dimensional metrology at various scales allows for the digitalization of the elements which locates it within the big data philosophy [4]. Increasingly often the manufactured element is analysed on the basis of a deviation map which enables to verify its quality holistically at various scales.

All presented aspects became a motivation for the research team to test how different measurement devices might be applied to digitize workpiece surface in meso scale, since choosing the right measurement technique might contribute to lower impact of the operator one the measured value [5].

2.1   Research Methods

Measurement results, presented further, were acquired using 3D scanners differing from one another – among others – by various methods of measurement data acquisition and various measuring strategies. Due to the symmetry of an object under study and the lack of geometrical characteristic shapes, the matching of respective scans is based solely on reference points. In order to achieve the best possible results for every single device, measurement parameters providing the best accessible resolution were selected with caution in each case. The Polytec TopMap TMS 500 coherence scanning interferometer was used as a reference device. Figure 4 illustrates the measuring instruments used in our study.

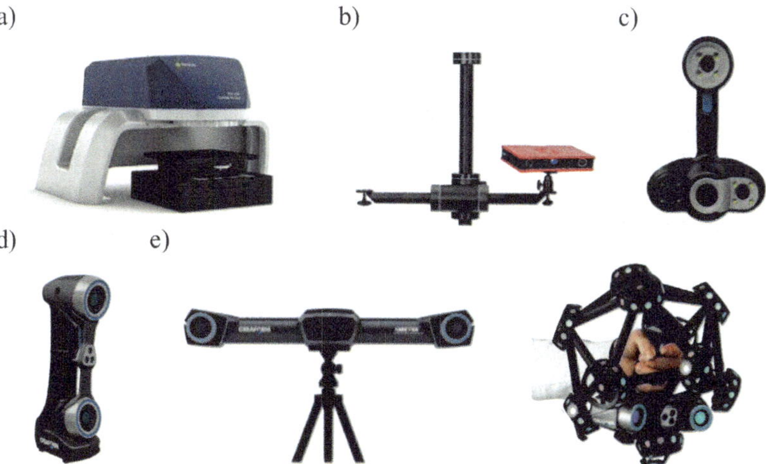

Fig. 4. Measuring instruments used in the study: (a) TMS-500 TopMap Pro.Surf - Polytec [5], (b) Atos Core 185 - GOM [6], (c) Go!SCAN 50 - CREAFORM [7], (d) HandySCAN 700 – CREAFORM [8], (e) MetraSCAN 750 - CREAFORM [9].

A reference measuring instrument (Fig. 4a) enables to perform a quick and precise determination of surface topography parameters characteristic for studied surfaces. The vertical measurement range of the interferometer is 70 mm, while measurement area dimensions reach 44 × 33 mm for a single scan. It is possible to broaden the area up to 220 × 230 mm if needed. Lateral measurement resolution is 28.2 μm, while the calculated optical resolution is 16 μm [6].

One of the devices used in surface topography comparative analysis (Fig. 4b) is a scanner with its action based on the projection of numerous and various structures of a high-bandwidth blue light. The aforementioned structures differ from one another in terms of density, number and location of fringes. A measurement head was used, with a projector that displayed a rectangular measurement area of 185 × 140 mm onto the studied area. Cameras, placed on both sides of the projector, acquired information about the location of points, based on the fringe distortion resulting from the shape of a studied object. This was the only study-implemented 3D scanner which did not allow for real-time data visualization [7].

One of the used hand scanners (Fig. 4c) is a device based on structural white light measurement technology. Its mode of action entails the projection of pattern in the form of rectangles onto the measurement area and the analysis of their deformities. The instrument acquires up to 550 thousand of points per second. The scanner functions, with regards to the acquisition of surface structure and colour, were not used for the experimental aims presented in this article. The measuring capacity of the scanner was 380 × 380 × 250 mm, while the maximum permissible error (MPE) was on the level of 0.1 + L/40000 mm. Resolution of this device was equal to 0.5 mm [8].

Also, scanners based on laser beam technology as a light source were used in our study. One of them was a device using 7 laser crosses (Fig. 4d) that allowed for data recording in the form of a cloud of points due to laser triangulation. The measurement volume of the presented device was 275 × 250 × 300 mm, while the maximum permissible error (MPE) was 0.02 + L/40000 mm. This scanner performs 480 thousand measurements per second, and its resolution goes down to 0.2 mm [9].

The last device (Fig. 4e) was a laser scanner equipped with a measuring head and an instrument tracking its location within a measurement volume. This solution enables scanning of objects without using any reference points. The measuring volume of the head projecting 7 laser crosses was 250 × 250 × 200 mm, and the resolution used in the study was 0.2 mm. The use of a tracking system allowed us to broaden the measurement space up to 16.6 m^3. The measuring velocity of the scanner was 480 thousand dots per second, while the maximum permissible error (MPE) was 0.044 + L/40000 mm [10].

3 Results

3.1 Qualitative Assessment

Two surfaces of two steel rings were selected for analysis, each measured with five different measurement systems. The qualitative assessment consists of comparison of the shape of workpieces. Each of the measurement systems uses different data acquisition technique and is featured by different resolution, which in turns has a significant impact on its ability to detect small surface asperities. This fact is clearly visible while evaluating valleys and peaks, since the lower the resolution is, the worse becomes the reproduction of the surface. One must always take into account, that, when operating within mesoscale, surface asperities have much higher values than commonly observed surfaces in industry applications.

Surfaces, that have undergone the analysis were created in the processes of hot-forging and hot-rolling. The analysis area was about 80 mm × 80 mm, which represents a much larger area, than conventionally considered. However, in this specific application, the use of a smaller area would result in loss of crucial information regarding surface characteristics of the steel ring.

In the first ring, the maximum difference between the peak height and valley depth is not greater than 1,9 mm. Each measurement system was able to reproduce the general characteristic of the surface, in which the regions close to edges are higher than central region (Fig. 5). In Fig. 5(a), where the result of the measurement from the handheld white-light 3D scanner is illustrated, multiple single asperities are noticeable. These may be the result of the light scattering and other measurement disturbance. This effect has not been seen for other measurement systems. One may also notice the effect of a much higher vertical resolution of a CSI, combined with consistent lateral resolution, which results in significantly more detailed surface reproduction. It may be assumed, that in every case, the gathered data allows for a qualitative surface structure assessment and its characteristic details are alike regarding both location and values.

Important part of Surface 3 is its forged marking which was executed during hot forging process (Fig. 6). Another characteristic feature is the hill located on both rounded sides of the element, with the smaller side (internal radius) being significantly higher. There is a radial valley in a central area.

Similarly to Surface 1, Surface 3 showed the greatest irregularities when measured with handheld white light 3D scanner. Its resolution seems to be not sufficient for precise reproduction of the markings shape and depth. A practical would be, that the localization of the marking may be derived from the measurement, but it is illegible. Other measurement methods obtained a legible marking, however the most detailed one was obtained using a CSI (Fig. 6e).

a)

b)

c)

d)

e)

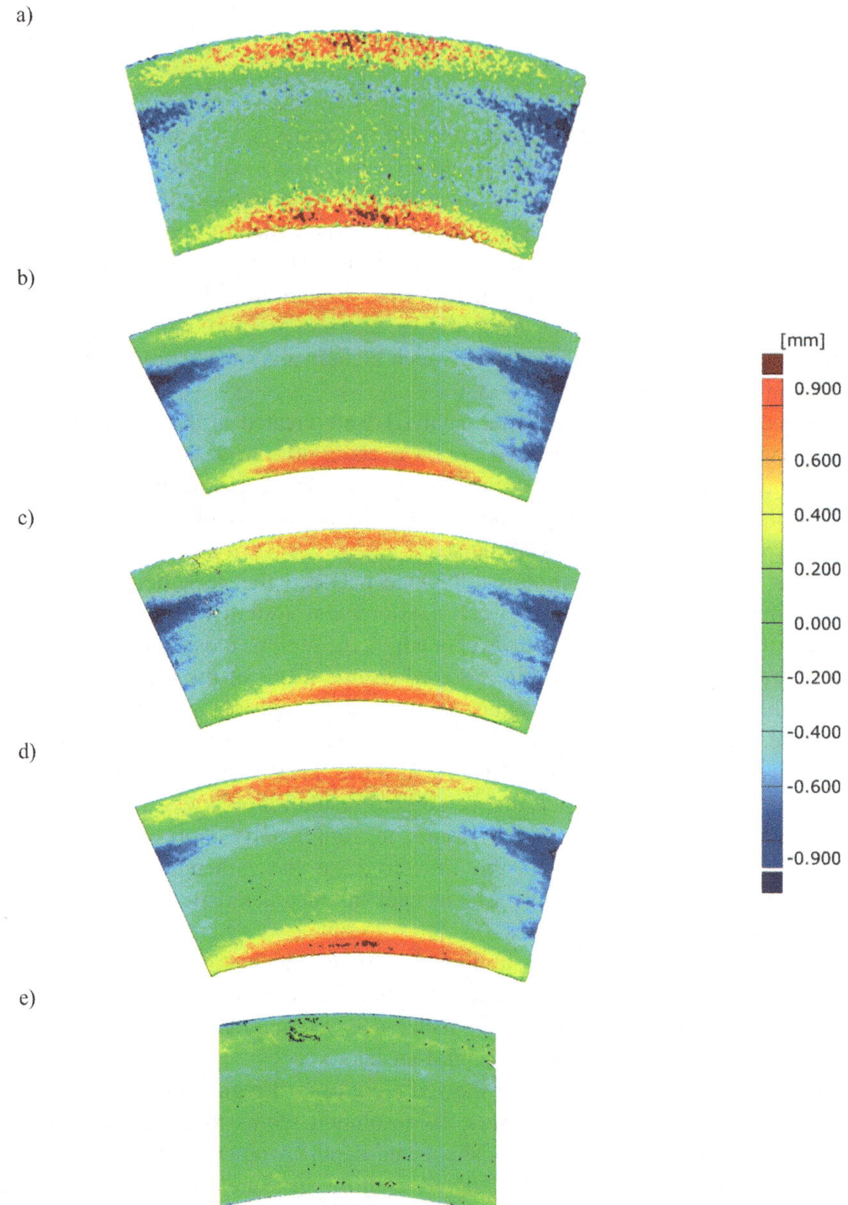

Fig. 5. A juxtaposition of Surface 1 measurement results: (a) handheld white light 3D scanner; (b) handheld laser scanner; (c) laser scanner with tracking system; (d) blue, structural light 3D scanner; (e) Coherent Scanning Interferometer (CSI).

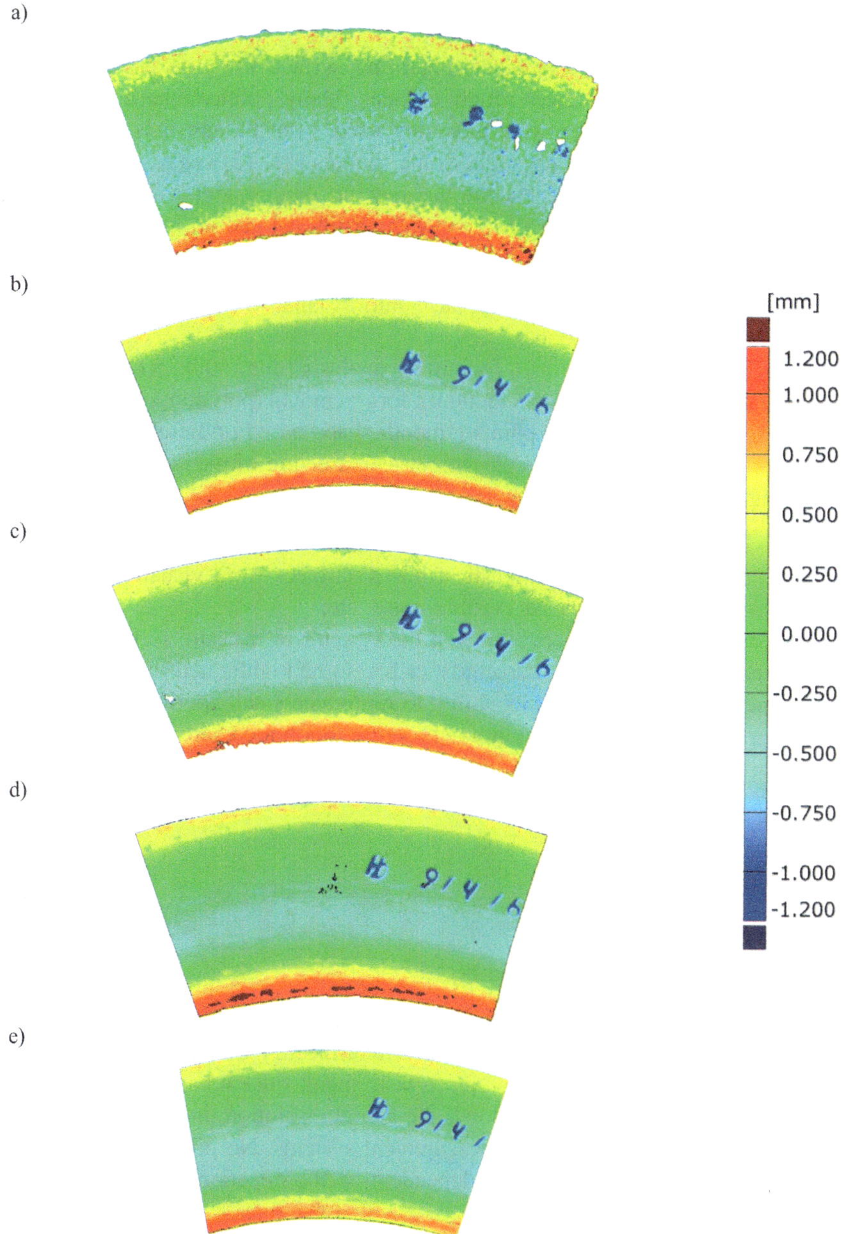

Fig. 6. A juxtaposition of Surface 3 measurement results: (a) handheld white light 3D scanner; (b) handheld laser scanner; (c) laser scanner with tracking system; (d) blue, structural light 3D scanner; (e) Coherent Scanning Interferometer (CSI).

3.2 Quantitative Assessment

A certain inconsistency in references, regarding surface filtration can be observed, particularly in case of extremely rough surfaces. Many publications do not mention which, or even whether there were any filter used, even though it is crucial for the means of reproducibility of a research. Some sources [11, 12] suggest that a filtration method should be entirely dependent the specific requirements of the application. Nevertheless the authors decided to analyse both filtered and unfiltered surfaces. For this purpose an algorithm was designed, partially based on previous publications [13]:

- Importing the files of measured surfaces into MountainsMap software.
- Initial levelling of the surface.
- Choice of a representative area for further analysis (min. 50 × 50 mm).
- Primary levelling (least squares method), form removal (polynomial of fifth degree).
- Filling in the non-measured points (smooth shape based on neighbouring points).
- Export of results and analysis of geometrical structure of the surface.
- Filtering, Gaussian robust filter, lc = 2,5 mm.
- Export of results and analysis of geometrical structure of the surface.

Before applying the algorithm, all acquired surfaces were visually examined for abnormal artefacts, such as peaks, visible noise, etc.

As far as estimating the capabilities of the systems to measure surface topography are concerned, certain surface characteristics had to be highlighted in order to analyse the efficiency of the system and its ability to reproduce small aperities on the surface. For this particular analysis, the authors chose the following characteristics:

- Bearing Curve.
- Texture direction.
- Functional Parameters Spk Svk Sk.
- Feature Parameters: Spc, Sdv, Shv.
- Skewness.
- Kurtosis.
- Areal Surface Roughness Sa and Sq.

The selected parameters will be used for the assessment for the capability of the systems to reproduce the asperities of a given surface. A Coherence Scanning Interferometer (CSI) will be used as a reference instrument, since it has significantly higher vertical resolution. The limiting factor is the lateral resolution, in case of CSI, and the volumetric resolution for all other measurement systems. No profile parameters were chosen, since their values are highly dependent on the direction of the profile.

3.2.1 Abbott – Firestone Curves

The obtained data were used to calculate the bearing area curves of the measured surfaces, since they provide information regarding both height and distribution of the peaks [14]. In Table 1, the bearing area curves for different devices are presented. Complementarily to the images of the curves, bearing ratio parameters *Sk, Spk, Svk* are also presented as *functional parameters*. The scanners generally represented the behaviour of bearing area curve in a similar way as CSI, still some differences are visible.

Table 1. Bearing area curves for Surface 1.

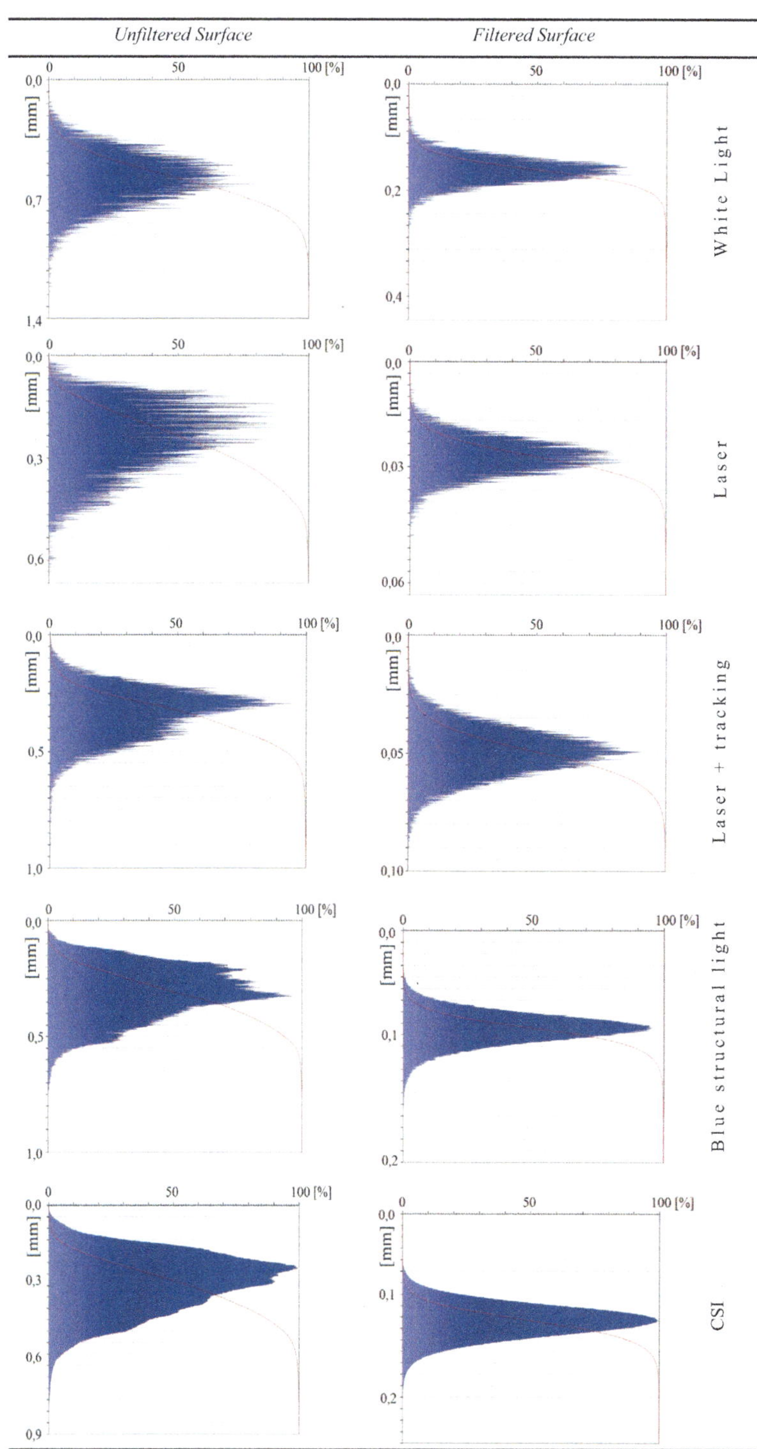

3.2.2 Texture Direction

The next table presents analysis of texture directionality. This was realized by means of texture direction plots showing the main and remaining directions of asperities. The differences for various scanners show how well they can represent a measured surface (Table 2).

Table 2. Texture direction of Surface 1.

Unfiltered Surface	*Filtered Surface*	
90° / 180° / 0°	90° / 180° / 0°	White Light
90° / 180° / 0°	90° / 180° / 0°	Laser
90° / 180° / 0°	90° / 180° / 0°	Laser + tracking
90° / 180° / 0°	90° / 180° / 0°	Blue structural light
90° / 180° / 0°	90° / 180° / 0°	CSI

3.2.3 Feature Parameters

The charts listed above were obtained using the data from each measurement system for both filtered and unfiltered surface. Arithmetic mean peak curvature does not differ significantly between both surfaces, therefore its chart appears relatively similar. Mean dale volume and mean hill volume decreased roughly of one order of magnitude.

Similarity of the corresponding charts (Surface 1 and 3, both filtered and unfiltered) is clearly visible (Fig. 7).

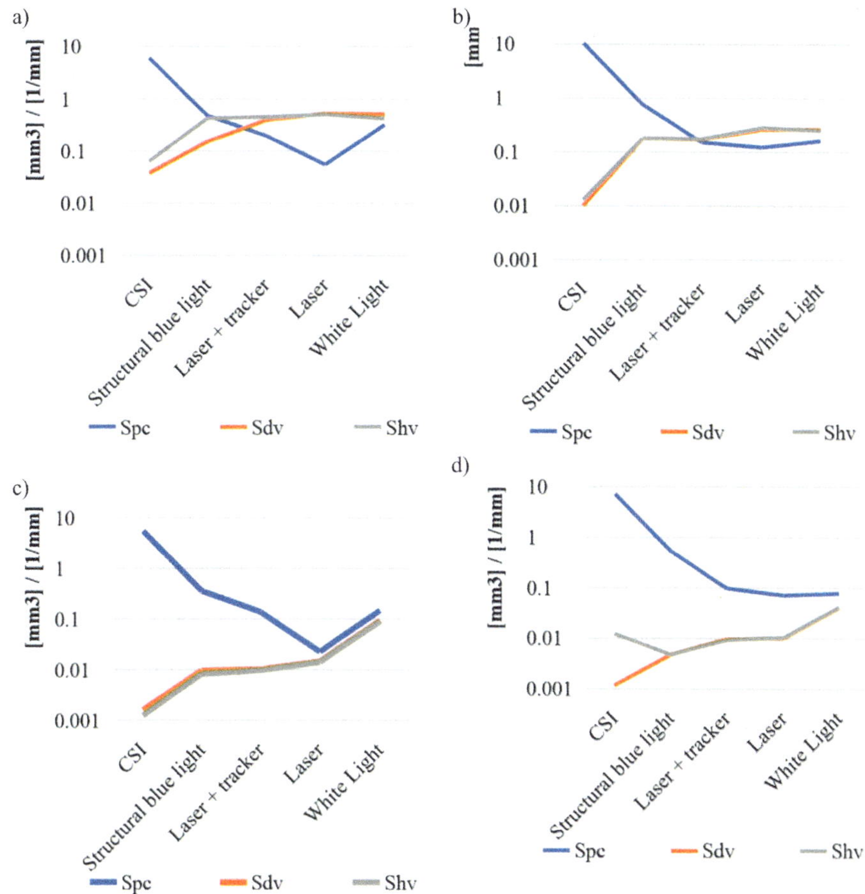

Fig. 7. Feature Parameters: (a) Surface 1, unfiltered; (b) Surface 3, unfiltered; (c) Surface 1, filtered; (d) Surface 3, unfiltered.

3.2.4   Kurtosis

The spread of the height distribution in Surface 1 (unfiltered) is nearly Gaussian (Kurtosis = 3). After applying the robust filter, Kurtosis increased in value (up to 6,52 for white light system, except for system c), where it decreased by 0,5. Surface 3 however, has shown much higher values of Kurtosis (unfiltered), which then decreased, after filtration, by more than half, with an exception of white light system, that only by the value of 0,84 (Fig. 8).

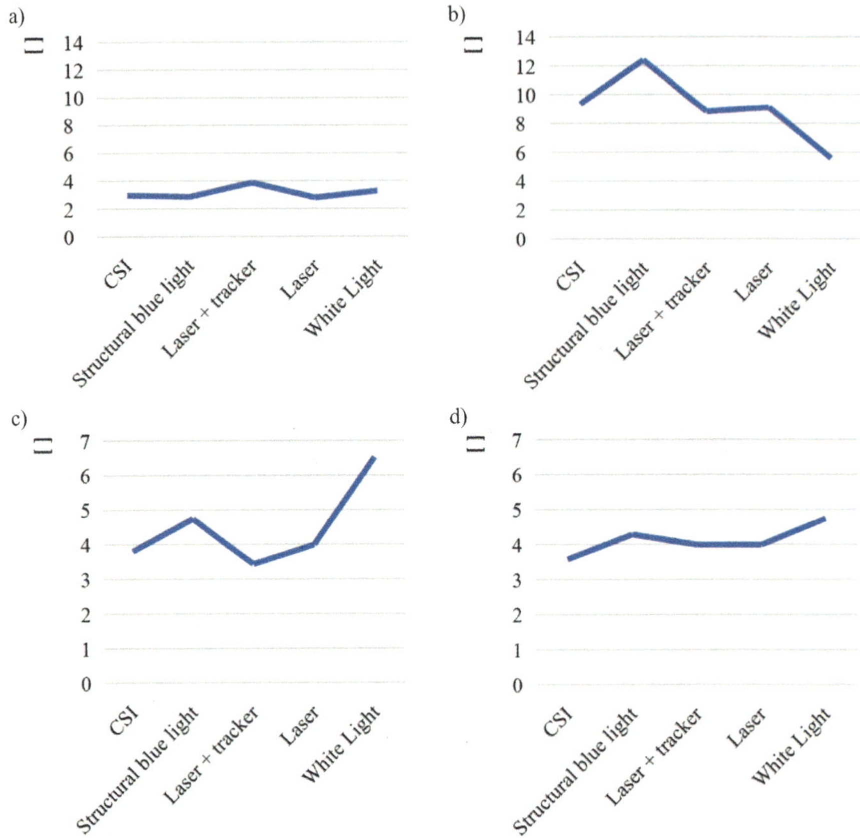

Fig. 8. Kurtosis of the measured surfaces: (a) Surface 1, unfiltered; (b) Surface 3, unfiltered; (c) Surface 1, filtered; (d) Surface 3, unfiltered.

3.2.5   Functional Parameters

All Functional parameters (Fig. 9), in both surfaces, show comparable trends, in which the values decrease for the first four consecutive measurement systems and rapidly increase for the latter one. The calculation process of these parameters already has a built-in filtration, therefore only the results from the unfiltered surfaces were used. Surprisingly, the measurement system (d) showed bigger differences from the reference values than (c) despite having nearly twice as good volumetric resolution.

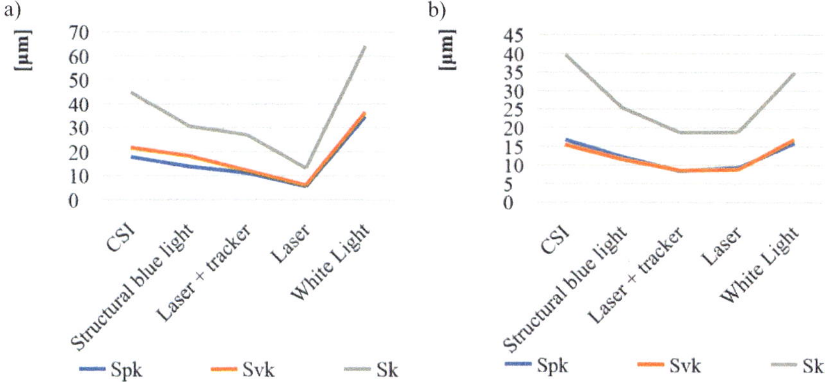

Fig. 9. Functional parameters: (a) Surface 1; (b) Surface 3.

3.2.6 Skewness – Ssk

The chart (Fig. 10) above illustrates the values of Skewness for different measurement methods. The skewness itself describes the shape of the topography height distribution and is very susceptible to isolated valleys and peaks. For Gaussian height distribution, Ssk = 0. Negative values of skewness indicate higher amount of peaks in the surface topography. One might assume, in terms of comparison of the measurement method, that skewness should remain negative or positive, depending on the reference value. That did not happen in two cases: in measurement of Surface 3, both unfiltered (measurement system c) and filtered (measurement system e).

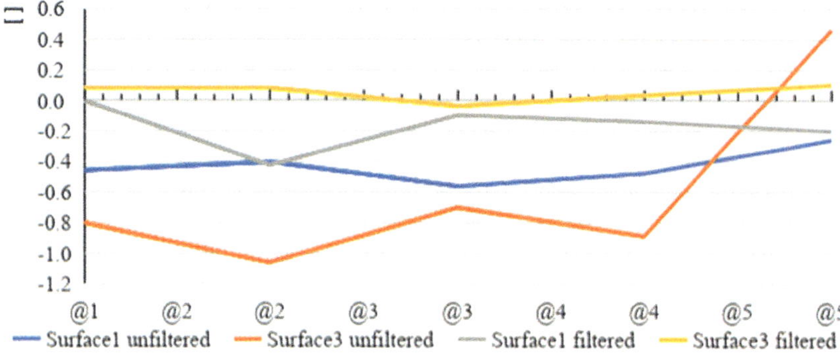

Fig. 10. A juxtaposition of the skewness values: 1 – CSI, 2 – structural blue light, 3 – Laser with tracker, 4 – Laser, 5 – White light.

3.2.7 Surface Topography

Two height parameters were chosen, Sa and Sq. The first one, because of its similarity to a very common profile parameter Ra. The choice of Sq parameter was based upon the fact that it's much more statistically significant. Moreover, some sources [15] state, that it can be related to the way that light scatters from a surface.

In three devices (a), (b), (d) the differences between Sa and Sq among the measurement systems are nearly identical, producing almost parallel chart lines. This is due to the fact, that these parameters are related to each other for each surface. In Fig. 11(c) the differences are noticeably different for systems (c) and (d).

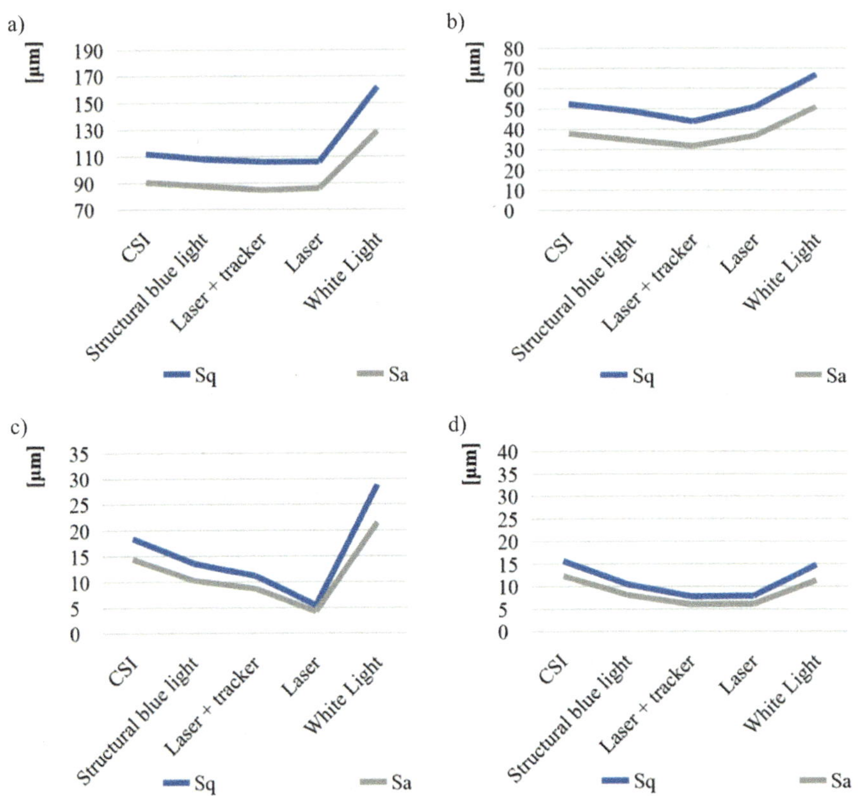

Fig. 11. A juxtaposition of Sa and Sq values of the measured surfaces: (a) Surface 1, unfiltered; (b) Surface 3, unfiltered; (c) Surface 1, filtered; (d) Surface 3, unfiltered.

4 Summary

This paper shows the possibility of using 3D scanners for mesoscale surface topography measurements. 4 different measurement systems were evaluated, in each case, a model of the measured surface was obtained, which allowed for its further quantitative and qualitative assessment. The derived surface parameters values were compared to the reference measuring device, in this case, a coherent scanning interferometer, which allowed for a much higher measurement resolution.

Qualitative assessment included evaluation of a bigger area on the surface (ca. 100 mm × 200 mm), which allowed for detecting major hills and valleys. These features appeared consistently at nearly the same locations. Scanners with lower resolution did not succeed in local asperities detection, nevertheless they enable for surface features evaluation, such as hills on the sides of the measured elements and a valley in its centre area.

Quantitative assessment was conducted on smaller area (min. 50 mm $\times$ 50 mm), out of which surface parameters were derived. These were later evaluated for deviations regarding reference values obtained with CSI.

Acknowledgments. The presented research was conducted within a task "Analysis of the influence of the surface preparation process on the measurement uncertainty of geometrical features of rings and rims" realized within a project No. POIR.01.01.01-00-0208/17, entitled: "Automated line for quality control and examination of rings and rims, with intelligent system of identification and measurement of internal defects using PA method, form measurement by means of 3D measurement heads, and inspection of mechanical properties SMART-HARD", realized by Huta Bankowa Sp. z o.o. within Smart Growth Operational Program 2014–2020. Priority Axis I: Support for R&D activity of enterprises. Investment Priority 1b: R&D projects of enterprises.

Research was cofinanced with grants for education allocated by the Ministry of Science and Higher Education in Poland No. 02/22/DSPB/1432.

References

1. Wieczorowski M (2018) Digitization of surfaces in micro, meso and macro applications. Mechanik 11
2. Senin N, Leach R (2018) Information-rich surface metrology. Procedia CIRP 75:19–26 ISSN 2212-8271
3. Marteau J, Wieczorowski M, Xia Y, Bigerelle M (2014) Multiscale assessment of the accuracy of surface replication. Surf Topogr Metrol Prop 2(2):044002
4. Youssra R, Sara R (2018) Big data and big data analytics: concepts, types and technologies. Int J Res Eng [S.l.] 5(9):524–528 ISSN 2348-7860
5. Cepova L, Kovacikova A, Cep R, Klaput P, Mizera O (2018) Measurement system analyses – Gauge repeatability and reproducibility methods. Measur Sci Rev 18(1):20–27
6. TopMap Family. https://www.polytec.com/eu/surface-metrology/products/large-area-measuring-systems/tms-500-topmap-prosurf/
7. Atos Core Optical 3D Scanner GOM. http://www.3dteam.pl/wp-content/uploads/2015/08/ATOS-Core.pdf
8. Go!SCAN 3D. https://www.creaform3d.com/en/metrology-solutions/handheld-portable-3d-scanner-goscan-3d
9. HandySCAN 3D. https://www.creaform3d.com/en/metrology-solutions/portable-3d-scanner-handyscan-3d
10. MetraSCAN 3D. https://www.creaform3d.com/en/metrology-solutions/optical-3d-scanner-metrascan
11. Deepak Lawrence K, Shanmugamani R, Ramamoorthy B (2014) Evaluation of image based Abbott-Firestone curve parameters using machine vision for the characterization of cylinder liner surface topography. Measurement 55:318–334 ISSN 0263-2241
12. Stout KJ, Blunt L, Mainsah E, Dong W, Mainsah E, Luo N, Mathia T, Sullivan P, Zahouani H (2003) Development of methods for the characterisation of roughness in three dimensions. Butterworth-Heinemann, Oxford

13. Tomkowski R, Kapłonek W, Kacalak W, Łukianowicz C, Lipiński D, Cincio R (2013) Digital filtration methods in the assessment of surface topography (in Polish). VI Kongres Metrologii, Kielce-Sandomierz
14. Laheurte R, Darnis P, Darbois N, Cahuc O, Neauport J (2012) Subsurface damage distribution characterization of ground surfaces using Abbott-Firestone curves. Opt Express 20:13551–13559
15. Petzing JN, Coupland JM, Leach RK (2010) The measurement of rough surface topography using coherence scanning interferometry. NPL Measurement good practice guide 116

Evaluation of the Usefulness
of the Measurement System in the Production
# of Surgical Instruments

Magdalena Diering[(✉)], Agnieszka Kujawińska, and Anna Olejnik

Poznan University of Technology, Poznań, Poland
Magdalena.Diering@put.poznan.pl

Abstract. The aim of the work undertaken in an enterprise producing surgical instruments, including needle holders and surgical scissors, was to improve the principles and methods of measurement for the critical features of selected tools. The quality control process for the production of surgical instruments was studied, statistical analysis of control and measurement systems for this process was performed and, on the basis of the %GRR value, their suitability for the measurement tasks was assessed. The conducted study shows how much influence the human factor as well as the organizational conditions and environment have on the result of the evaluation of the usefulness of the measurement system.

Keywords: Measurement system analysis (MSA) · ARM method · %GRR · Repeatability · Reproducibility · Surgical instruments

1 Introduction

One of the main criteria for the success of manufacturing enterprises is the quality of the products they offer. Quality is understood as the degree of fulfillment of the requirements agreed with the client and/or customarily accepted, resulting from the needs and expectations of buyers and users of the product [1–3]. Managing the quality of products is possible, among others, thanks to control, i.e. an action aimed at verifying the compliance of the state of the achieved process or product with the assumed state [4, 5]. In relation to technological processes, quality control mainly refers to key and critical characteristics of products. Most often these are dimensions, shape, appearance and product surface quality, features which are assessed organoleptically. The confirmation of the compliance of characteristics assessed in a measurable manner is based on the comparison of the measurement result with a given compliance criterion [4, 6]. The appropriate number and distribution of control operations in the manufacturing process helps to ensure the required product quality. Control is also a tool for identifying errors and potential discrepancies that occur during production, and thus – contributes to the minimization of losses (including material, time). Data obtained as a result of the measurement translate into information about the perception of the quality of a given product. It can be said that the quality of measurement results affects the

M. Diering et al. (Eds.): *Advances in Manufacturing II* - Volume 5, LNME, pp. 257–269, 2019.
https://doi.org/10.1007/978-3-030-18682-1_20

quality of decisions made. Therefore, it is necessary to ensure that the measurement system fulfils its purpose and the obtained results are reliable and credible.

The measurement system is all elements (object of measurement, measuring instruments and devices, measurement method, norms, standards and tests as well as the operator and environment) occurring in the measurement process that may affect its result [7–10]. The set of operations whose purpose is to determine the numerical value of the measured characteristic is called measurement. The result of the measurement is an approximation of the true value of the measured quantity because it is always burdened with some error. The most frequently determined parameters to assess the properties of measurement systems are repeatability and reproducibility. You can also include the resolution of the measuring instrument and the ndc indicator [7, 11]. The authors of the article present the analysis of these parameters and statistical properties, based on the tests carried out in the selected process of quality control of surgical instruments.

Surgical instruments are specialist instruments used in medicine during operations and procedures. Due to their function and contact with the human body, surgical instruments must meet very rigorous requirements: durability, resistance to abrasion and corrosion, construction ergonomics, aesthetics of workmanship and safety of use for the patient and the operator. Committing the first type of error, i.e. qualifying the non-compliant product as the one that meets the requirements and sending it to the client may have, in the worst case, tragic consequences for the life and health of the patient. The authors of the article, as a result of observations and measurements carried out and based on statistical analysis of the control and measurement system, developed a set of principles and measurement methods for the critical characteristics of surgical instruments for the needs of a selected manufacturing enterprise.

2 Research Object

The aim of the work undertaken in an enterprise producing needle holders and surgical scissors, i.e. selected surgical instruments, was to develop a set of principles and measurement methods for the critical characteristics of these instruments. Surgical instruments are used in medicine during operations and treatments. Needle holders are used to hold the needle while sewing and binding the seams on the tool. The main task of surgical scissors is preparation of soft tissues or cutting of sutures and surgical threads. Due to the function and contact with the human body, surgical instruments are made with a very high requirements, including durability, abrasion resistance and corrosion resistance.

Already the preliminary tests conducted by a quality engineer from the organization showed that the results of measurements of a given characteristic performed by several controllers differ from each other, and the measurement systems require detailed validation. Therefore, the process of quality control of surgical instruments was performed, statistical analysis of collected results was carried out and – by determining %GRR and ndc – the usefulness of control and measurement systems used in this process was evaluated. The research results were to indicate (confirm) the correctness of applying these systems to the objectives adopted by the organization or to develop new principles and methods of control. The analysis of the measurement systems was carried out according to the ARM procedure, which is the most frequently chosen procedure for characteristics assessed in a measurable way [7, 12, 13]. This procedure makes it possible to assess both the repeatability and the reproducibility of the measurement system, also taking into account the change of the operator. Among others, the %GRR and ndc values were determined. Criteria for the evaluation of measurement systems when the reference value (RF) is part of the tolerance field (in this case 1/6 T) is presented in Table 1.

Table 1. Criteria for the evaluation of the measurement system when the reference value is tolerance (own elaboration based on [7]).

%GRR value	ndc value	Measurement system
<10%	5	Acceptable
10%–30%	$2 \geq$ ndc ≥ 4	Conditionally acceptable
>30%	ndc < 2	Unacceptable

Measurements for the analysis of measurement systems were made for the critical features of needle holders and surgical scissors. The critical feature of needle holders is the thickness of the jaws. In the construction of scissors, except for one of the critical dimensions – the width of the jaws, it is important to deflect the working part of the scissors. For each feature, a separate measurement system is specified, because a different measurement method is assigned to each feature. Due to the design of the working parts of these instruments, the company developed a special method of measuring the characteristics at a set distance from the tip of the instrument. For this purpose, the caliper was modified. After milling the lower jaw of the fixed caliper, a metal plate supporting the measurement of the width and thickness at a specific point of the instrument was attached to its movable jaw (Fig. 1).

In accordance with to the assumptions contained in the control drawings, the measurement of the thickness of the jaws of needle holders is made using a metal plate with the possibility of measuring at a distance of 3 mm from the tip of the instrument. When measuring with a caliper with metal plates, sliding the movable jaw along the vernier scale, the caliper's jaws are brought into contact with the instrument, which is inserted between the jaws from below (Fig. 2a). Due to rounded edges, uneven surface, pointed tips and various thicknesses and deflections of instruments, it is difficult to correctly position the measuring instrument in relation to the tested element.

Measurements of the deflection of scissors are made using a depth gauge, as shown in Fig. 2b. The surface for the depth measurement is brought into contact with the tip of the scissors lying flat on the measuring table.

The elements of the discussed measurement systems, in addition to the measuring instruments already described, are the parts subject to measurements (medical instruments), methods and patterns, personnel (quality controllers) and the environment.

The final quality control takes place in a room separated from the production hall. The influence of such factors as temperature and humidity, lighting or vibrations on the course of measurements was not noted. The measurements were made at tables covered

Fig. 1. Electronic caliper with metal plates and visible milling of the jaws (own elaboration).

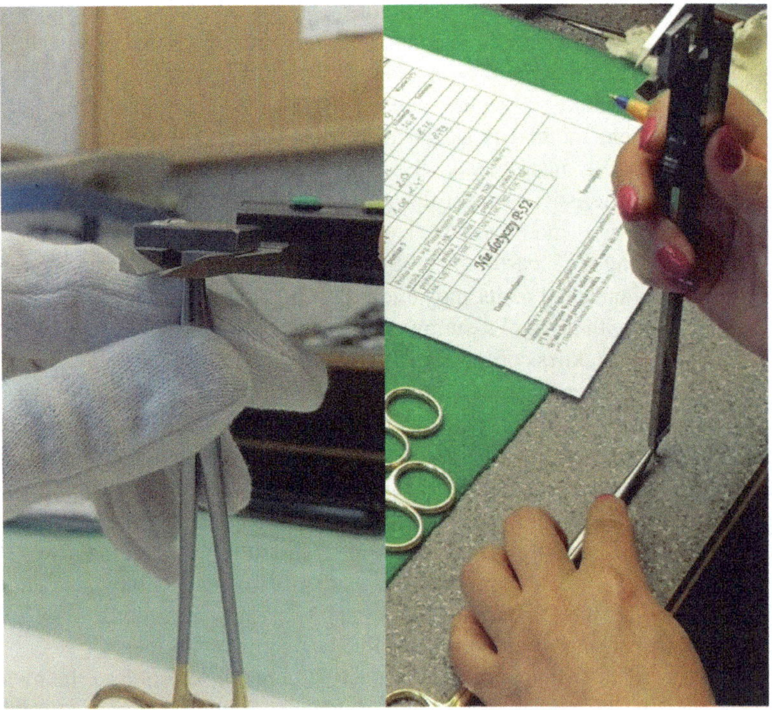

Fig. 2. a. Application of a caliper to the product - measurement of the width of the jaws of needle holders, b. Application of a depth gauge to the product - measurement of the scissors deflection value (own elaboration).

with a flexible, rubber mat. It is unacceptable to base the tested instrument or measuring device during the measurement on a soft surface as it falsifies the measurement results. When measuring the deflection of the scissors with a depth gauge, as shown in Fig. 2b, the pin is stuck in the soft ground.

The final control of measurements is made by authorized quality controllers. They have various levels of experience and employment periods. During the observation of the controllers, it was noticed that they perform measurements in a similar manner, but with different pressure of the jaws. Controllers pay attention to the setting of the jaws of the caliper at the right angle – perpendicular to the object being tested and try to make the vernier movement uniform. Operators are unable to clearly determine whether the individual elements of the caliper are worn out and damaged, and whether the caliper is suitable for use and fulfills its functions. In addition, in other departments, it was noted that the measurements are made at the wrong measuring point (incorrect application), "hanging" measurements are made or the caliper is moved during measurement and "the desired result is looked for". Errors made by operators result, among others, from the operators' habits acquired in the course of work, inadequate measurement methodology and lack of concentration and haste.

3   Research Methodology

In order to understand the actual impact of the measurement system on the evaluation of finished products, the tests for statistical measurement system analysis (MSA) were carried out under real conditions, i.e. in the natural environment of controls carried out on a daily basis, at quality control stations and by four authorized quality controllers. Due to the fact that it was not advisable to leave workstations by employees and that tests were performed during work, all controllers stayed in the room. Before each series of measurements made by each operator, the parts were mixed (randomization of results), and their numerical markings were invisible to the controllers. In this way, the appearance of the interaction effect was minimized. Each operator performed three series of measurements for a given feature, according to the assumption that on one day the controller could perform only one series of measurements of a given feature. All measurements were made with the electronic caliper Mitutoyo ABSOLUTE AOS Digimatic with a range of 0–150 mm, with a resolution of 0.01 mm and an accuracy of ±0.02 mm. Plates for measurements at 1.5 mm and 3 mm were screwed on the caliper's jaws. Every day before testing, the surface condition of the jaws was checked and the measuring instrument was calibrated.

For testing the measurement systems, 11 needle holders and 10 scissors were taken from the production process. In order to check the dimensional accuracy of the instruments, reference measurements of selected three critical features of the instruments were made. The measurements were performed in the Division of Metrology and Measurement Systems at Poznan University of Technology. The measurement results confirmed that the selected parts are good, i.e. their dimensions are within tolerance.

4 Results and Discussion

After collecting the measurement results, they were analyzed and calculations were made, as a result of which the values of selected indicators and information on statistical properties of the assessed measurement systems were obtained. Tables 2a, 2b, 3a, 3b, 4a and 4b shows the results of measurements, and Table 5 presents the results of tests for three measurement systems.

Table 2a. MSA study – measurement results of thickness of needle holder jaws (1) (own elaboration).

	1	2	3	4	6	7	8
1	Name of feature:		Thickness of needle holder jaws (1)				
2	Dimension:		$2,6 \pm 0,15$ mm				
3	Parts:		11				
4	Trials:		3				
5	Appraisers:		4				
6	*APPRAISER*	*TRIAL*	*PART*				
7			1	2	3	4	5
8	A	1	2,49	2,52	2,51	2,48	2,55
9		2	2,49	2,54	2,54	2,49	2,57
10		3	2,48	2,50	2,51	2,47	2,53
11		AVE	2,49	2,52	2,52	2,48	2,55
12		R	0,01	0,04	0,03	0,02	0,04
13	B	1	2,50	2,51	2,52	2,44	2,55
14		2	2,45	2,49	2,50	2,43	2,52
15		3	2,46	2,48	2,51	2,42	2,52
16		AVE	2,47	2,49	2,51	2,43	2,53
17		R	0,05	0,03	0,02	0,02	0,03
18	C	1	2,46	2,50	2,51	2,42	2,52
19		2	2,49	2,52	2,51	2,45	2,55
20		3	2,45	2,48	2,52	2,45	2,54
21		AVE	2,47	2,50	2,51	2,44	2,54
22		R	0,04	0,04	0,01	0,03	0,03
23	D	1	2,52	2,56	2,54	2,50	2,59
24		2	2,48	2,54	2,53	2,48	2,57
25		3	2,50	2,54	2,56	2,50	2,58
26		AVE	2,50	2,55	2,54	2,49	2,58
27		R	0,04	0,02	0,03	0,02	0,02
28	PART AVE (x_p) =		2,48	2,52	2,52	2,46	2,55
29							
30	$\bar{\bar{R}} = (\bar{R}_A + \bar{R}_B + \bar{R}_C + \bar{R}_D)/4 = (0{,}046 + 0{,}036 + 0{,}029 + 0{,}035)/4 = 0{,}037$						
31	$\bar{x}_{Diff} = \mathrm{Max}\,\bar{x} - \mathrm{Min}\,\bar{x} = 2{,}563 - 2{,}513 = 0{,}050$						
32							
33	$UCL_R = \bar{R} * D4 = 0{,}037 * 2{,}58 = 0{,}095$						

Table 2b. MSA study – measurement results of thickness of needle holder jaws (1) (own elaboration).

	1	2	9	10	11	12	13	14	15	16
1	Name of feature:		Thickness of needle holder jaws (1)							
2	Dimension:		2,6 ± 0,15 mm							
3	Parts:		11							
4	Trials:		3							
5	Appraisers:		4							
6	*APPRAISER*	*TRIAL*	*PART*						*AVERAGE*	
7			6	7	8	9	10	11		
8	A	1	2,57	2,60	2,66	2,54	2,52	2,44	2,54	
9		2	2,58	2,52	2,61	2,58	2,48	2,41	2,53	
10		3	2,63	2,58	2,62	2,60	2,51	2,49	2,54	
11		AVE	2,59	2,57	2,63	2,57	2,50	2,45	$\bar{x}_A = 2{,}53$	
12		R	0,06	0,08	0,05	0,06	0,04	0,08	$R_A = 0{,}05$	
13	B	1	2,52	2,59	2,62	2,56	2,48	2,45	2,52	
14		2	2,57	2,54	2,58	2,58	2,48	2,42	2,51	
15		3	2,55	2,54	2,67	2,58	2,47	2,43	2,51	
16		AVE	2,55	2,56	2,62	2,57	2,48	2,43	$\bar{x}_B = 2{,}51$	
17		R	0,05	0,05	0,09	0,02	0,01	0,03	$R_B = 0{,}04$	
18	C	1	2,55	2,57	2,61	2,56	2,48	2,46	2,51	
19		2	2,59	2,55	2,61	2,60	2,48	2,42	2,53	
20		3	2,56	2,54	2,62	2,59	2,47	2,42	2,51	
21		AVE	2,57	2,55	2,61	2,58	2,48	2,43	$\bar{x}_C = 2{,}52$	
22		R	0,04	0,03	0,01	0,04	0,01	0,04	$R_C = 0{,}03$	
23	D	1	2,62	2,60	2,67	2,63	2,55	2,50	2,57	
24		2	2,60	2,57	2,65	2,61	2,51	2,46	2,55	
25		3	2,66	2,60	2,72	2,62	2,53	2,50	2,57	
26		AVE	2,63	2,59	2,68	2,62	2,53	2,49	$\bar{x}_D = 2{,}56$	
27		R	0,06	0,03	0,07	0,02	0,04	0,04	$R_D = 0{,}04$	
28	PART AVE (x_p) =		2,58	2,57	2,64	2,59	2,50	2,45	$\bar{x}_p = 2{,}53$	
29									$R_p = 0{,}04$	
30									$\bar{\bar{R}} = 0{,}04$	
31									$\bar{x}_{Diff} = 0{,}05$	
32									$D_4 = 2{,}58$	
33									$UCL_R = 0{,}10$	

Based on the criteria for the evaluation of measurement systems adopted from [7] and [8], the measurement systems under investigation for selected features assessed in a measurable way are not useful. The measuring instrument and operators performing the measurements influenced the scattering of the measurement results. The number of different categories of data (ndc) for each of the three assessed systems is below 2, which also indicates their low suitability for their control and measurement task. These systems are not suitable for controlling the quality of the indicated critical features of needle holders (thickness and width) and surgical scissors (deflection).

Analyzing the results gathered in Tables 2a, 2b, 3a, 3b, 4a and 4b it can be noticed that for the thickness of needle holders jaws (1) and the width of scissors jaws (2), similar values were obtained for repeatability and reproducibility. For the first feature (1), the %GRR values were equal to 62%, while for the second feature (2) they were equal to 69%. This is most likely due to the fact that the measurements of these features

Table 3a. MSA study – measurement results the width of scissors jaws (2) (own elaboration).

	1	2	3	4	6	7	8
1	Name of feature:		Width of scissors jaws (2)				
2	Dimension:		$1{,}6 \pm 0{,}15$ mm				
3	Parts:		10				
4	Trials:		3				
5	Appraisers:		4				
6	*APPRAISER*	*TRIAL*	*PART*				
7			1	2	3	4	5
8	A	1	1,60	1,62	1,70	1,65	1,59
9		2	1,54	1,62	1,67	1,60	1,58
10		3	1,56	1,62	1,62	1,66	1,63
11		AVE	1,57	1,62	1,66	1,64	1,60
12		R	0,06	0	0,08	0,06	0,05
13	B	1	1,58	1,63	1,67	1,52	1,66
14		2	1,52	1,61	1,62	1,50	1,63
15		3	1,53	1,59	1,64	1,50	1,61
16		AVE	1,54	1,61	1,64	1,51	1,63
17		R	0,06	0,04	0,05	0,02	0,05
18	C	1	1,55	1,61	1,62	1,65	1,57
19		2	1,53	1,63	1,64	1,59	1,59
20		3	1,52	1,59	1,65	1,59	1,56
21		AVE	1,53	1,61	1,64	1,61	1,57
22		R	0,03	0,04	0,03	0,06	0,03
23	D	1	1,52	1,62	1,64	1,56	1,58
24		2	1,56	1,59	1,60	1,60	1,56
25		3	1,55	1,61	1,63	1,57	1,60
26		AVE	1,54	1,61	1,62	1,58	1,58
27		R	0,04	0,03	0,04	0,04	0,04
28	PART AVE (x_p) =		1,55	1,61	1,64	1,58	1,60
29							
30	$\overline{\overline{R}} = (\bar{R}_A + \bar{R}_B + \bar{R}_C + \bar{R}_D)/4 = (0{,}060 + 0{,}034 + 0{,}043 + 0{,}035)/4 = 0{,}043$						
31	$\bar{x}_{Diff} = \mathrm{Max}\,\bar{x} - \mathrm{Min}\,\bar{x} = 1{,}639 - 1{,}585 = 0{,}054$						
32							
33	$\mathrm{UCL}_R = \bar{R} * 0{,}043 * 2{,}58 = 0{,}111$						

were made in the same way, i.e. using plates screwed to the caliper for measurement at 3 mm and 1.5 mm respectively. Due to the insufficient knowledge about the impact of the quality of metal plates on the accuracy of measurements made by electronic calipers and based on the obtained test results, one gets the impression that measurements with a caliper with a metal blade are not reliable. The metal plate may be improperly milled and may become deformed and damaged during use. The numerical analysis for the

Table 3b. MSA study – measurement results the width of scissors jaws (2) (own elaboration).

	1	2	9	10	11	12	13	14	15	16
1	Name of feature:		Width of scissors jaws (2)							
2	Dimension:		$1,6 \pm 0,15$ mm							
3	Parts:		10							
4	Trials:		3							
5	Appraisers:		4							
6	*APPRAISER*	*TRIAL*	*PART*						*AVERAGE*	
7			6	7	8	9	10			
8	A	1	1,59	1,72	1,72	1,58	1,72		1,649	
9		2	1,56	1,66	1,75	1,62	1,70		1,630	
10		3	1,62	1,61	1,68	1,67	1,71		1,638	
11		AVE	1,59	1,66	1,72	1,62	1,71		$\bar{x}_A = 1,639$	
12		R	0,06	0,11	0,07	0,09	0,02		$\bar{R}_A = 0,060$	
13	B	1	1,59	1,57	1,61	1,59	1,59		1,601	
14		2	1,58	1,60	1,60	1,60	1,58		1,584	
15		3	1,60	1,60	1,57	1,59	1,57		1,580	
16		AVE	1,59	1,59	1,59	1,59	1,58		$\bar{x}_B = 1,588$	
17		R	0,02	0,03	0,04	0,01	0,02		$\bar{R}_B = 0,034$	
18	C	1	1,57	1,61	1,65	1,56	1,63		1,602	
19		2	1,57	1,62	1,62	1,58	1,67		1,604	
20		3	1,56	1,55	1,59	1,59	1,60		1,580	
21		AVE	1,57	1,59	1,62	1,58	1,63		$\bar{x}_C = 1,595$	
22		R	0,01	0,07	0,06	0,03	0,07		$\bar{R}_C = 0,043$	
23	D	1	1,57	1,58	1,61	1,58	1,63		1,589	
24		2	1,56	1,55	1,58	1,58	1,56		1,574	
25		3	1,58	1,56	1,60	1,59	1,63		1,592	
26		AVE	1,57	1,56	1,60	1,58	1,61		$\bar{x}_D = 1,585$	
27		R	0,02	0,03	0,03	0,01	0,07		$\bar{R}_D = 0,035$	
28	PART AVE (x_p) =		1,58	1,60	1,63	1,59	1,63		$\bar{x}_p = 1,602$	
29									$R_p = 0,095$	
30									$\bar{\bar{R}} = 0,043$	
31									$\bar{x}_{Diff} = 0,054$	
32									$D_4 = 2,580$	
33									$UCL_R = 0,111$	

critical feature of needle holders was slightly better than that of the critical feature of scissors. This dependence may be related to the fact that the measurement of the width of needle holders jaws is much easier due to the design of the instruments and the need to measure them at a height of 3 mm, and not like with scissors at a height of 1.5 mm.

The percentage value of reproducibility and repeatability for the third feature (3), i.e. for scissors, is estimated at 79%. This is the highest result obtained in the research carried out for the three critical features of the selected instruments. The measurements results performed with a depth gauge could have been influenced by the fact that they were performed on the soft ground. During the measurements, the pin was driven into the rubber at various depths. The sharp ending of the scissors blades and a small surface for measuring the depth (ending of the caliper) may have made it difficult to maintain a stable and perpendicular setting of the caliper relative to the tested instrument. These are potential causes of systematic errors that appear in the measurement of this feature.

Table 4a. MSA study – measurement results of scissors deflection (3) (own elaboration).

	1	2	3	4	6	7	8
1	Name of feature:		Scissors deflection (3)				
2	Dimension:		$0{,}6 \pm 0{,}5$ mm				
3	Parts:		10				
4	Trials:		3				
5	Appraisers:		4				
6	*APPRAISER*	*TRIAL*	*PART*				
7			1	2	3	4	5
8	A	1	6,17	5,78	5,95	5,94	5,72
9		2	6,19	5,47	5,83	5,90	5,62
10		3	6,00	5,66	5,93	5,80	5,74
11		AVE	6,12	5,64	5,90	5,88	5,69
12		R	0,19	0,31	0,12	0,14	0,12
13	B	1	6,23	5,63	5,97	5,91	5,83
14		2	6,15	5,74	5,96	6,09	5,82
15		3	6,18	5,79	6,02	5,90	5,61
16		AVE	6,19	5,72	5,98	5,97	5,75
17		R	0,08	0,16	0,06	0,19	0,22
18	C	1	6,28	5,76	6,13	6,03	5,97
19		2	6,32	5,79	6,22	6,07	5,74
20		3	6,35	5,69	6,07	6,01	5,88
21		AVE	6,32	5,75	6,14	6,04	5,86
22		R	0,07	0,1	0,15	0,06	0,23
23	D	1	6,21	5,60	5,82	5,91	5,76
24		2	6,10	5,55	5,94	5,91	5,68
25		3	6,42	5,67	6,15	6,08	5,88
26		AVE	6,24	5,61	5,97	5,97	5,77
27		R	0,32	0,12	0,33	0,17	0,2
28	PART AVE (x_p) =		6,22	5,68	6,00	5,96	5,77
29							
30	$\bar{\bar{R}} = (\bar{R}_A + \bar{R}_B + \bar{R}_C + \bar{R}_D)/4 = (0{,}203 + 0{,}137 + 0{,}114 + 0{,}188)/4 = 0{,}161$						
31	$\bar{x}_{\text{Diff}} = \text{Max}\,\bar{x} - \text{Min}\,\bar{x} = 6{,}041 - 5{,}832 = 0{,}210$						
32							
33	$\text{UCL}_R = \bar{R} * D4 = 0{,}161 * 2{,}58 = 0{,}414$						

As can be seen from the analysis of the results for each of the features, the most significant impact on the %EV index had the results of controller A measurements. The analysis, omitting these measurement results, allows to draw a new conclusion. For example, when analyzing the width of scissors jaws – after removing controller A the biggest change was noted for the %AV indicator, whose value changed from 47% to 9%. Based on the %GRR and ndc values after eliminating controller A, the measurement system for the discussed feature could conditionally be considered as acceptable.

Table 4b. MSA study – measurement results of scissors deflection (3) (own elaboration).

	1	2	9	10	11	12	13	15	16
1	Name of feature:		Scissors deflection (3)						
2	Dimension:		$0{,}6 \pm 0{,}5$ mm						
3	Parts:		10						
4	Trials:		3						
5	Appraisers:		4						
6	*APPRAISER*	*TRIAL*	*PART*					*AVERAGE*	
7			6	7	8	9	10		
8	A	1	5,89	5,96	6,02	5,51	5,67	5,861	
9		2	6,00	5,89	6,34	5,62	5,55	5,841	
10		3	5,65	6,07	5,98	5,48	5,62	5,793	
11		AVE	5,85	5,97	6,11	5,54	5,61	$\bar{x}_A = 5{,}832$	
12		R	0,35	0,18	0,36	0,14	0,12	$R_A = 0{,}203$	
13	B	1	5,89	6,18	6,28	5,90	5,83	5,965	
14		2	6,05	6,10	6,19	5,82	5,86	5,978	
15		3	6,08	6,16	6,15	5,67	5,85	5,941	
16		AVE	6,01	6,15	6,21	5,80	5,85	$\bar{x}_B = 5{,}961$	
17		R	0,19	0,08	0.13	0,23	0,03	$R_B = 0{,}137$	
18	C	1	5,95	6,32	6,27	5,90	5,93	6,054	
19		2	6,01	6,30	6,40	5,78	5,87	6,050	
20		3	6,00	6,23	6,20	5,89	5,88	6,020	
21		AVE	5,99	6,28	6,29	5,86	5,89	$\bar{x}_C = 6{,}041$	
22		R	0,06	0,09	0,2	0,12	0,06	$R_C = 0{,}114$	
23	D	1	5,90	6,17	6,11	5,95	5,83	5,926	
24		2	5,88	6,20	6,18	5,68	5,79	5,891	
25		3	6,03	6,27	6,26	5,79	5,86	6,041	
26		AVE	5,94	6,21	6,18	5,81	5,83	$\bar{x}_D = 5{,}953$	
27		R	0,15	0,1	0,15	0,27	0,07	$R_D = 0{,}188$	
28	PART AVE (x_p) =		1,58	1.60	1,63	1,59	1,63	$\bar{x}_p = 5{,}947$	
29								$R_p = 0{,}539$	
30								$\bar{\bar{R}} = 0{,}161$	
31								$\bar{x}_{Diff} = 0{,}210$	
32								$D_4 = 2{,}580$	
33								$UCL_R = 0{,}414$	

Table 5. MSA study – results of statistical analysis (own elaboration).

Property/characteristics of the measurement system	Result		
	Thickness of needle holder jaws (1)	Width of scissors jaws (2)	Scissors deflection (3)
Repeatability (EV) - device variability (impact of the instrument)	0.022 mm	0.025 mm	0.095 mm
Reproducibility (AV) - evaluator variability (operator impact)	0.022 mm	0.024 mm	0.092 mm
Repeatability and reproducibility (GRR)	0.031 mm	0.035 mm	0.132 mm
Reference value (RF)	0.050 mm	0.050 mm	0.167 mm
Part volatility (PV) (manufacturing process)	0.039 mm	0.036 mm	0.102 mm
%EV	43%	51%	57%
%AV	44%	47%	55%
%GRR	62%	69%	79%
ndc	1.78	1.46	1.09

This experience shows how much impact on the assessment of the measurement system can be made by a human factor and direct evaluators (controllers, operators) – their learned methods of measurement, habits or susceptibility to stress and fatigue (Table 6).

Table 6. Comparison of the analysis results for the width of scissors jaws with and without the participation of controller A (own elaboration).

Indicator	Indicator value	
	Original version (with the participation of controller A)	Version after modification (without the participation of controller A)
%EV	51%	33%
%AV	47%	9%
%GRR	69%	34%
ndc	1.462	3.864

5 Conclusions and Summary

The conducted study shows how much impact on the assessment of the measurement system can be exerted by a human factor and direct evaluators (controllers, operators) – their learned methods of measurement, habits or susceptibility to stress and fatigue. When performing measurements, special attention should be paid to the organizational conditions and environment (including the base of the measuring table, the point of application of the measuring instrument to the measured element, etc.).

In order to improve the ability of correct measurements by employees, it is necessary to constantly improve in this area and to indicate the validity of measurements in the assessment of the quality of products. Training should be of a theoretical and practical nature as well (what is more, improving the efficiency of measurement processes should be related to knowledge management in the organization [14]). Authors, based on study in enterprise producing surgical instruments, prepared an instruction to measuring with digital caliper. It includes notes on the storage and use of the caliper, the proper setting of the caliper in relation to the medical instrument and a description of the correct course of the measurement. The instruction also proposes a method of unifying the rules of approximation and rounding of measurement results.

Acknowledgments. The results presented in the paper come from a scientific statutory research conducted at the Chair of Management and Production Engineering, Faculty of Mechanical Engineering and Management, Poznan University of Technology, Poland, supported by the Polish Ministry of Science and Higher Education from the financial means in 201802/23/DSPB/7716.

References

1. Araújo AF, Varela MLR, Gomes MS, Barreto RCC, Trojanowska J (2018) Development of an intelligent and automated system for lean industrial production adding maximum productivity and efficiency in the production process. In: Hamrol A, Ciszak O, Legutko S, Jurczyk M (eds) Advances in manufacturing. Lecture notes in mechanical engineering. Springer, Cham, pp 131–140

2. Skołud B, Krenczyk D, Zemczak M (2015) Multi-assortment rhythmic production planning and control. IOP Conf Ser Mater Sci Eng 95:012133. https://doi.org/10.1088/1757-899X/95/1/012133

3. Goliński M, Spychała M, Szafrański M, Graczyk-Kucharska M (2016) Competency management as the direction of the development of enterprises – based on research. In: Proceedings of 2016 3rd international conference on social science, Shanghai, China. Published by DEStech Publications, Inc., pp 391–399

4. Hamrol A (2015) Strategies and practices of efficient operations of Lean Six Sigma and other (in Polish). PWN

5. Starzyńska B, Klembalska A (2017) A digital repository of science resources of research institute as a source of knowledge from the area of production engineering for SMEs. ICPR DEStech Trans Eng Technol Res 525–530

6. Gwet KL (2012) Handbook of inter-rater reliability. Definitive guide to measuring the extent of agreement among mulitple raters, 4th edn. Advanced Analytics, LLC, Gaithersburg

7. AIAG-Work Group (2010) Measurement system analysis, 4th edn., Reference manual, AIAG-Work Group, Daimler Chrysler Corporation, Ford Motor Company, General Motors Corporation

8. VDA 5.2 (2013) Capability of measurement processes for the torque inspection on bolted joints

9. ISO-10012 (2003) Measurement management system – requirements for measurement processes and measuring equipment

10. Wojciechowski J, Suszynski M (2017) Optical scanner assisted robotic assembly. Assembly Autom 37(4):434–441

11. Barrentine LB (2003) Concept for R&R studies, 2nd edn. ASQ Quality Press, Milwaukee

12. Perez-Wilson M (2007) Gauge R&R studies for destructive and non-destructive testing

13. Perez-Wilson M (2003) Gauge R&R studies. ASC Press

14. Patalas-Maliszewska J, Kłos S (2017) A study on improving the effectiveness of a manufacturing company in the context of knowledge management - research results. Found Manage Int J 9(1):149–160

NDT Porosity Evaluation of AlSi10Mg Samples Fabricated by Selective Laser Sintering Method

Joanna Maszybrocka[1], Bartosz Gapiński[2(✉)], Andrzej Stwora[3], and Grzegorz Skrabalak[3]

[1] Institute of Materials Science, University of Silesia, Chorzów, Poland
[2] Faculty of Mechanical Engineering and Management, Institute of Mechanical Technology, Poznan University of Technology, Poznan, Poland
bartosz.gapinski@put.poznan.pl
[3] Institute of Advanced Manufacturing Technology, Kraków, Poland

Abstract. The orientation of the sample during selective laser sintering (SLS) process is one of the factors that affect quality of the final product made from the AlSi10Mg powder. Most of the properties of AlSi10Mg bulk fabricated by SLS are strongly related to porosity, therefore one of the stages of quality inspection should enable for quick and precise analysis of the porosity without destruction the manufactured part. The paper shows that an effective tool in this area is computed microtomography. The presented methodology enabled not only the detection and visualization of pores in the whole volume of the samples, but, above all, allows for a full quantitative analysis, which included description of such features as: number, volume, shape and arrangement of the pores.

Keywords: Additive manufacturing · Computed tomography · Selective laser sintering (SLS) · AlSi10Mg alloy · Porosity

1 Introduction

Selective laser sintering (SLS) currently is one of the most dynamically developing branches associated with additive manufacturing technologies. With technological progress, a significant increase in the market share of this segment has been observed [1]. Almost unlimited possibilities for producing complex shapes and the ability to make internal channels with complex geometry cause, that SLS technology is used inter alia in: in the tool industry (injection molds with conformal cooling channels), in the medical industry (parts of knee and hip implants), in the aerospace industry (turbine blades and fuel injectors for aircraft engines) [2]. The SLS process is illustrated by the schematic diagram presented in Fig. 1.

During the first stage of the selective laser sintering (SLS) process, a thin layer of metallic powder is spread over the preheated plate using a scraper (20–50 μm). Then, using a laser beam, the powder particles are sintered together to form a homogeneous layer. At the next stage of the process, the table is lowered to a predetermined height

© Springer Nature Switzerland AG 2019
M. Diering et al. (Eds.): *Advances in Manufacturing II* - Volume 5, LNME, pp. 270–284, 2019.
https://doi.org/10.1007/978-3-030-18682-1_21

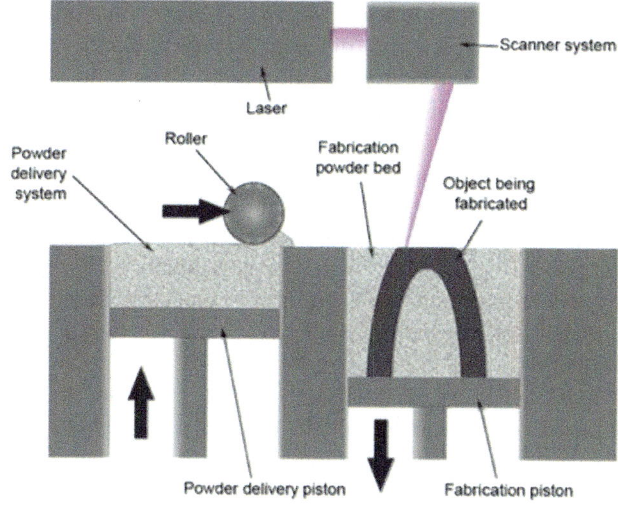

Fig. 1. Selective laser sintering process scheme [3].

corresponding to the thickness of the layer, the next thin layer of powder is distributed and the grains are re-bonded. The process is repeated until the whole object is sintered [4, 5].

The growing demand for strong, lightweight construction, often with complex geometry, results in more and more focus on the use of selective laser sintering for the manufacture of aluminum alloy components [6–9]. Among them special attention is paid to AlSi10Mg alloy. It belongs to the family of alloys that have good casting properties i.e. good ductility or low shrinkage. This alloy is also characterized by relatively low melting point and good thermal conductivity. Its high strength combined with low weight and flexible end-of-life processing (e.g. welding, polishing) makes it suitable for use in the aerospace or automotive industry [10].

Most of the properties of AlSi10Mg fabricated by SLS are strongly related to porosity. The presence of pores in the solid structure of the material may significantly reduce its mechanical properties, i.e. tensile strength, fatigue resistance, as in the case of conventional die-cast aluminum [11, 12]. The pores give rise to local stress concentrations in addition to reducing the effective load-bearing cross section. The effect of pores on mechanical properties is complex and depends not only on their number and distribution but also on shape, size and spatial orientation [13]. One of the challenges for the selective laser sintering technology, is the minimization of the degree of porosity of the sintered component. This is achievable, inter alia, by optimizing the parameters of the sintering process, among which the most important ones are included: scanning speed, hatch spacing, laser power and layer thickness. AlSi10Mg powder is characterized by low density, poor flowability, high reflectivity and high thermal conductivity relative to other materials which makes the selection of optimal parameters of the SLS process a difficult task [14]. The matter is further complicated by the fact that AlSi10Mg powders from different manufacturers may differ in chemical composition, particle size, distribution and flowability. The same process parameters may occur at the same time with noticeable difference in density, what Kempen et al. [15] research has shown.

In order to ensure the required quality of selectively sintered AlSi10Mg alloy components, it is also necessary, in addition to optimizing sintering parameters, to choose the appropriate direction of manufacturing of the model. Studies have shown that AlSi10Mg samples sintered at different angles exhibit anisotropy of mechanical properties. This may be due, among other things to greater borderline porosity, which makes the Z-oriented tensile parts more sensitive to crack initiation, compared to XY oriented tensile samples.

Bearing in mind the above, it can be stated that one of the key elements of the quality control process of SLS powdered AlSi10Mg powders should be to control their porosity. In practice, most often such an assessment is made on the basis of microscopic examination of metallographic inscriptions. Metallographic specimens are usually made on a cross-sectional or longitudinal section of a specimen in the direction of its manufacturing. Sometimes, the method of computer image analysis is included in the quantitative description of porosity, which gives the ability to determine the type of porosity based on such features as: location, shape, size, number. Unfortunately, microscopic studies generally allow for the assessment of local porosity and are not always representative for the entire volume of sinter e.g. due to the non-uniform distribution of pores in the material structure. In addition, the result of the evaluation of porosity features may also be influenced by the method of preparation of metallographic samples (e.g. grinding and polishing processes may have impact on the size and shape of the observed pores due to metal smearing) [16]. Thus, it seems appropriate to include other methods of quantifying the total porosity of an element e.g. computed microtomography [17–20]. Computed microtomography is one of the most effective methods of three-dimensional analysis of all structural elements as long as they exhibit density differences. This method has been increasingly used for optimization of manufacturing processes. [21–24]. An undoubted advantage of computed microtomography is the possibility of detailed analysis of pores in the entire volume of the examined object. Data is obtained in a relatively short time without the need to destroy the sample and its special preparation [25]. Taking this into account, computed microtomography can be an effective step in the quality control of SLS sintered samples, especially in case of complex structures with closed spaces or hard to access. The purpose of this paper is to evaluate the effect of sample orientation during the SLS process on its porosity using computed microtomography.

2   Research Material

The test samples were manufactured from Renishaw AlSi10Mg powders. The chemical compositions of the AlSi10Mg powder were shown in Table 1.

Table 1. Chemical composition of AlSi10Mg powder used in the studies (data from Renishaw - powder producer).

Element	Mass (%)	Element	Mass (%)
Aluminum	Balance	Zinc	<0.10
Silicon	9.00 to 11.00	Manganese	<0.10
Magnesium	0.25 to 0.45	Nickel	<0.05
Iron	<0.25	Copper	<0.05
Nitrogen	<0.20	Lead	<0.02
Oxygen	<0.20	Tin	<0.02
Titanium	<0.15		

SEM observations have shown that the powder has a large variation in shape and particle size (Fig. 2). Both spherical particles of $30 \div 50$ µm in diameter and strongly elongated particles of up to 100 µm in length are visible. Such geometry of the powder particles is not preferable for the SLS technique, as it can significantly affect the homogeneity of the obtained structure and also the porosity of the part.

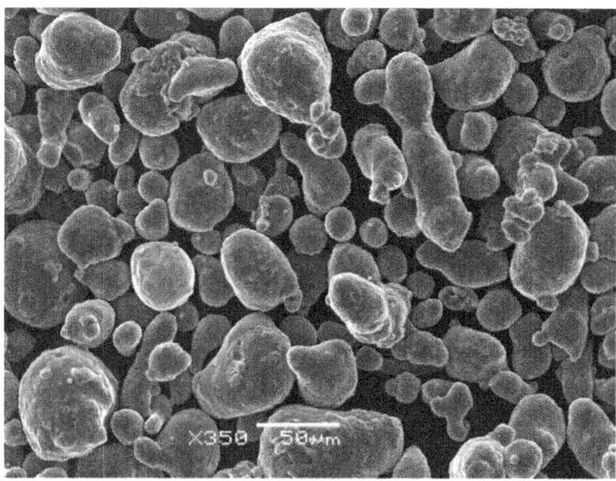

Fig. 2. AlSi10Mg powder (SEM).

2.1   Selective Laser Sintering

The samples were fabricated using a selective laser sintering device AM 250 Renishaw equipped with a fiber laser Yb-Fiber ($\lambda = 1064$ nm) with a power of 400 W. During manufacturing of the element in the working chamber of the device the inert gas atmosphere (oxygen content below 100 ppm) is kept. Sintering parameters are summarized in Table 2, The shape and orientation of the samples during the sintering process are shown on Fig. 3. The shape of the samples corresponds to the round samples that are used during the static tensile test [26]. To determine the effect of the manufacturing direction on porosity, the samples were oriented at angles 0°, 45° and

90° relative to the base plate. They are denoted by the symbols P00, P45 and P90 respectively. Technological parameters and model preparation for the SLS process were performed in Autofab software.

Table 2. Parameters used during the SLS process for AlSi10Mg powders.

Parameters		
Constant	Point distance	75 [μm]
	Exposure time	75 [μs]
	Laser power	400 [W]
	Scan speed	150 [mm/s]
Variable	The angle of inclination of the samples	0–90 [°]

After the sintering process, the samples were cut off from the base plate on which they were made, using Electrical Discharge Machining (EDM) wire cutting. Then, the support elements of the individual samples were removed.

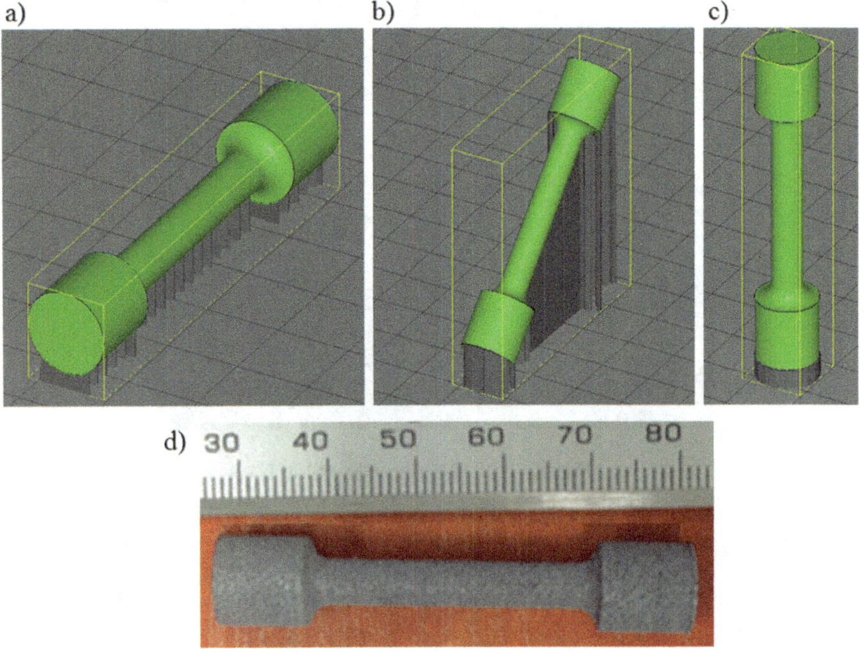

Fig. 3. Sample orientation in AutoFab software: (a) P00, (b) P45, (c) P90, (d) sample after selective laser sintering.

3   Methodology of Research

The test was performed at General Electric phoenix v|tome|x s 240 equipped with 180 kV nanofocus x-ray transmission target tube (Fig. 4).

An image in computed tomography is obtained by overlaying multiple X-ray images taken from different angular positions with respect to a common axis of rotation. The final result is a spatial image containing information about visible and invisible surfaces of the examined element and internal structure i.e. cracks, porosity, inclusions or deliberately introduced additives such as fibers.

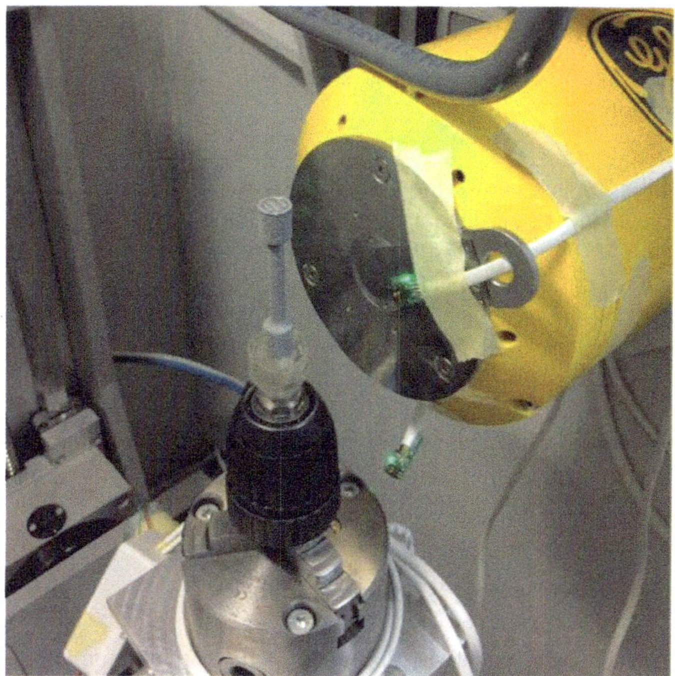

Fig. 4. Sample placed in the tomograph measuring chamber.

Investigations of the discussed samples were made with a lamp current of 90 kV and a current of 150 µA. A panel detector with a resolution of 1000 × 1000 pixels was used to record the image. The exposure time of a single measurement shot was 200 ms. In order to obtain the correct image resolution, it was necessary to place the sample close to the x-ray source. In this case, only a part of the sample may be registered. During the reconstruction of the image, a multiscan function was used to combine five scans in a single image with different positions relative to the sample axis (Fig. 5).

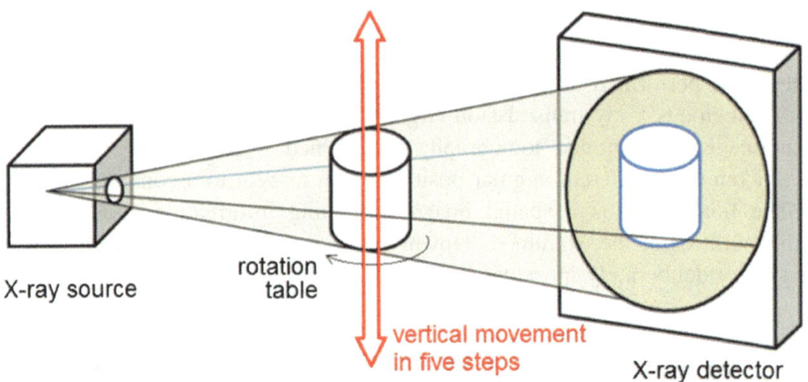

Fig. 5. Vertical movement of the sample allowing to combine measurements from five measurement positions.

The resulting image allows for full geometric analysis of the samples and their porosity. Calculations were done using Volume Graphics 2.2. This type of research allows not only to observe porosity without destroying the samples, but also to calculate their volume, position, orientation or dimension. This is a qualitative assessment, which is based on the visual evaluation of the examined element. Unfortunately, the disadvantage is constituted by the subjectivity of the researcher. In this aspect, it is too imprecise to unequivocally determine whether the components of the structure are the same or not. However, the effect of the tomographic measurement is also the ability to quantify the results, which allows for a precise and unambiguous assessment of porosity of the studied samples. It includes the designation of characteristic features of objects such as their number, size, volume, shape or distribution.

4   Research Results and Their Analysis

Figures 6, 7 and 8 show spatial visualization of pore distribution in P00, P45 and P90 samples. A summary of the results of the analysis is presented in Table 3.

Pore sizes were calculated as the equivalent diameter for a spherical representation of pores. The distributions shown in Fig. 9 illustrate the quantity and volume of pores in different size classes. The diameter of the majority of pores in the analyzed samples is in the range of 0.1–1 mm. The largest dispersion of the equivalent diameters was found for sample P00. There are a few pores with diameter exceeding 1 mm (Fig. 9a), but they represent 0.39% of the total porosity of the sample (Fig. 9b) which was 1.15% (Table 3).

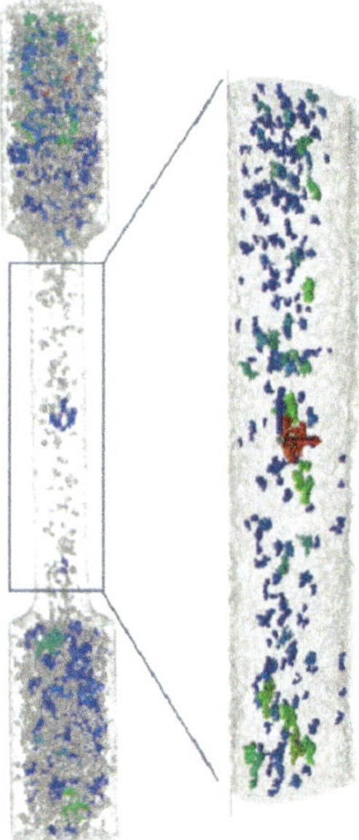

Fig. 6. Spatial visualization of pore distribution in P00 sample.

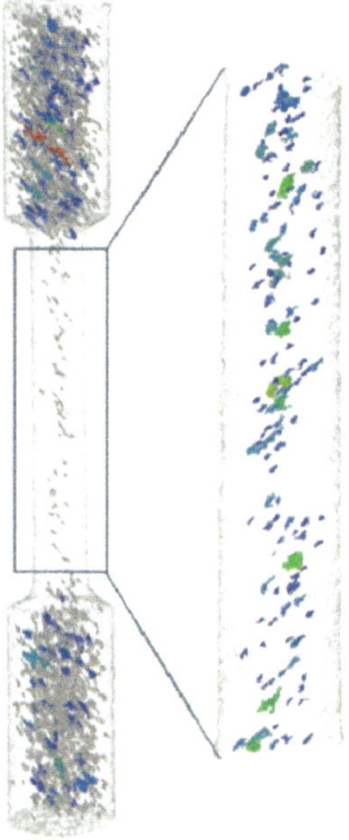

Fig. 7. Spatial visualization of pore distribution in P45 sample.

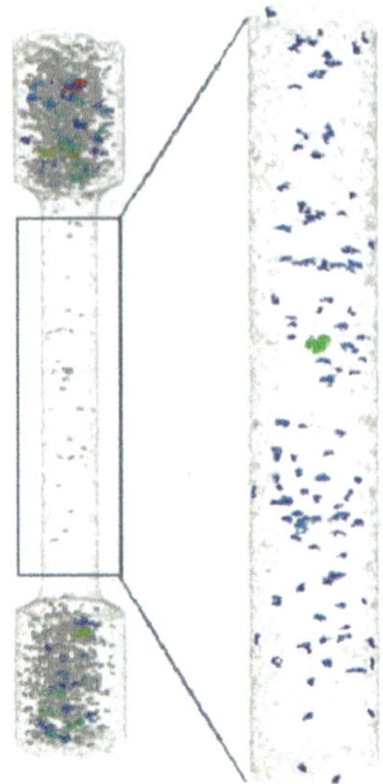

Fig. 8. Spatial visualization of pore distribution in P90 sample.

Table 3. Determined parameters of the pores in the individual samples.

P00	Min	Max	Mean	Std.dev.
Diameter [mm]	0.19	2.31	0.45	0.25
Surface [mm^2]	0.096	8.24	0.48	0.74
Volume [mm^3]	0.0013	0.1646	0.0078	0.0151
Compactness	0.02	0.39	0.12	0.06
Sphericity	0.18	0.62	0.41	0.07
P45	Min	Max	Mean	Std.dev.
Diameter [mm]	0.18	1.45	0.37	0.19
Surface [mm^2]	0.091	4.30	0.31	0.48
Volume [mm^3]	0.0013	0.1091	0.0053	0.0105
Compactness	0.03	0.51	0.16	0.07
Sphericity	0.23	0.65	0.48	0.08
P90	Min	Max	Mean	Std.dev.
Diameter [mm]	0.17	1.31	0.35	0.13
Surface [mm^2]	0.091	4.44	0.22	0.21
Volume [mm^3]	0.0013	0.0884	0.0032	0.0042
Compactness	0.01	0.50	0.14	0.08
Sphericity	0.22	0.66	0.48	0.07
	P00	P45	P90	
Vv [%]	1,15	0,63	0,47	
Nv [1/ mm^3]	1,47	1,16	1,44	

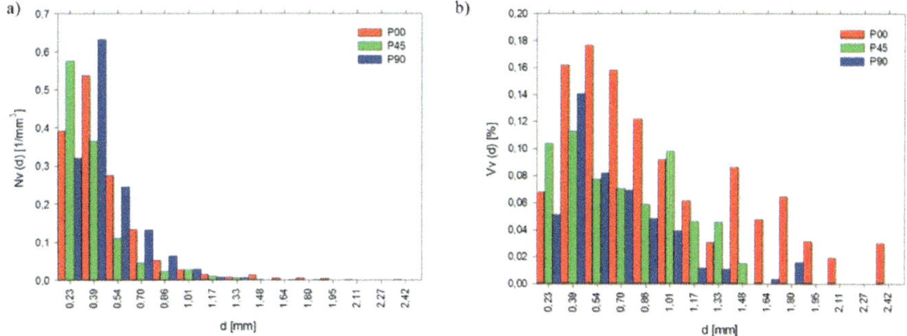

Fig. 9. Quantity and volume of the pores in different size classes.

Taking into account that spatial visualization of pores (Figs. 6, 7 and 8) performed on individual samples shows their non-spherical shape, the analysis was supplemented by determining the spherical shape coefficients (α) and compactness (β).

Sphericity is a measure of how spherical a 3D object is and it can be calculated according to the following formula (1):

$$= \frac{\sqrt[3]{(6V)^{\frac{2}{3}}}}{S} \tag{1}$$

where V and S are the object volume and surface area, respectively.

For complex, non-spherical objects the surface area of the volume equivalent sphere will be much smaller than the particle surface area, thus α will be low. The maximum value possible is 1, which would be obtained for a sphere [27]. The results of the sphericity analysis for pores and aggregate are presented in Fig. 10.

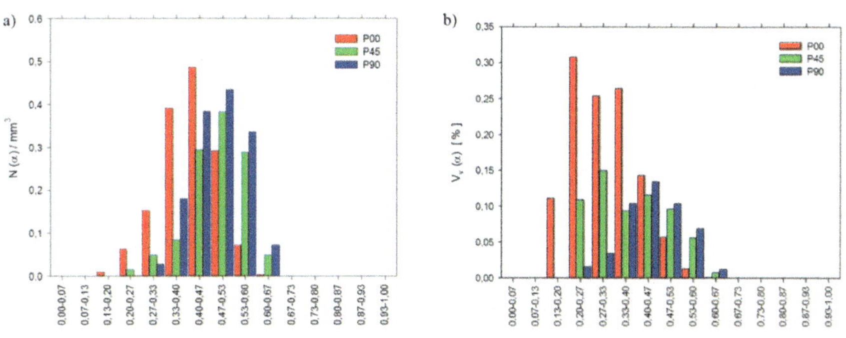

Fig. 10. Quantity and volume fraction of the pores with shape factor α.

Low sphericity coefficients values mean that the pore shape in all the analyzed samples is distinctly different from spherical. The sphericity coefficients of most of the pores in the samples P45 and P90 is in the range of 0.4–0.6. A slight shift of the spherical index to the lower values is noted for the pore population in the sample P00, indicating the most irregular shape of these pores.

Compactness is a measure expressing how compact a feature is. It was calculated as ratio between the volume of the defect and the volume of the circumscribed sphere (2).

$$= \frac{V_{defect}}{V_{sphere}} \tag{2}$$

Only for the sphere, the compactness coefficient has the value 1. In other cases it is less than 1. The results of the compactness analysis was shown in the Fig. 11. In all the analyzed samples, the compactness coefficient was found to be less than 0.4, indicating irregular pore shape.

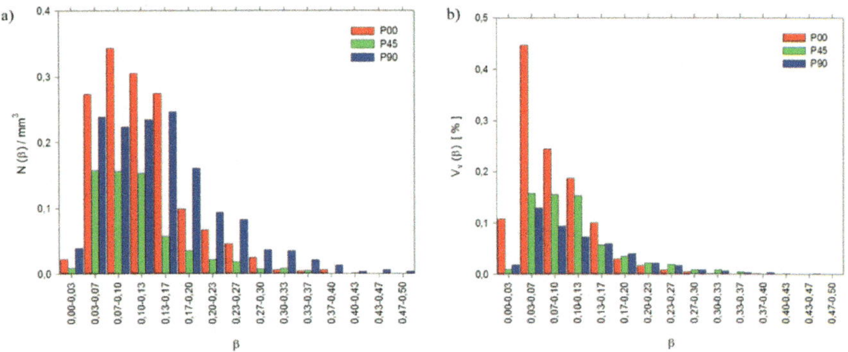

Fig. 11. Quantity and volume fraction of the pores with the shape factor β.

Analysis of cross sectional images of samples made in three perpendicular directions indicates the characteristic arrangement of the pores in each sample. This is especially evident when observing a section made in the XZ plane (Fig. 12). With this in mind, the quantity and volume of pores oriented at different angles to the Z axis in the XZ plane were determined. They were determined on the basis of the size of the bounding box surrounding the defect in the scene coordinate, calculating the ratio of the dimensions $\gamma = SX/SZ$ (Fig. 12a).

The obtained results, presented in the form of histograms in Fig. 13, confirm that in each sample there is a dominant direction of the pores. Correspondingly, in case of sample P00 the largest quantity and volume fraction belongs to the pores whose coefficient γ belongs to the interval $\gamma = 1.63 \div 1.87$ in case of the sample P45 $\gamma = 0.93 \div 1.17$ and in case of the sample P90 $\gamma = 0.47 \div 0.70$. Referring the determined coefficients γ to the build direction, it can be stated, that the voids are mainly oriented with their long axis perpendicular to building direction.

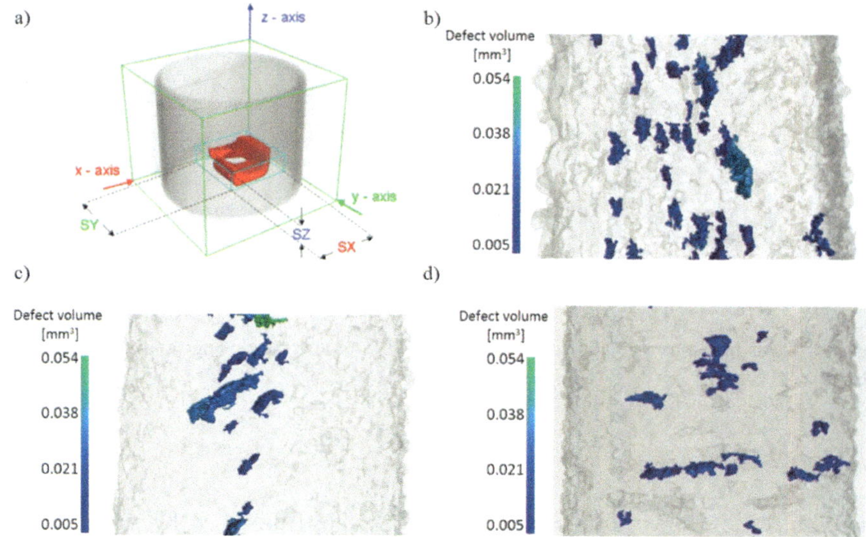

Fig. 12. Idea of dimensioning pores in the XYZ system (a) and visualization of cross-section of pores in XZ plane for the samples: (b) P00, (c) P45 and (d) P90.

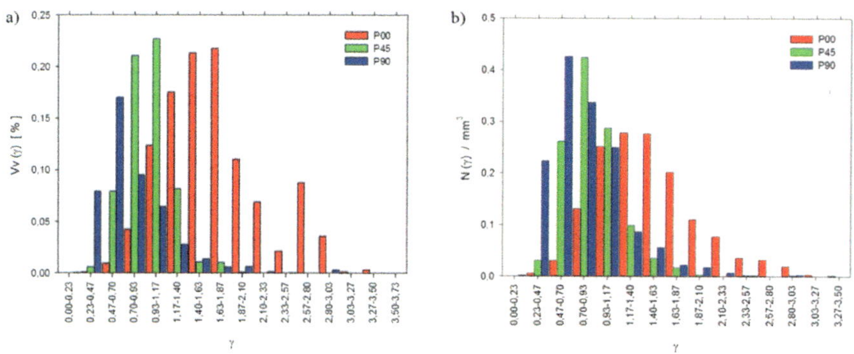

Fig. 13. Volume fraction and quantity of the pores with the direction factor γ.

The observed directionality of the pores is also confirmed by images from scanning microscope (Fig. 14). Cross-sectional observation of sample P90 showed, that the pores have a clearly elongated shape and are perpendicular to the manufacturing direction. Inside the pores there are visible loose or partially molten powders of the base material. Both the shape and arrangement of the pores suggest that defects are formed between the layers of sintered powder.

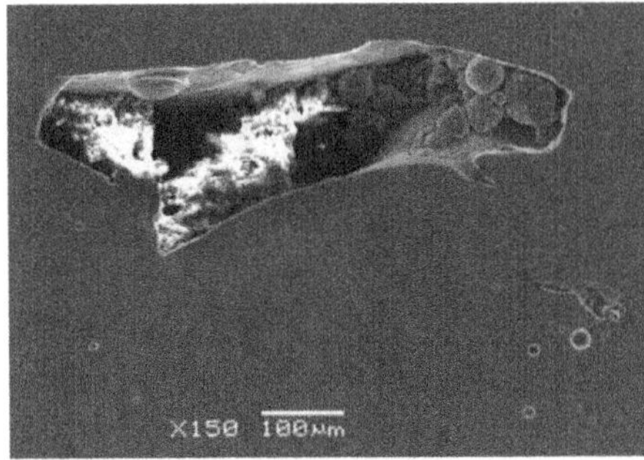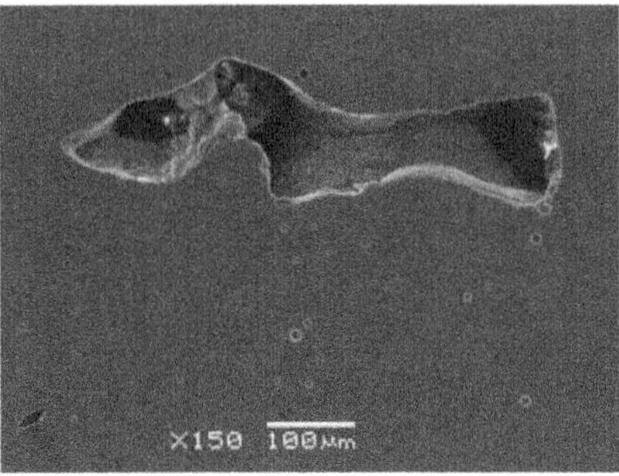

Fig. 14. Clear pore directionality in the sample P00, the pores enclosing non-molten powders (SEM).

5 Conclusions

The presence of pores in aluminum AlSi10Mg alloys produced by selective laser sintering is one of the most common problems that can, as a consequence, affect the deterioration of the strength properties of the manufactured parts. The study confirmed that computed X-ray microtomography allows a rapid and precise analysis of porosity without destroying the already prepared component and may be an effective step of controlling the quality of parts produced by SLS. It also allows to evaluate the anisotropy of the manufactured element and to choose the right print strategy to achieve the desired properties of the final product. The presented methodology allowed to detect and visualize the pores in the entire volume of the examined part and allowed for full quantitative analysis. Referring the description of such pore characteristics as number, size, volume ratio, shape and distribution to the model orientation during manufacturing process, it has been shown that:

- The sample whose direction of manufacturing was perpendicular to the base plate (P90) was characterized by the smallest porosity $Vv = 0.47\%$. In this case also the smallest mean pore diameter of $d = 0.35$ mm was recorded.
- The sample whose manufacturing direction was parallel to the base plate (P00) was characterized by the highest porosity $Vv = 1.15\%$. In this case, the largest average pore diameter $d = 0.45$ mm and the largest diameter spread was found.
- The shape of the pores in all the analyzed samples is distinctly different from the spherical one and the particles have a "non compact" shape, as demonstrated by the values of the shape coefficients $\alpha < 0.6$ and $\beta < 0.6$.
- Based on the histograms showing the distribution of volume and the volume fraction of pores with a shape factor α or β it was found that the pore population in the P00 sample, whose direction of manufacturing was parallel to the base plate, was of the most irregular shape.
- In each sample, there is a population of pores whose orientation is perpendicular to the direction of sample sintering.

Acknowledgments. Research was cofinanced with grants for education allocated by the Ministry of Science and Higher Education in Poland No. 02/22/DSPB/1432.

References

1. Wohlers T, Caffrey T (2014) 3D printing and additive manufacturing state of the industry annual worldwide progress report. Wohlers Report
2. Guo N, Leu MC (2013) Additive manufacturing technology, applications and research needs. Front Mech Eng 8:215–243. https://doi.org/10.1007/s11465-013-0248-8
3. https://www.tth.com/3d-printing/sls-prototyping. Accessed Nov 2018
4. Frazier WE (2014) Metal additive manufacturing: a review. J Mater Eng Perform 23:1917–1928. https://doi.org/10.1007/s11665-014-0958-z
5. Brandt M (2016) Laser additive manufacturing: materials, design, technologies, and applications. Laser Mater Process. https://doi.org/10.1016/b978-0-08-100433-3.01001-0
6. Olakanmi EO, Cochrane RF, Dalgarno KW (2015) A review on selective laser sintering/melting (SLS/SLM) of aluminum alloy powders: processing, microstructure, and properties. Prog Mater Sci 74:401–477. https://doi.org/10.1016/j.pmatsci.2015.03.002
7. Thijs L, Kempen K, Kruth JP, Van Humbeeck J (2013) Fine-structured aluminum products with controllable texture by selective laser melting of pre-alloyed AlSi10Mg powder. Acta Mater 61:1809–1819. https://doi.org/10.1016/j.actamat.2012.11.052
8. Louvis E, Fox P, Sutcliffe CJ (2011) Selective laser melting of aluminum components. J Mater Process Technol 211:275–284. https://doi.org/10.1016/j.jmatprotec.2010.09.019
9. Olakanmi EO (2013) Selective laser sintering/melting (SLS/SLM) of pure Al, Al-Mg, and Al-Si powders: Effect of processing conditions and powder properties. J Mater Process Technol 213:1387–1405. https://doi.org/10.1016/j.jmatprotec.2013.03.009
10. Kaufman JG, Rooy EL (2004) Aluminum alloy castings: properties, processes, and applications. Alum Alloy Cast Prop Process Appl 340 https://doi.org/10.1017/cbo9781107415324.004
11. Brandl E, Heckenberger U, Holzinger V, Buchbinder D (2012) Additive manufactured AlSi10Mg samples using Selective Laser Melting (SLM): microstructure, high cycle fatigue, and fracture behavior. Mater Des 34:159–169. https://doi.org/10.1016/j.matdes.2011.07.067
12. Anwar AB, Pham QC (2017) Selective laser melting of AlSi10 Mg: effects of scan direction, part placement and inert gas flow velocity on tensile strength. J Mater Process Technol 240:388–396. https://doi.org/10.1016/j.jmatprotec.2016.10.015
13. Tang M, Pistorius PC (2017) Oxides, porosity and fatigue performance of AlSi10Mg parts produced by selective laser melting. Int J Fatigue 94:192–201. https://doi.org/10.1016/j.ijfatigue.2016.06.002
14. Aboulkhair NT, Everitt NM, Ashcroft I, Tuck C (2014) Reducing porosity in AlSi10Mg parts processed by selective laser melting. Addit Manuf 1:77–86. https://doi.org/10.1016/j.addma.2014.08.001
15. Kempen K, Thijs L, Van Humbeeck J, Kruth J-P (2015) Processing AlSi10Mg by selective laser melting: parameter optimisation and material characterisation. Mater Sci Technol 31:917–923. https://doi.org/10.1179/1743284714Y.0000000702
16. Tyczynski P, Siemiatkowski Z, Rucki M (2018) Analysis of the drill base body fabricated with additive manufacturing technology. In: Proceedings of 18th international euspen conference & exhibition, Venice, Italy, 4–8 June 2018, pp 287–288

17. Plessis A, Yadroitsev I, Yadroitsava I, Le Roux S (2018) X-Ray microcomputed tomography in additive manufacturing: a review of the current technology and applications. 3D Printing Addit Manufact 5(3). https://doi.org/10.1089/3dp.2018.0060

18. Maskery I, Aboulkhair NT, Corfield MR et al (2016) Quantification and characterization of porosity in selectively laser melted Al-Si10-Mg using X-ray computed tomography. Mater Charact 111:193–204. https://doi.org/10.1016/j.matchar.2015.12.001

19. Cai X, Malcolm AA, Wong BS, Fan Z (2015) Measurement and characterization of porosity in aluminium selective laser melting parts using X-ray CT. Virtual Phys Prototyp 10:195–206. https://doi.org/10.1080/17452759.2015.1112412

20. Gapinski B, Wieczorowski M, Grzelka M, Alonso PA, Tome AB (2017) The application of micro computed tomography to assess quality of parts manufactured by means of rapid prototyping. Polimery 62(1):53–59. https://doi.org/10.14314/polimery.2017.053

21. Maire E, Buffière JY, Salvo L et al (2001) On the Application of X-ray Microtomography in the Field of Materials Science. Adv Eng Mater 3:539–546. https://doi.org/10.1002/1527-2648(200108)3:8%3c539:AID-ADEM539%3e3.0.CO;2-6

22. De Chiffre L, Carmignato S, Kruth JP et al (2014) Industrial applications of computed tomography. CIRP Ann Manuf Technol 63:655–677. https://doi.org/10.1016/j.cirp.2014.05.011

23. Tammas-Williams S, Zhao H, Léonard F et al (2015) XCT analysis of the influence of melt strategies on defect population in Ti-6Al-4 V components manufactured by selective electron beam melting. Mater Charact 102:47–61. https://doi.org/10.1016/j.matchar.2015.02.008

24. Ziolkowski G, Chlebus E, Szymczyk P, Kurzac J (2014) Application of X-ray CT method for discontinuity and porosity detection in 316L stainless steel parts produced with SLM technology. Arch Civ Mech Eng 14:608–614. https://doi.org/10.1016/j.acme.2014.02.003

25. Gapinski B, Wieczorowski M, Bak A, Pereira Dominguez A, Mathia T (2018) The assessment of accuracy of inner shapes manufactured by FDM. In: AIP conference proceedings, vol 1960. https://doi.org/10.1063/1.5035001

26. PN-EN ISO 6892-1:2016-09

27. Zhao B, Wang J (2016) 3D quantitative shape analysis on form, roundness, and compactness with μCT. Powder Technol 291:262–275. https://doi.org/10.1016/j.powtec.2015.12.029

Evaluation of the Longitudinal Roughness of the Thin-Walled Cooler for the Robot Control System Made Using CAM Programming

Peter Tirpak, Peter Michalik[(✉)], Jozef Zajac, Vieroslav Molnar, Dusan Knezo, and Michal Petruš

Faculty Manufacturing and Technologies, Technical University of Kosice,
Sturova 31, 08001 Presov, Slovakia
`peter.michalik@tuke.sk`

Abstract. The article deals by design, programming production, measuring and evaluation of the longitudinal roughness of the contact surface of the thin-walled cooler for the robot control system. The dimensional design was made on the basis of the output of the power supply. A software model Autodesk Inventor 2019 was used to model the 3D model of the heat sink and generate the NC code for the program needed for its production. The milling and drilling was done on a Pinacacle 2100 Vertical CNC Machining Center with Fanuc Control System. The surface roughness was measured using a Mitutoyo SJ 400 whereby was measured with a maximum value of Ra = 0,69 µm and Rz = 4,1 µm.

Keywords: Longitudal roughness · Thin-walled cooler · Robot ·
CAM programming

1 Introduction

The performance of processors of robot control systems is steadily increasing, and there are also increased requirements for their execution and, in particular, for more serious problems in the design of processors - cooling. Integrated circuits (e.g., CPU and GPU) are the main generators of heat in modern computers. Heat generation can be reduced by efficient design and selection of operating para meters such as voltage and frequency, but ultimately, acceptable performance can often only be achieved by managing significant heat generation. The dust buildup on this laptop CPU heat sink after three years of use has made the laptop unusable due to frequent thermal shutdowns. In operation, the temperature of a computer's components will rise until the heat transferred to the surroundings is equal to the heat produced by the component, that is, when thermal equilibrium is reached. For reliable operation, the temperature must never exceed a specified maximum permissible value unique to each component. For semiconductors, instantaneous junction temperature, rather than component case, heatsink, or ambient temperature is critical. Cooling can be changed by:

© Springer Nature Switzerland AG 2019
M. Diering et al. (Eds.): *Advances in Manufacturing II* - Volume 5, LNME, pp. 285–296, 2019.
https://doi.org/10.1007/978-3-030-18682-1_22

- Dust acting as a thermal insulator and impeding airflow, thereby reducing heat sink and fan performance.
- Poor airflow including turbulence due to friction against impeding components such as ribbon cables, or incorrect orientation of fans, can reduce the amount of air flowing through a case and even create localized whirlpools of hot air in the case. In some cases of equipment with bad thermal design, cooling air can easily flow out through "cooling" holes before passing over hot components; cooling in such cases can often be improved by blocking of selected holes.
- Poor heat transfer due to poor thermal contact between components to be cooled and cooling devices. This can be improved by the use of thermal compounds to even out surface imperfections, or even by lapping.

Because high temperatures can significantly reduce life span or cause permanent damage to components, and the heat output of components can sometimes exceed the computer's cooling capacity, manufacturers often take additional precautions to ensure that temperatures remain within safe limits. A computer with thermal sensors integrated in the CPU, motherboard, chipset, or GPU can shut itself down when high temperatures are detected to prevent permanent damage, although this may not completely guarantee long-term safe operation. Before an overheating component reaches this point, it may be "throttled" until temperatures fall below a safe point using dynamic frequency scaling technology. Throttling reduces the operating frequency and voltage of an integrated circuit or disables non-essential features of the chip to reduce heat output, often at the cost of slightly or significantly reduced performance. For desktop and notebook computers, throttling is often controlled at the BIOS level. Throttling is also commonly used to manage temperatures in smartphones and tablets, where components are packed tightly together with little to no active cooling, and with additional heat transferred from the hand of the user. The first processors did not require any special cooling, but the heat sinks were increasing over time, and their construction today relies on research into the thermal conductivity of metals and aerodynamics. At present, passive and active radiators are used.

The passive coolers of Fig. 1 are made of aluminum and are made of a ribbed section that is attached to the processor by a base. Primarily used aluminum for the production of paints is increasingly common in copper, which has even better thermal conductivity [1].

Passive heat-sink cooling involves attaching a block of machined or extruded metal to the part that needs cooling. A thermal adhesive may be used. More commonly for a personal-computer CPU, a clamp holds the heat sink directly over the chip, with a thermal grease or thermal pad spread between. This block has fins and ridges to increase its surface area. The heat conductivity of metal is much better than that of air, and it radiates heat better than the component that it is protecting (usually an integrated circuit or CPU). Fan-cooled aluminium heat sinks were originally the norm for desktop computers, but nowadays many heat sinks feature copper base-plates or are entirely made of copper. Dust buildup between the metal fins of a heat sink gradually reduces efficiency, but can be countered with a gas duster by blowing away the dust along with any other unwanted excess material. Passive heat sinks are commonly found on older CPUs, parts that do not get very hot (such as the chipset), and low-power computers.

Fig. 1. Passive cooler [1].

Usually a heat sink is attached to the integrated heat spreader (IHS), essentially a large, flat plate attached to the CPU, with conduction paste layered between. This dissipates or spreads the heat locally. Unlike a heat sink, a spreader is meant to redistribute heat, not to remove it. In addition, the IHS protects the fragile CPU. Passive cooling involves no fan noise as convection forces move air over the heatsink [1].

The active cooler is a fan that blows the air into the processor to remove excess heat. With this cooling, there is a greater risk of fan failure and thus damage the processor. Improved ventilators use rolling bearings. Ventilators with sliding bearings are cheaper but more faulty [1] (Fig. 2).

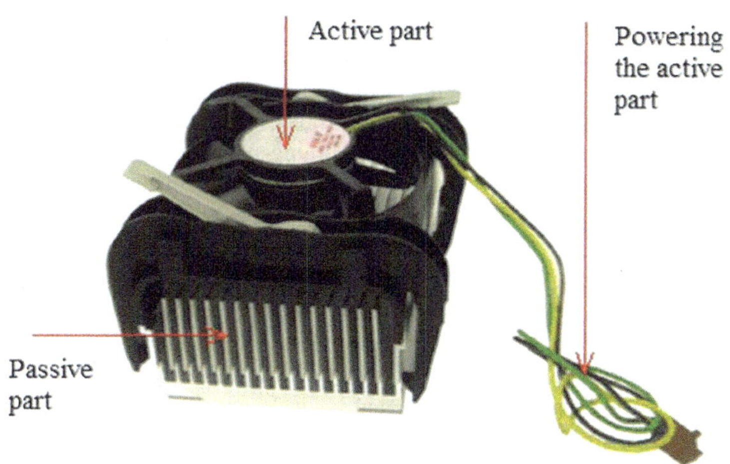

Fig. 2. Active cooler with individual parts [1].

The fan propellers are in a different form and are positioned to minimize aerodynamic noise while maintaining maximum cooling efficiency. A more sophisticated solution for high-performance processors is a combination of a passive and active heat sink that provides sufficient heat gain. Typically, the bottom of the heat sink is a passive section and the upper part is an active part that supplies a sufficient amount of air between the cooling coil rebate. An equally important factor is the choice of the right material for the contact surfaces of the heat sink and processor - the heat transfer paste. It fills microscopic inequalities and ensures maximum heat transfer from the processor body to the chiller block. Quality coolers have their entire surface, in addition to their lower part, covered by inequalities. This increases the heat transfer area from the passive section. In addition to cooling by air, it is possible to cool with water, liquid nitrogen or special electrical equipment - Peltier element. However, these alternatives are construction-intensive and are not yet in common desktop computers. The processor is not the only heat source in the robot control system. Almost every installed component emits heat in the on-off state, which is not only a part of the control system but of every electronic component. Another fan is part of the power supply. Another growing trend due to the increasing heat density of computer, GPU, FPGA, and ASICs is to immerse the entire computer or select components in a thermally, but not electrically, conductive liquid. Although rarely used for the cooling of personal computers, liquid immersion is a routine method of cooling large power distribution components such as transformers. It is also becoming popular with data centers. Personal computers cooled in this manner may not require either fans or pumps, and may be cooled exclusively by passive heat exchange between the computer hardware and the enclosure it is placed in. A heat exchanger (i.e. heater core or radiator) might still be needed though, and the piping also needs to be placed correctly. The coolant used must have sufficiently low electrical conductivity not to interfere with the normal operation of the computer. If the liquid is somewhat electrically conductive, it may cause electrical shorts between components or traces and permanently damage them. For these reasons, it is preferred that the liquid be an insulator (dielectric) and not conduct electricity. A wide variety of liquids exist for this purpose, including transformer oils, synthetic single-phase dielectric coolants such as Engineered Fluids' ElectroCooll, and 2-phase coolants such as 3M Fluorinert. Non-purpose oils, including cooking, motor and silicone oils, have been successfully used for cooling personal computers. Some fluids used in immersion cooling, especially hydrocarbon based materials such as mineral oils, cooking oils, and organic esters, may degrade some common materials used in computers such as rubbers, PVC, and thermal greases. Therefore it is critical to review the material compatibility of such fluids prior to use. Mineral oil in particular has been found to have negative effects on PVC and rubber-based wire insulation. Thermal pastes used to transfer heat to heat sinks from processors and graphic cards has been reported to dissolve in some liquids, however with negligible impact to cooling, unless the components were removed and operated in air. Evaporation, especially for 2-phase coolants, can pose a problem, and the liquid may require either to be regularly refilled or sealed inside the computer's enclosure [1].

Liquid cooling is a highly effective method of removing excess heat, with the most common heat transfer fluid in desktop PCs being (distilled) water. The advantages of water cooling over air cooling include water's higher specific heat capacity and thermal conductivity. The principle used in a typical (active) liquid cooling system for computers is identical to that used in an automobile's internal combustion engine, with the water being circulated by a water pump through a waterblock mounted on the CPU (and sometimes additional components as GPU and northbridge) and out to a heat exchanger, typically a radiator. The radiator is itself usually cooled additionally by means of a fan. Besides a fan, it could possibly also be cooled by other means, such as a Peltier cooler (although Peltier elements are most commonly placed directly on top of the hardware to be cooled, and the coolant is used to conduct the heat away from the hot side of the Peltier element). A coolant reservoir is often also connected to the system. Besides active liquid cooling systems, passive liquid cooling systems are also sometimes used. These systems often discard a fan or a water pump, hence theoretically increasing the reliability of the system, and/or making it quieter than active systems. Downsides of these systems however are that they are much less efficient in discarding the heat and thus also need to have much more coolant -and thus a much bigger coolant reservoir- (giving more time to the coolant to cool down). Liquids allow the transfer of more heat from the parts being cooled than air, making liquid cooling suitable for overclocking and high performance computer applications. Compared to air cooling, liquid cooling is also influenced less by the ambient temperature. Liquid cooling's comparatively low noise-level compares favorably to that of active cooling, which can become quite noisy [1].

Disadvantages of liquid cooling include complexity and the potential for a coolant leak. Leaked water (or more importantly any additives added to the water) can damage any electronic components with which it comes into contact, and the need to test for and repair leaks makes for more complex and less reliable installations. (Notably, the first major foray into the field of liquid-cooled personal computers for general use, the high-end versions of Apple's Power Mac G5, was ultimately doomed by a propensity for coolant leaks.) An air-cooled heat sink is generally much simpler to build, install, and maintain than a water cooling solution, although CPU-specific water cooling kits can also be found, which may be just as easy to install as an air cooler. These are not limited to CPUs, however, and liquid cooling of GPU cards is also possible. While originally limited to mainframe computers, liquid cooling has become a practice largely associated with overclocking in the form of either manufactured kits, or in the form of do-it-yourself setups assembled from individually gathered parts. The past few years have seen an increase in the popularity of liquid cooling in pre-assembled, moderate to high performance, desktop computers. Sealed ("closed-loop") systems incorporating a small pre-filled radiator, fan, and waterblock simplify the installation and maintenance of water cooling at a slight cost in cooling effectiveness relative to larger and more complex setups. Liquid cooling is typically combined with air cooling, using liquid cooling for the hottest components, such as CPUs or GPUs, while retaining the simpler and cheaper air cooling for less demanding components. Sometimes the power is insufficient to remove the heat from the cabinet [1, 2]. The surface quality of thin-walled parts of C45 steel was studied Michalik et al. [3, 4]. Milling of thin-walled parts and evaluation of machining quality for the aerospace industry was dealt with by Meshreki [5]. The prediction of the machining of the machined thin walled parts was performed Kuram et al. [6, 7]. The wear of cutting materials during the C45 steel was evaluated Duplak et al. [8], Fedorko et al. [9, 10]. Comparison of the quality of the surface quality of the raw materials in the water treatment was carried out Lehocka et al. [11]. Selected diagnostic methods for assessing the condition of the gearbox of machining center accrued from collaboration applied Baron et al. [12]. CAM programming for the production of thin-walled parts has been applied Fabian et al. Michalik et al. Kral et al. [13–18]. Testing of composite materials has been evaluated Hutyrova et al. [19].

2 Manufacturing Diagram and Machined Materials

The dimensions of the thin-walled cooler Fig. 3 for the robot's control system were adapted to the size of the processor where the size of the cooling surface was 40 × 40 mm. Other dimensions have been adapted to the structure of the carrier box and the liquid cooling system.

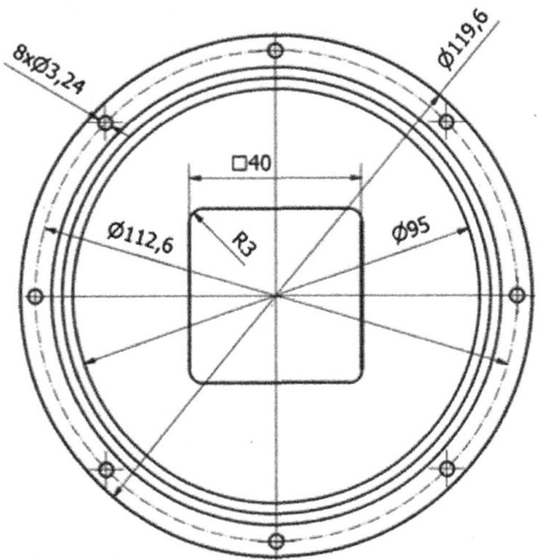

Fig. 3. Manufacturing diagram of thin-walled cooler.

The thin-walled cooler material for the robot control system selected from the available AlCu4PbMgMn sources with Table 1. The workpiece was a 120 mm diameter circular.

Table 1. Data of material AlCu4PbMgMn.

Component	Si	Fe	Cu	Mn	Mg	Cr	Ni	Zn	Ti	Pb
Element content	0,8	0,8	3,3	0,5	1,8	0,1	0,2	0,8	0,2	1,5

The milling tool used with a diameter of 12 mm and clamped into a short clamp holder using a pinch of Fig. 4. The tool speed was 1200 min and a feed rate of 600 mm/min.

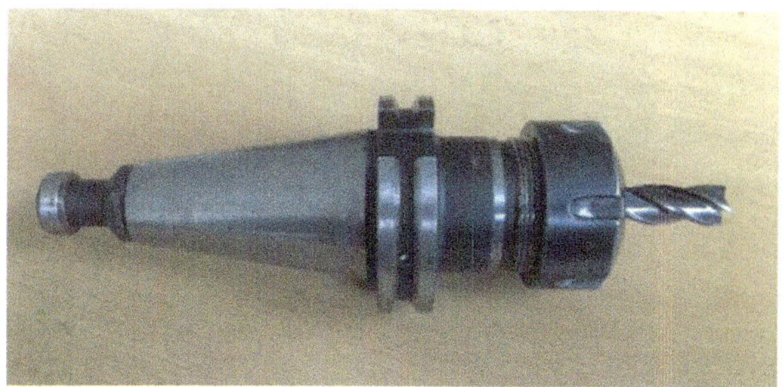

Fig. 4. Short clamp milling tool.

3   Programming of Manufacturing Cooler

The NC code for programming the thin-walled cooler for the robot control system generated using the Autodesk Inventor 2019 software. The first step in the CAM environment after loading the cooler model is the placement of the zero point of the finished component Fig. 5.

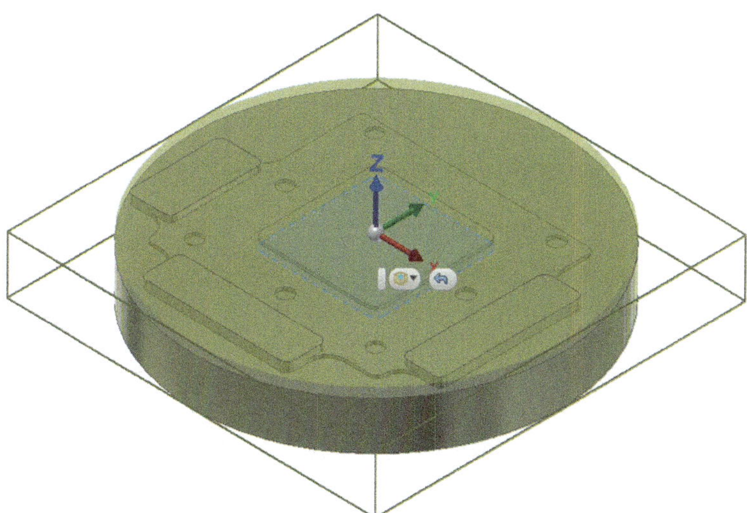

Fig. 5. Selection position of coordinate system.

Other step is selection tool and cutting conditions Fig. 6.

When roughing, the shape of the component was made with an addition of 0.2 mm in the axes x, y and z. In this machining, a carbide cutter with a diameter of Ø 12 mm used. After selecting the technological machining conditions, simulation of the roughing machining Fig. 7 and finishing of Fig. 8 follows.

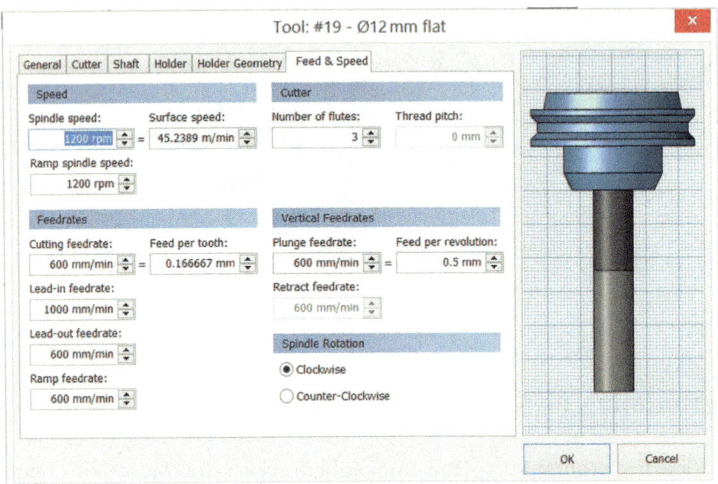

Fig. 6. Selection cutting tool and cutting conditions.

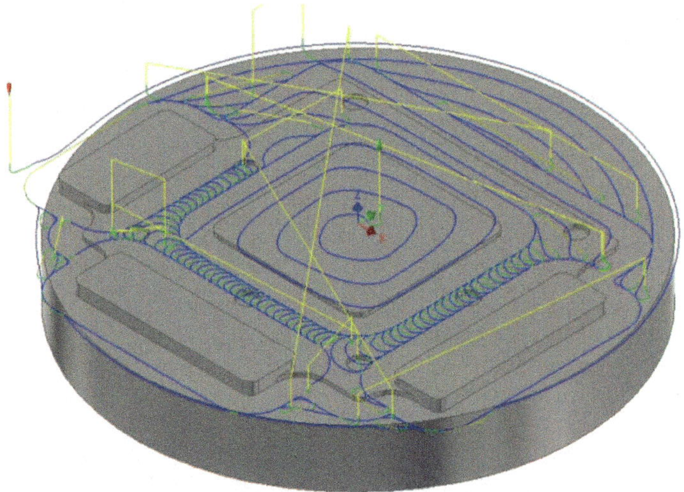

Fig. 7. Simulation of roughing machining.

4   Machining and Measuring Thin-Walled Cooler

Thin-walled radiator for the robot control system produced by the Pinnacle 2100 Vertical CNC Machining Center with control system Fanuc Fig. 9.

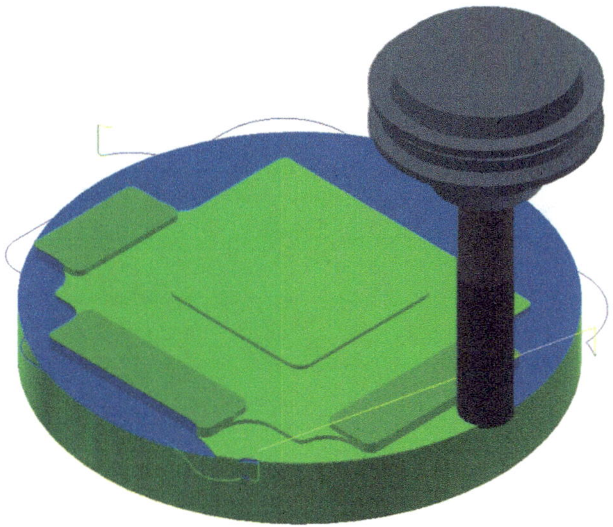

Fig. 8. Simulation of finishing machining.

Fig. 9. Machining thin-walled component on the vertical CNC milling centre Pinnacle 2100.

The semi-fabric clamped in the tri-chuck chuck at the adjacent worktable with clamps. Longitudinal roughness measured on the Mituoyo SJ 400 Fig. 10. The tool paths were labeled A; B; C; D; A1; B1; C1; D1. Sampling was followed at these locations and the Ra and Rz values were written into tabs and resulting graphs for the tool A path; B; C; D is on Fig. 11, for the A1 track path; B1; C1; D1 Fig. 12. Measurement at each location repeated ten times.

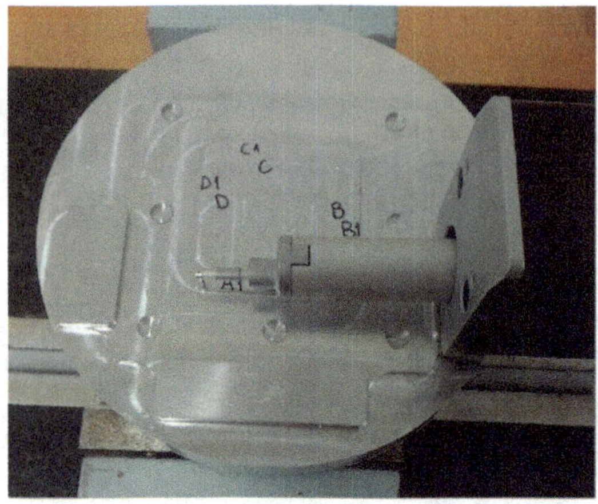

Fig. 10. Simulation of finishing machining.

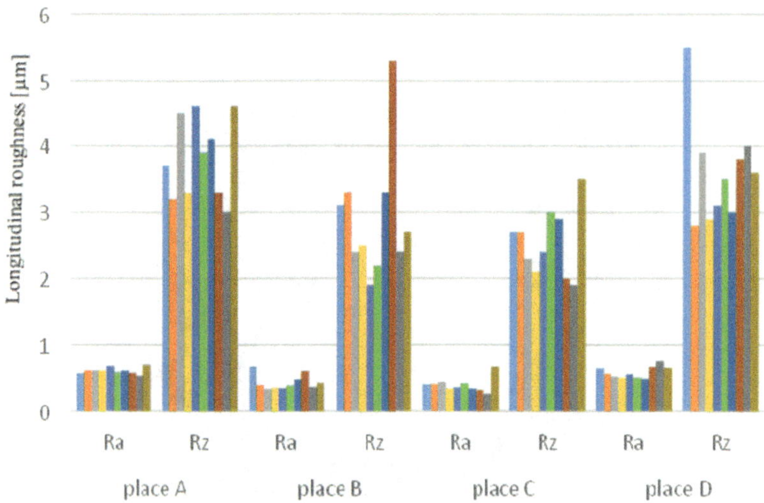

Fig. 11. Longitudinal roughness on the place A; B; C; D.

5 Discussion

Article deals with the experimental testing of longitudinal roughness focused on surface of the thin wall component. Milling considered as an untraditional way of manufacturing thin wall components – ribs. Materials used for manufacturing of thin wall components are another area of research, which has to solve. Three dimension views overall surface can be next aspect of research.

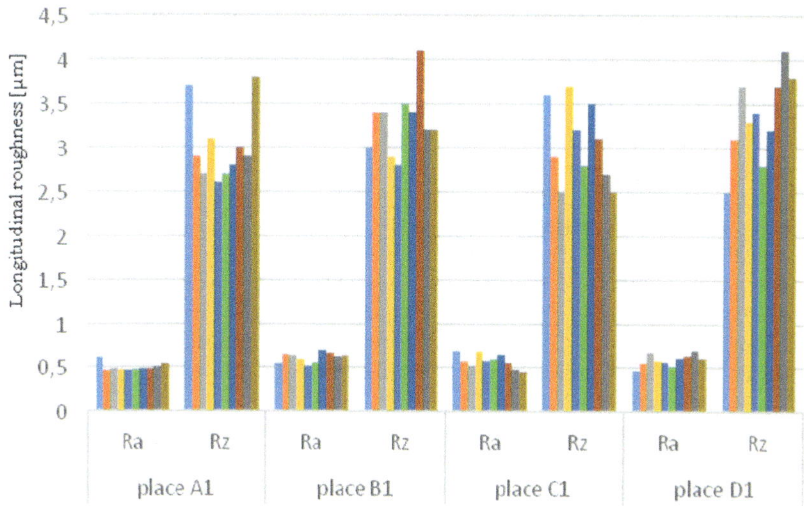

Fig. 12. Longitudinal roughness on the place A1; B1; C1; D1.

6   Conclusion

Today are used Heat Pipes. These are copper tubes filled with a special liquid that drain heat from the processor of the robot control system to the radiator block. As a result of physical processes, this fluid flows freely. Generally, the more the cooler of these tubes is, the better it will cool down. The quality of the heat sink area also affects the heat transfer itself with the processor itself. Based on the measurements, the roughness values were found. The lowest value Ra = 0.26 µm, Rz = 1.9 µm was measured at the location C. The maximum value Ra = 0.69 µm Rz = 4.1 µm was measured at the D1 site.

Acknowledgments. This work is a part of research project VEGA 1/0045/18, VEGA 1/0492/16, VEGA 1/0403/18

References

1. https://en.wikipedia.org/wiki/Computer_cooling
2. Mudawar I (2001) Assessment of high-heat-flux thermal management schemes. IEEE Trans Compon Packag Tech 24(2):122–141
3. Michalik P et al (2014) Monitoring surface roughness of thin-walled components from steel C45 machining down and up milling. Measurement 58:416–428
4. Straka L et al (2016) Properties evaluation of thin microhardened surface layer of tool steel after wire EDM. Metals, 95–103
5. Meshreki M (2009) Dynamics of Thin-Walled Aerospace Structures for Fixture Design in Multi-axis Milling (2009)

6. Kuram E et al (2013) Multi-objective optimization using Taguchi based grey relational analysis for micro-milling of Al 7075 material with ball nose end mill. Measurement 46:1849–1864
7. Miko E et al (2012) Analysis and verification of surface roughness constitution model after machining process. Procedia Eng 39:395–404
8. Duplak J et al Evaluation of T-vc dependence for the most commonly used cutting tools. Key Eng Mater 663:278–285
9. Fedorko G et al (2011) The influence of Ni and Cr-content on mechanical properties of Hadfield's steel. In: Met., Tanger Ostrava, pp 1–6
10. Fedorko G et al (2014) Failure analysis of irreversible changes in the construction of rubber–textile conveyor belt damaged by sharp-edge material impact. Eng Fail Anal 39
11. Lehocka D et al (2017) Comparison of the influence of acoustically enhanced pulsating water jet on selected surface integrity characteristics of CW004A copper and CW614 N brass. Measurement 110:230–238
12. Baron P et al (2016) Research and application of methods of technical diagnostics for the verification of the design node. Measurement 94:245–253
13. Fabian M et al (2014) Influence of the CAM parameters and selection of end-mill cutter when assessing the resultant surface quality in 3D milling. Appl Mech Mater 474:267–272
14. Michalik P et al (2013) Programming CNC machines using computer-aided manufacturing software. Adv Sci Lett 19:369–373
15. Kral J et al (2008) Creation of 3D parametric surfaces in CAD systems. Acta Mech Slovaca, 223–228
16. Mantic M et al (2016) Influence of selected digitization methods on final accuracy of 3D model, 475–480
17. Michalik P et al (2011) CAM software products for creation of programs for CNC machining. Lect Notes Electr Eng 141:421–425
18. Peterka J et al (2013) Optical 3D scanning of cutting tools. Appl Mech Mater 421:663–667
19. Hutyrova Z et al (2013) Non-destructive testing of inhomogeneity of composite material. In: ICET 2013, pp 133–136

Author Index

Made in the USA
Monee, IL
07 July 2026

56544704R00175